ANCIENT
AUSTRALIA

THE THIRD EDITION OF
CHARLES LASERON'S

ANCIENT AUSTRALIA

The Story of its Past Geography and Life

Revised by

RUDOLF OSKAR
BRUNNSCHWEILER, Ph.D.

Angus & Robertson Publishers

ANGUS & ROBERTSON PUBLISHERS
London · Sydney · Melbourne

This book is copyright. Apart from any fair dealing for the purposes
of private study, research, criticism or review, as permitted under the
Copyright Act, no part may be reproduced by any process without
written permission. Inquiries should be addressed to the publishers.

First published by Angus & Roberston Publishers, Australia, 1954
Reprinted 1955
Revised edition 1969
Revised 1984

© 1984
Rudolf Oskar Brunnschweiler
The Laseron estate

National Library of Australia
Cataloguing-in-publication data.

Laseron, Charles Francis.
  Ancient Australia: the story of its past geography and life.

  Rev. ed. / by Rudolf Oskar Brunnschweiler.
  (Australian natural science library).
  Index
  First published, Sydney: Angus & Robertson, 1969.
  Bibliography
  ISBN 0 207 14181 9

  1. Paleography—Australia. 2. Paleontology—Australia.
  I. Brunnschweiler, Rudolf Oskar. I. Title. (Series)

551.7

Typeset in 11 pt Baskerville by Asco Trade Typesetting Limited, Hong Kong
Printed in Hong Kong

PREFACE
to Third Edition

"I HAVE stood on the bed of the ocean just as it was 200 000 000 years ago. It was at Ulladulla, a short distance from Sydney on the South Coast of New South Wales. There, at the foot of the cliffs, erosion had stripped the upper layers of rock to expose a flat rocky platform some hundreds of yards in extent. At one stage in the great Permian Period this platform had literally been the bed of the sea. When I visited it, and for a great length of time before, the sea no longer covered it and the sand and mud on the bottom had hardened into rock, but otherwise it was exactly the same as it had been in that remote age. Scattered about were the remains of the creatures which had lived at the time, now embedded in the hard rock. There were large scallop-shells, innumerable lamp-shells, delicate lace-corals, and the plates of extinct sea-lilies. Here and there large boulders protruded above the surface, just where they had been dropped from melting icebergs floating overhead. Here, it seemed to me, was the nucleus of a book, a story of Australia with its changing geography and life from the beginning of geological time to the present."

Three decades have passed since the late Charles F. Laseron proceeded from idea to action and wrote the first edition of this book. Published in 1954, and reprinted in 1955, it turned out to be a very successful piece of Australian popular-science writing and, thanks to Charles Laseron's initial effort, the revised and much enlarged second edition of 1969 enjoyed similar popularity, so that it too has now been out of print for several years.

Originally, the book was intended for three main groups of readers. Firstly, for those who, without any special predilection for science, are interested from a cultural point of view in any general survey of a scientific subject if it is readable and avoids technical jargon. The second group contains the amateur naturalists, and particularly those among them who would make geology their hobby. The third is composed of students in the early stages of their studies who have to familiarize themselves with a multitude of new terms and concepts.

With the enlarged second edition I aimed at broadening the readership to include a fourth group—the professionals. Not so much the Australian, perhaps, than the many geologists from overseas who came to participate in the search for petroleum and minerals in Australia. For them it was an inexpensive and easily readable reference book which, in addition, contained a number of new ideas and interpretations to help them on their way into the intricacies of this continent's geological history.

The present third edition continues in the same vein and hopes to find the same general readership. As mentioned in the preface to the 1969 edition, a dozen or so years, in our times of rapid progress in the sciences, is "old age" for a book of this kind. Although some of the contents here remain unchanged, or but slightly updated, others needed considerable revision and re-writing. Apart from updating it I have also endeavoured to improve the index by introducing geographic co-ordinates for localities mentioned in the text, an omission frequently complained about by overseas readers. The bibliographic references have also been complemented by adding pertinent studies published up to mid-1979. As it is already twelve years too since the only modern and comprehensive professorial description of Australia's geology appeared in print (*Brown et al., 1968*), this book may now also prove useful to the Australian professional geologist for quick reference.

Although my work as a geologist has taken me across the length and breadth of Australia and well beyond, and although keeping up to date in my profession requires voracious reading, some sections of this book could hardly have been written without additional information obtained from discussions with many professional colleagues, such as those of the Commonwealth Bureau of Mineral Resources, Geology and Geophysics in Canberra, whose activities extend over all States and the Territories, as well as such neighbouring countries as Niugini, Irian Jaya and the Solomon Islands, through arrangements with the respective governments. Pleasant memories of similarly fruitful conversations with many colleagues in the Universities, State Geological Surveys and Museums, private mining and oil companies across the land and through the years are equally gratefully acknowledged. I trust my friends will forgive me that I cannot mention them all by name.

I am also indebted to various organizations for allowing me to reproduce numerous illustrations, especially palaeontological ones, from their scientific publications, and to some governmental agencies which made available photographs of Australian scenery. The sources of all this illustrative material are acknowledged in the appropriate places.

R.O.B

CONTENTS

Preface to Third Edition — v

(I) INTRODUCTION
1 The Changing Face of the Land — 3
2 Fossils and the Geological Record — 19

(II) BIRTH AND GROWTH OF AUSTRALIA IN THE PRECAMBRIAN EONS
3 The Precambrian Eons — 33
4 Life and Evolution — 68

(III) ADOLESCENCE AND ADJUSTMENT IN THE PALAEOZOIC ERA
5 The Cambrian Period — 83
6 The Ordovician Period — 99
7 The Silurian Period — 113
8 The Devonian Period — 124
9 The Carboniferous Period — 144
10 The Permian Period — 157

(IV) THROUGH CONSOLIDATION TO MATURITY IN THE MESOZOIC ERA
11 The Triassic Period — 193
12 The Jurassic Period — 213
13 The Great Flood in the Cretaceous Period — 229

(V) THE LAST 65 MILLION YEARS—SERENE ISOLATION IN THE CAINOZOIC ERA
14 From the Palaeocene to the Present — 253

References — 297
Index — 311

ILLUSTRATIONS AND MAPS

PALAEONTOLOGICAL PLATES *Following page 150*
- 1–2 Proterozoic Fossils
- 3–4 Cambrian Fossils
- 5–6 Ordovician Fossils
- 7–8 Silurian Fossils
- 9–10 Devonian Fossils
- 11 Carboniferous Fossils
- 12–15 Permian Fossils
- 16–17 Triassic Fossils
- 18 Jurassic Fossils
- 19–25 Cretaceous Fossils
- 26–28 Cainozoic Fossils

GEOLOGICAL PLATES
1. Lower Precambrian Willyama Series
2. Chewings Range, a quartzite ridge in the Arunta Complex
3. Tilted Mount Isa Shales
4. An outcrop of Precambrian rocks, Mount Isa
5. The Flinders Ranges
6. Weathered granite on Kangaroo Island
7. Gosses Bluff
8. The Grand Arch at Jenolan Caves
9. Cathedral Rocks
10. Westerly dipping sandstones of the Grampian Range
11. Bathurst granite intrusion—Carboniferous
12. Permian fossil shells

ILLUSTRATIONS AND MAPS ix

13 Triassic sandstones at North Avalon, New South Wales
14 Cradle Mountain
15 The Warrumbungles
16 Sink-hole in the Nullarbor
17 Tertiary rocks at Port Campbell
18 Foliated granite tors near Tennant Creek
19 Gibber Plains, near Woomera
20 Bangemall Basin
21 Devonian reef complex, Fitzroy Trough
22 Windjana Gorge

PALAEOGEOGRAPHICAL MAPS

1	Later Archaeozoic Era—2700 to 2600 millions years ago	41
2	Late Archaeozoic Kalgoorlie-Rum Jungle Orogeny—2500 million years ago	43
3	Early Proterozoic Era—2300 to 2200 million years ago	46
4	After sediments of Halls Creek-Pine Creek-Hatches Creek geosynclines were folded—1900 to 1700 million years ago	49
5	Mid-Proterozoic archipelago—1600 million years ago	51
6	After the mountain-building event—1500 million years ago	54
7	During the deposition of Bitter Springs Dolomite—1300 to 1200 million years ago	55
8	Early Upper Proterozoic Period—1000 million years ago	59
9	Late Proterozoic Eon—650 to 550 million years ago	65
10	End of the early Middle Cambrian Delamerian Orogeny	89
11	Early part of the Late Cambrian—500 million years ago	91
12	Early Ordovician—460 million years ago	100
13	After the Centralian Orogeny and the Benambran phase—420 to 410 million years ago	114
14	Lower Devonian time—380–370 million years ago	127

ILLUSTRATIONS AND MAPS

15	Late Devonian time—360 million years ago	128
16	Mid-Carboniferous time—310 million years ago	147
17	Glaciers of early Permian Period—275 to 260 million years ago	159
18	Early Triassic, after the Hunter-Bowen Orogeny—220 million years ago	194
19	During the Mesozoic Era—150 million years ago	215
20	Early Cretaceous times—130 to 120 million years ago	230
21	In the Albian Epoch at the end of the Middle Cretaceous—110 million years ago	232
22	At the end of the Maryburian Orogeny, in the Santonian Epoch of the Cretaceous—85 to 75 million years ago	234
23	The early part of the Miocene Epoch during the Tertiary Period—21 million years ago	260
24	The Pleistocene Epoch—20 000 to 15 000 years ago	272

OTHERS

25	Geological Map	290
26	Sedimentary Basins—Australia and Papua New Guinea	292
27	Mineral Map	294

Geographical Map of Australia—Endpapers

TEXT FIGURES

1	Plate-tectonic map of the earth's surface	12
2	The major exposure areas of Precambrian rocks	35
3	Record of life in Australian rocks	79
4	Correlation of Middle Cambrian formations	86
5	Reconstruction of the Australian Jurassic dinosaur	224
6	Diagrammatic section across the Great Artesian Basin	238

I
INTRODUCTION

1 The CHANGING FACE of the LAND

CLUES TO the past lie in the present. In nature as in human affairs, history repeats itself, and the forces operating on the surface of the earth have so operated for countless ages. Measured by the span of a man's life, geographical changes are almost imperceptibly slow, yet it needs little imagination to realize the potency of the forces at work before our eyes. Destruction and construction go on unceasingly and eventually balance each other in an endless cycle. We will consider these forces separately, taking destruction first.

Let us begin with a rain-storm, such a storm as comes only a few times a year. For days it has been raining, and every creek is running a banker. The flooded rivers pour billions of tonnes of muddy water into the sea. For the most part the mud consists of fine surface soil, but in every cubic metre of muddy water there will also be some solid matter. Depending on the carrying power of a river, such solids may range from a few grains of sand to pebbles and even boulders. One river system in a period of flood can transport many millions of tonnes of solid matter. Multiply the one river system by the number of river systems in a country, the single rain-storm by the number of rain-storms in a year; consider the years that make a century, and the countless centuries of the past—and then it will be realized how tremendous is the bulk of material involved.

Whence comes this mass of earth and soil which is forever being washed out to sea?

In any given locality, and particularly on the hills, the soil is rarely more than a few metres thick. If it were not renewed every vestige would soon disappear, and only the bare rocks of the hills would remain. However, it is renewed, for the rocks themselves are not stable, but are constantly being broken up to form fresh soil.

This happens in many ways. First come the numerous agents of direct physical destruction. Frost in the high land is one of the most potent. Moisture seeps into cracks and between the grains of the rock, and when it freezes expands and forces the grains apart, even to the extent of shattering the whole rock. The roots of plants penetrate the finest fissures and exert a similar pressure. Wind-borne grains of sand act as a sand blast and abrade the exposed surfaces. In the streams grains and pebbles carried or rolled along by the running water act as chisels, and gouge out and rapidly deepen the channel. Even burrowing animals play their part in the endless work of destruction.

In the high mountainous regions, where the summits are above the snow-line, destruction is most rapid. Here moving ice is particularly destructive. Snow which accumulates on the summits gradually consolidates into ice, and by its own weight begins to move down the sides in the form of rivers of ice or glaciers. They carry with them masses of loose rock already broken by frost, and these act as gouges which grind and excavate with comparative rapidity. Glaciers thus cut out for themselves deep U-shaped valleys, with steep sides, down which avalanches carry still more material to the surface of the ice beneath. Along the sides, at the front, and underneath the glaciers, part of this material accumulates in what are known as moraines. The rest is carried away by streams born of the melting ice. Moraines, being made of loose material, are rather easily removed by subsequent erosion cycles, so that only large cobbles and blocks are left behind. These are known as glacial erratics; they occur sporadically in groups or singly and indicate the location of former moraines.

In the polar and sub-polar regions, where the snow-line is at sea-level or only slightly above it, glaciers do not melt on land but push into the sea. As they melt in the water, their burden of rock is dropped, the finer material being carried by currents and distributed over a considerable area. When large masses of ice break off as icebergs they may drift for hundreds of kilometres before they finally melt. In this way huge pieces of rock are transported great distances from their place of origin to be dropped finally on the bed of the ocean. These are known as iceberg-erratics, and later we shall see how they are found in the marine rocks of past ages as well as at the present time.

In arid desert regions wind is the chief carrier of eroded material. Strong winds, bearing grains of sand, are also very potent erosive agents. While seepage water and great daily temperature differences tend to break up the rock vertically, wind erosion works in the horizontal, cutting into and undermining rock masses rapidly, and carving them into bizarre shapes. Topography of arid lands is very characteristic, particularly where the rock is sandstone. Huge rock masses are often undercut to become top-heavy, or they may form needle-pointed pinnacles, standing isolated from the main outcrops.

Holes and caves are worn into their sides, natural arches and bridges are common. Even the hardest rocks are not immune to constant abrasion by sand-laden wind.

The great sand-storms of the Sahara transport countless tonnes of sand hundreds of kilometres from the areas where it was eroded. A proportion of this is borne out to sea to be deposited on the sea floor; much of it accumulates on the low land in the form of dunes. Wind-deposited sand may remain and harden into new beds of sandstone. Such aeolian deposits, as they are called, belonging to past ages, are to be found in many parts of the world. Even in less arid countries, if there is a drought, winds play a large part in removing the surface soil. Dust-storms, familiar in the back country of Australia, may be of great violence and carry away millions of tonnes of soil. The finer dust is carried right up into the stratosphere, and has been known to obscure the sky not only as far as the coast, but for hundreds of kilometres out to sea. Dust from the heart of Australia is, in fact, frequently deposited across the Tasman Sea in New Zealand. In the course of ages the total amount of material so transported is colossal.

Apart from physical destruction there is the chemical decomposition of the rocks. Many volcanic rocks—lavas such as basalt, for instance—decompose fairly rapidly when exposed to the atmosphere and biosphere. The molecular structure of the various rock minerals is broken down by purely chemical attack as well as by the concerted and often symbiotic biochemical action of micro-organisms such as fungi, algae, and bacteria. The resulting products are most commonly clays which are easily softened and removed by running water. A certain proportion of the decomposed material is in the form of soluble salts. These may be removed in solution, but they can also be redeposited by chemical precipitation as well as biochemical influences in favourable places on and below the surface. Many important ore deposits of copper, lead, zinc and uranium have been formed in this manner.

Even such an apparently solid and permanent rock as granite decomposes into a mixture of clay and sand. Limestone does not break up as readily as many other rocks, but it is directly soluble in rain-water charged with carbonic acid from the atmosphere, and so disappears in solution fairly rapidly. In some districts the hardness of water is frequently due to the presence of carbonate of lime derived from the corrosion of limestone rocks in the vicinity. All rocks, in fact, are to a varying degree destructible when exposed to the air, and so may form new soil to replace that continually removed by wind and running water.

Another agent of destruction, but only on the fringe of the land, is the sea itself. Breakers on the seashore hurl grains of sand and pebbles against the cliffs and gradually undermine them. Where the rocks are soft, marine erosion can be rather rapid. Within the history of man large areas of land, including villages and towns, have in some parts of the world disappeared beneath the sea.

So great and ceaseless is this work of destruction that even the highest mountains must in time inevitably disappear. The land would then be worn down and lowered to a nearly flat and monotonous plain, were it not for other compensating factors which will be dealt with presently. Eventually the stage is reached at which the rocks are covered and protected by surface soil and the fall, or gradient, of the drainage is so slight that it can neither erode nor carry away the soil. Such a level is called base-level, and the resulting surface is a peneplain. Some parts of the earth's surface have been reduced to peneplains, or very nearly so, many times in the course of geological history.

Let us now consider nature's constructive work. In lakes, on the flood-plains of rivers, at the mouth of rivers in the form of deltas, on the sea-bottom, there is continual accumulation of material derived from the destruction of the land. This consists not only of sand and mud, but also of the hard parts of animals and plants, of microscopic organisms of various kinds, of larger forms such as corals, shells, starfish, even the bones and teeth of larger animals. The material of which these organisms are composed is largely carbonate of lime secreted from the sea-water, but derived in the first place from limestone rocks on land.

Lakes absorb a comparatively small portion of the total amount of sediment and they are soon filled; nevertheless there are many lake deposits, some of large area and thickness, such as those formed in low-lying regions subject to periodical flooding by rivers. The great alluvial flood-plains of central Victoria, New South Wales and Queensland cover many thousands of square kilometres, and the sediments deposited are in places very thick. Here alone is sufficient material to form the bulk of a considerable mountain range.

It is the sea, however, which is the depository of the greater part of the earth's detritus. Here again, layer by layer, sediments thousands of metres in thickness are built up and eventually hardened into new rocks, similar to those from which they were derived. Mud is hardened into beds of claystone and shale, sand into sandstone, pebbles and cobbles into conglomerate, lime mud, corals and shells into limestone.

So is completed a cycle of destruction and construction.

One might expect that with the land wearing to base-level and the sea clogged with sediments, equilibrium would be attained and further action cease. However, there is another factor, rejuvenation, whereby mountains are built up again by an entirely different set of forces.

THE BUILDING OF MOUNTAINS

There is abundant evidence that in every period of the world's history, from the very earliest of which there is knowledge, high mountains similar to those of the present day have been in existence. The process of their destruction has been a continuous one but, to compensate, forces which built new mountain ranges have always been at work too.

It must be remembered that the crust of the earth is anything but stable. There have been many theories as to the exact nature and composition of the centre of the earth, but all agree that the stony portion with which we are familiar is merely an extremely thin crust. On a globe of one metre radius this crust is on the average less than one centimetre thick! Evidently, it must be subject to tremendous stresses and strains caused by continual adjustment to the forces operating on it from within and without. There is no need to consider the merits of the various hypotheses of the precise cause of those forces; it is enough for our purposes to observe the effects of their work.

In a broad sense, crustal movements are of two main types, vertical and horizontal. Vertical movements may elevate large areas into tablelands or plateaux—the Blue Mountains in New South Wales, for example—or they may depress land areas to a level that may be even lower than that of the sea. The sea may flow into these depressions, or they may become lakes, or remain dry land. Examples of these are Lake Eyre in South Australia, the Dead Sea in Israel, the Red Sea, the Qattara Basin in the Libyan Desert and, on a smaller scale, the Lake George Basin in New South Wales.

Such movements have often been on a vast scale, but they are gradual and rather slow, causing very few abrupt dislocations in the areas affected. Particularly along the margins of a continent large areas of sea may thus become land, and vice versa.

In many places piled-up layers or strata of rocks thousands of metres thick have been accurately measured and, despite their thickness, found to be of such uniformity that there is little doubt that they were laid down under the same conditions throughout. From the lowermost to the uppermost strata they were all deposited in shallow water, far shallower than the total thickness of rocks. This is contrary to expectation: in a sea of, say 3000 metres depth, which was gradually filled with sediments, it would be thought that only the topmost beds would be of shallow-water origin, and that the lowermost layers would be similar to those now being formed in the abyssal depths of the ocean. The explanation is that the sea was always shallow, but that subsidence continued at more or less the same rate as the deposition of sediment, so that there was very little change in the depth of water. Slight variations in the rate of subsidence, its temporary cessation, or even short periods of elevation, would produce corresponding variations in the strata deposited. The floor of the sea might even for a time become dry land, or it might be converted into freshwater lakes and swamps, in which successive generations of plants accumulated to be eventually turned into seams of coal. It is the record of these successive changes in the past which can be read in the rocks today.

An interesting feature of marine rocks now exposed on the land throughout the world is that true deep-sea deposits are not at all common in them. The seas which have at one time or another covered the whole of the existing land surface have mostly been shallow and of a temporary nature. In geological language they are known as epicontinental, transgressive seas, as distinct from the main oceans and the deep-sea trenches with their abysses several kilometres deep. This situation is one of the main reasons for assuming that in spite of the many changes in the geographical distribution of land and sea, such changes have not affected the relative positions of the main continents and oceans. In other words, that the great land masses of the earth, though they have changed in shape and size, have always been affixed where they are now, and that the great depths of ocean have also been in their present places since the dawn of geological time. This is known as the theory of the permanence of oceans and continents. However, because ancient deep-sea deposits nevertheless do occur also within continental regions—and rather more frequently than was once thought—this theory does not have many adherents today.

Much more spectacular than the simple vertical movements of the earth's crust, and on a grander scale, are the horizontal movements which are the chief cause of the formation of folded mountain ranges.

A number of hypotheses have been advanced to explain the reasons for and the dynamic mechanics of these colossal horizontal displacements. One of them is based on the analogy with a shrinking apple. When an apple dries, the skin has to accommodate itself to a shrinking centre; it is squeezed sideways until it puckers into numerous wrinkles and furrows. As recently as 60 years ago most geologists believed that something of this kind has happened and is still happening to the surface rocks of the earth. It seems a reasonable enough theory—provided our globe was and is cooling and shrinking. Yet, as we know now, this is not and has not been the case for some 2000 or more million years. Rather the contrary may have happened, namely expansion (*Carey, 1977*).

What other causes for horizontal movements are there? In fact, several. Three of the most important forces straining the earth's crust are gravitation, rotation, and the effects of heat-convection currents in the molten masses beneath the stony layer.

The formation of new rocks to the thickness of thousands of metres on a subsiding sea-bed, for example, imposes an enormous and increasing pressure on the formations beneath. At the same time the removal of this material from the neighbouring land causes a corresponding release of pressure there. To restore the balance it is necessary for the molten viscous masses below the outer crust to flow from areas of high pressure sideways to areas of low pressure. This helps to deepen the subsiding depression and at the same time to bulge up the crust in the neighbourhood, thereby maintaining or even increasing the rate of erosion on land, so that the depression continues to receive large amounts of sediment not only as masses of small particles of sand and mud, but also in the form of contiguous parcels and sheets of entire rock sequences gliding down the slopes of the "geo-tumor" into the bottom of the nearby marine depressions. This in turn increases the load and pressure there again. These processes then repeat themselves many times over. The movements are accelerated by heat-convection currents in the subcrustal masses because continuing gravitational imbalance upsets the normal heat exchange from the earth's interior through the crust to the atmosphere. In such disturbed areas the normal convection currents which effect the heat exchange will be deflected and accelerated. They work overtime, so to speak, and become an additional force in the dynamic confusion because they create increased friction on the underside of the outer crust.

Ultimately the process would either exhaust itself or move elsewhere, leaving behind a colossal scar in the crust, along the flanks of which the originally horizontal beds are now tilted and distorted, pushed on end, or even overturned, older rock sequences now resting on top of younger ones, and so on.

Here then we have a theory which makes the forces of gravity and convection alone responsible for the building of folded mountain ranges. Alas, this vivid theory too has its flaws. Apart from the underlying assumption of the permanence of the oceans and continents—the crust as a whole remains laterally immovable, only superficial sediment parcels are sliding off the "geo-tumor"—the theory has, for example, a symmetry problem. Both aspects of this gravity-and-convection theory, that is, both the downward "geo-suction" and the upward compensatory geo-tumor, are intrinsically symmetrical tectonic processes. Yet it is well known that all major folded mountain systems are, in cross section, patently asymmetrical. This, of course, indicates that tangential, horizontal, forces play a very important role. In other words, the theory is at variance with our observations or, at best, explains only part of what has really happened.

What then about horizontal forces caused, for instance, by the earth's rotation? We know that the average specific gravity of the rocks which make up the continents is rather less than that of rocks under the deep oceans. One can say the continents are lighter than the ocean floors, and this would tend to make them react differently to the centrifugal forces working on the earth's surface layers. It was the German geophysicist and polar explorer Alfred Wegener (*1920*) who first suggested the crust would break up and the continents go adrift—from equatorial regions towards the poles, for instance, or westward because of drag and tidal effects, or both. A few years later the Swiss, Rudolf Staub (*1927*), added another interesting idea to Wegener's hypothesis. He proposed that the movements of the continents would probably be cyclical. For instance, with their centrifugal drift towards the poles they would drag a part of the molten subcrustal masses with them, but these would eventually have to flow back towards the equator. This return flow would then also drag the continents at least part of the way back to the equatorial zone. After a time the whole dynamic process would repeat itself—until such time as the outer crust becomes too thick and stiff to allow further drifting; a time which has not arrived yet!

Here then we have a theory which proposes the utter non-permanence of the major features on earth, the oceans and continents. Moreover, if one accepts that continents can move at all, it is of course also likely that their movements may create folded mountain ranges. The theory of continental drifting thus envisages spectacular mountain-building as a result of collisions between two or more continents, whereby the larger continent usually overrides the opposing margin of the smaller one, such as Asia over India in the Himalayas, or Africa over Europe in the Alps.

It must be said, though, that for decades such revolutionary concepts were rather too much for the vast majority of the world's geologists. Except for the geology schools at the Swiss universities, and for a few courageous individualists like Professor A. L. Du Toit in South Africa and Professor S. W. Carey in Tasmania, the idea of drifting continents was treated with disdain and even ridicule as recently as fifteen years ago.

Nowadays, of course, few doubt that the crust is thoroughly broken up into moving fragments of various dimensions. The largest fragments—or "plates" as they are now called, although, as parts of the shell of the globe, they are really of convex spherical shape—tend to be built around the thick crust of a continent as their core. Thus one speaks of an African, a North American, Australian, Antarctic plate, but there are also many thinner crusted oceanic plates. Evidently, if these plates are movable—and there is no reason why they should not be—they will jostle each other along their margins. They may collide head-on with or slide and squeeze sideways past one another. One plate margin may mount the opposing one, or be sucked or "subducted" beneath it. All this becomes a rather entertaining game of jigsaw puzzles if, as is believed by most advocates of this new "plate tectonics" theory, the plates are rigid fragments of the crust, which cannot be folded or collapsed except along their margins—a rather dubious postulate. Mobility of crustal plates implies still another assumption: crust subducted, that is, disappearing downward—where it will eventually melt—away from the surface in one area, must be replaced by newly formed crust somewhere else. The latter phenomenon is believed to take place along circum-global zones which are volcanically and seismically very active and located in the depths of the oceans (Fig. 1). Along such oceanic "spreading centres" the adjacent plates appear to move apart to make room for the new basaltic lava crust which is forced up by convection currents from below. It is this process, rather than passive drifting due to the earth's rotation, which is now claimed to be the dynamic mechanism for the changes to and displacements of continents and oceans, and for the building of folded mountain ranges.

Plate tectonics has become a popular pastime in the 1970s, and the ideas of the pioneer drifters (*Wegener 1920, Argand 1924, Staub 1927, Du Toit 1937*) are now regarded as the respectable forerunners of modern concepts. These, of course, have the advantage of being based on new observations by methods and means, such as bathyscaphs, deep-sea submarines, deep-sea drilling, palaeomagnetism, highly refined rock age determinations, and the like, which the old pioneers could at best think of only in their wildest dreams. Yet even so there remain serious reservations (*Carey, 1977*) as to certain assumptions and implications of the current theory of plate tectonics. In the end no single theory of mountain-building can claim general and unreserved recognition. All our theories are, in fact, working hypotheses which have to be constantly adjusted to newly found facts of observation. The deeper problem is that we know not nearly enough about the forces acting from the interior on the earth's crust; we can only suspect, surmise and interpret, not observe and measure them.

Even the hardest rocks react to and fold under stress somewhat like plasticine, if the pressure is applied unceasingly over long periods of time such as we are dealing with in geological history. Nevertheless, we know and have observed that rocks can also simply break under strain, and that this results in a displacement, perhaps of tens of centimetres, perhaps of dozens of metres. The whole of the earth's

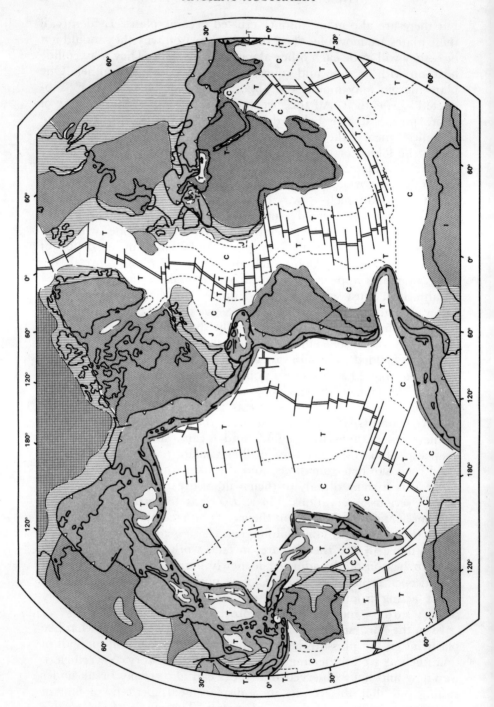

CONTINENTAL CRUST
(Age of Consolidation)

- MESO-CAINOZOIC
- PALAEOZOIC
- PRECAMBRIAN

OCEANIC CRUST
(Age of Formation)

- T TERTIARY
- C CRETACEOUS
- J JURASSIC
- NOT KNOWN

Overthrust Side / Subducted Side — COLLISION SUTURE BETWEEN CONTINENTAL MASSES

Overthrust Side / Subducted Side — COLLISION SUTURE BETWEEN CONTINENTAL AND OCEANIC CRUST

MID-OCEANIC RIDGES WITH TRANSVERSE SLIP FAULTS (Central zone where the spreading or widening of the oceans is said to originate)

Fig. 1
Plate-tectonic map of the earth's surface — a modern working-hypothesis (amended from Bally & Snelson, 1979). The more or less meridionally trending fat white lines mark — with one exception — the currently active oceanic spreading centres (zones). The exception lies between Australia and New Zealand, where the spreading was active only during Cretaceous times. Note that the oceans have grown drastically since their current expansion began late in the Triassic. The somewhat thinner white lines crossing the spreading zones represent trans-current (sideways) adjustment faults. Such transverse movements slice the oceanic crust because of local and regional stress and strain differences which interfere with the overall movement of the main crustal plates.

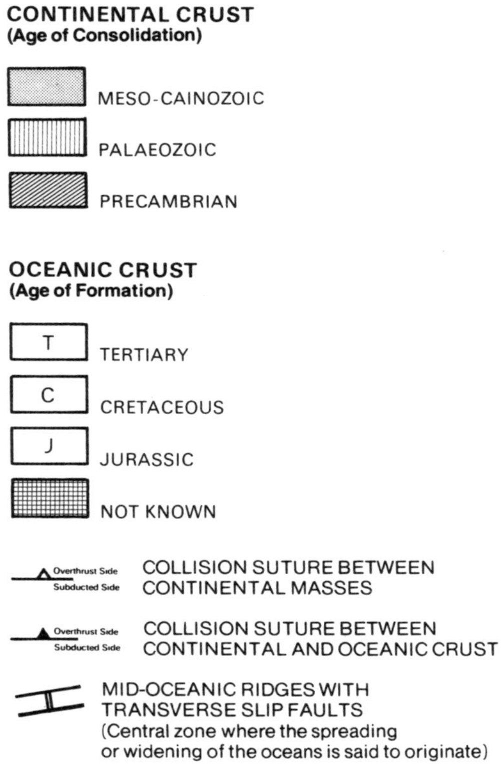

crust shivers slightly, and these shivers are recorded on seismographs throughout the world. Locally there is an earthquake—from the human point of view perhaps a major catastrophe, but merely a minor adjustment in the vast natural process which is forever going on. Thus earthquakes are the result rather than the cause of earth movements.

Mountain-building movements may go on for a vast period of time but the altitude of the mountains is limited by continuous erosion of the summits. Parallelling the high ranges on land are great troughs and depressions in adjacent seas. Into these is poured the debris from the destruction of the mountains, and thus the formation of new rocks keeps pace with the destruction of the old. It happens at the present time all around the Pacific Ocean—an area of earthquakes and volcanic eruptions with high young mountain ranges flanking deep seas. When ultimately some stability is reached and movements die down, and earthquakes become rare, there remains little more than the progressively slower wearing-down of the high land to base-level.

Australia is at present enjoying such a period of relative stability. For many millions of years there have been no more than minor adjustments of level caused by vertical earth movements. Yet the records of our geological history, which we read from the rocks, show that there were many great periods of mountain-building on the Australian continent, but the last of these ceased long ago, long before the rocks which were folded into the Alps and the Himalayas were formed.

Can evidence of such movements be seen? Railway and road cuttings, especially in eastern Australia, reveal a great deal. In some the rocks may lie undisturbed and horizontally bedded, but more often than not the exposed slates, sandstones and limestones are dipping in one direction or another, or stand on end, or are bent, folded or even double-folded. The beds may be broken by the outcrop of more or less vertical planes along which up or down displacements of a few metres took place long ago. One can also see that folded and fractured rocks are commonly traversed by veins of quartz which fill the fissures and cracks caused by the breaking of the beds under the strain. All these observations can, of course, also be made on coastal cliffs, on hillsides and mountain slopes, in quarries and in mines. A more systematic and detailed examination will link many of these beds over considerable areas, and show them to be the remnants of major folds, the upper parts of which have long since been removed by erosion. Yet all were once horizontal beds of sand and mud laid down beneath the waters of ancient seas.

Throughout the world there is evidence of an endless cycle of events. Seas turn into land, land turns into sea, mountains are born and grow and die. The records which tell the story are contained in

THE CHANGING FACE OF THE LAND

the rocks themselves, and in later chapters we endeavour to tell this story of our ancient continent in its proper sequence by showing what these records are, and how to read them.

CHANGES OF SEA-LEVEL

So far only changes in the level of the land have been considered, but there is evidence that the level of the sea is also subject to change. This is partly caused by the variation in the shape of the land. Elevation of a large area of land would restrict the area occupied by the sea and raise its general level. The accumulation of sediment on the floor of the sea would also raise its level and cause it to spill on to the low-lying land. Conversely the subsidence of the land below sea-level in one place or the subsidence of the sea-bed itself would increase the water-carrying capacity of the ocean basins and reduce sea-level throughout the world. In tracing the sequence of alternating sea and land in any one country it is not always possible to say whether the changes have been produced by local movements of the land or by changes in sea-level caused by cataclysms on the other side of the world.

Another important cause of a change in sea-level is the alternate melting and freezing of the polar ice-caps. In geological history there is evidence that there were periods of refrigeration when the climate was much colder than it is now. At these times ice accumulated in the polar regions as well as on mountain ranges in lower latitudes to an enormous extent, causing a diminution of the water in the sea. Such a period or, rather, such periods—for there were several interludes of warmer climate—occurred comparatively recently, and the last overlapped the advent of man upon the earth. It was only about 10 000 years ago that vast ice-sheets, many thousands of metres thick and covering almost the whole of Canada, the northern United States, much of northern and central Europe, and the mountains of central Asia, finally melted.

The water derived from the melting of the ice raised the sea-level throughout the world by about 90 metres, flooding river valleys, inundating coasts and submerging many areas of low-lying country. Port Jackson is such an inundated river valley, as are many other harbours and bays on the Australian coast. Tasmania became an island instead of a peninsula, many islands in the Pacific Ocean were separated from the mainland, and Torres Strait came into existence. A renewal of glacial conditions, with the re-formation of continental ice-caps—an event just as likely to occur in the future as it did in the past—would lead to a corresponding withdrawal of water from the sea and the consequent lowering of its level.

Apart from this comparatively recent glaciation, known as the Great Ice Age or Pleistocene Epoch, there have been several much more ancient ice ages, though their extent and duration are not so well known. Three are known in Australian geological history, and these will be discussed when the periods in which they occurred are considered.

SOMETHING ABOUT ROCKS

Rocks are the bricks of nature, the material which builds up the crust of the earth. They need not be hard; geologically, sand and mud are as much rock as is granite. Sedimentary rocks, such as sandstone, shale and limestone, have already been spoken of, but other types of rock will be mentioned in later chapters, and some knowledge of the common varieties is necessary to make the subject intelligible. For a more detailed knowledge, a text-book of geology or petrology must be consulted, petrology being the very specialized science of the origin and constitution of rocks.

Apart from sedimentary rocks, which are deposited as sediments in the sea, in lakes or rivers, or even on land, two classes of rocks will be frequently referred to, igneous and metamorphic. Igneous rocks are those which have consolidated by cooling from a molten state; metamorphic rocks are either sedimentary or igneous rocks which have subsequently become altered, sometimes by direct chemical action but more often by subjection to pressure and heat.

The most familiar igneous rock is lava, occurring as flows from volcanoes, both recent and extinct, and in all ages. There are many rock types included in the general term "lava"—dark coloured basalt, dark to grey andesite, light coloured trachyte, rhyolite and others. Of these basalt is perhaps the best known because of its wide use as road metal. Since lavas have generally cooled quickly they have had insufficient time to crystallize, and as a result nearly all are fine-grained, with a residue of natural glass between such small crystals as have developed. Some, such as obsidian and tachylite, are entirely glassy, and are most commonly found on the surface or edges of a lava flow. Lava rocks often show streaky wavy lines in the direction of flow, known as flow texture.

A second class of igneous rocks, though now often exposed in rock sections, failed originally to reach the surface, having cooled at varying depths within the earth. These filled the necks of ancient volcanoes, or forced their way through fissures or lines of weakness in other rocks as dykes, or between bedding planes as sills. These intrusions may be of considerable size or may be only about two centimetres or so across. Dyke rocks are generally coarser than lavas, the included crystals are larger, with clear outlines, and the texture of the matrix between them is microscopically crystalline and not glassy. Basalts may occur as dyke rocks; but the commonest type, particularly in the larger intrusions, is porphyry, of which there are many species. Porphyries were greatly prized by the ancients, not so much because of their rarity as because of the great labour involved in grinding them and their retention of a high polish. Excellent examples of intrusive porphyries are the tall rock pinnacles at the entrance to Port Stephens in New South Wales. Harder than the rocks into which they were originally forced, they have resisted erosion better, and now stand isolated, well above the level of the surrounding country.

Deep below the dykes and smaller intrusions are the plutonic rocks—intrusive masses, sometimes of vast extent, and derived from still greater reservoirs below. Plutonic masses are most commonly found forced upwards into folded mountain ranges, and are revealed on the surface only when the overlying rocks have been removed by erosion. Having cooled very slowly, they are coarse in texture, and consist of an interwoven mass of crystals of such minerals as quartz, various species of feldspar and mica, hornblende, augite, olivine, and many others. The most familiar plutonic rock is granite, but there are many others, including diorite—well known to miners—and syenite, gabbro and pyroxenite.

Metamorphic rocks are those which have been changed or metamorphosed from their original state. Under great pressure shale is changed to slate; it becomes harder and more lustrous, with the fissile grain or cleavage at right angles to the direction of pressure, and often cutting directly across the original laminae of which the shale was composed. Under intense heat and pressure shale and slate become further altered; they partially fuse and recrystallize while still retaining their fissile structure. The resulting, mostly fine-grained rock is a schist, containing such minerals as mica and garnet. Coarse-grained metamorphic rocks are known as gneisses. Sandstone, like shale, may be partly fused; or silica from heated solutions may be deposited between the grains to bind them into a hard homogeneous mass. This is metaquartzite. Limestone, and magnesian limestone or dolomite, may be completely recrystallized into true marble. (Marble, incidentally, is a term often loosely used, for, commercially, any limestone which will take a polish is so called.)

Sedimentary rocks when completely fused and recrystallized may behave and look like igneous rocks. Igneous rocks themselves are further altered by great heat and pressure. Gneisses, which are very common in the oldest known formations in the world, are often altered granites or other plutonic rocks which, among other things, have developed a foliated structure. When metamorphosis has been very strong, or repeated several times over, it may become virtually impossible to distinguish what was once sedimentary rock from what was igneous.

There is still one type of rock to be mentioned, which should perhaps be classified with the lavas, for it, too, results from volcanic action. This is tuff—the "u" pronounced as in "pudding"—which is composed of volcanic ash. The explosive action of volcanic eruptions produces great amounts of very fine crystalline dust, which is distributed over the neighbouring country. It may accumulate on the land—as did, for instance, the ash from Vesuvius which buried Pompeii and Herculaneum in historic times—or it may fly far out to sea and be redistributed there by currents, when it becomes a true sedimentary rock. When the material is coarse and composed of angular fragments of country rock exploded by the eruption, it forms a volcanic breccia or agglomerate. Such deposits often choke the vents of volcanoes and are finally consolidated into a hard mass. An excellent example of a volcanic breccia forms the breakwater at Port Macquarie in New South Wales, and the quarry from which it was taken, once the neck of an ancient volcano, lies just behind the town.

2 FOSSILS and the GEOLOGICAL RECORD

INTERWOVEN WITH the story of geographical change is that of the life of the past, of plants which once grew and animals which once roamed on the land, of birds, insects and reptiles which flew in the air, of fish and other organisms which swarmed in the sea. And so we come to fossils.

Literally, a fossil is "something dug up", an unearthed record of the past. In general, any record of past events is a "fossilium", and in that sense all rocks are fossiliferous. However, here we shall speak only of such fossils as are the remains or traces of a plant or animal which lived in previous periods of the earth's history. The term is not confined to the actual remains of the animal or plant, though these may sometimes be preserved. The mere cast or impression of an organism in a rock is a fossil, even a footprint impressed in what was once soft sand or mud.

The presence in rocks of objects which simulated the appearance of living things aroused the curiosity of people in the very early days of civilization. It was a long time before the real nature of these objects was understood, and longer still before it was realized what an important part they play in interpreting the past.

Ancient Greek scientists were a remarkable and learned body of men, but much of their time was taken up with mathematics and with philosophical conceptions of origins. Some, however, found time for actual observation of their natural surroundings, including fossils. Amongst the first was Xenophanes of Colophon, who was born about 614 B.C. He noticed shells on mountains in the middle of the land, impressions of leaves in the rocks of Paros, and recorded evidence of a former sea on Malta. These he attributed to great cataclysms of the ocean, during which men and cities were submerged. A little later, about 500 B.C., Xanthus of Sardis also found shells in Armenia, Phrygia, and Lydia, and concluded that these areas had formerly been the bed of the ocean, and that the land and the sea were constantly being interchanged. Herodotus, born 484 B.C., recorded fossil shells from the mountains of Egypt, and he considered that Lower Egypt had at one time been covered by the sea, and had eventually been filled in by the growing delta of the Nile.

Strabo, another Greek, who was born about 63 B.C., is often cited as the father of modern geology. He not only noted fossils, but taught that elevation and subsidence were often on a scale sufficient to affect whole continents, and that there was a continual cycle of change. A Roman, Seneca, was a clever observer, and his writings on the destructive effects of running water have quite a modern flavour.

These are only a few of the many scientists and writers of ancient times who went to nature for facts on which to base their theories, and there is no doubt that their keen and logical reasoning would have led to greater discoveries if it had not been for the downfall of the Roman Empire. For many centuries there was a general collapse of scientific inquiry, and geology, with other sciences, lapsed into a state of coma. Theology seemed to be the one controversial subject, and wars were waged over differences of opinion on some obscure point of dogma. That the whole of the work of the Greek and Roman scientists was not for ever lost is due largely to the Arabs, who had a profound respect for the classics. They did little new work themselves, but did their best to preserve the ancient writings and translated many of them into Arabic.

It was the Moorish invasion of Europe that ushered in the Renaissance and led to the revival of scientific research. Fossils again came under notice, but in the early part of the Renaissance the fine work of the classical writers was ignored. From the fifteenth century onwards there was a great deal of controversy as to the nature of these strange formations in the rocks. Avicenna, an Arabian commentator on Aristotle, suggested that they were unsuccessful attempts by nature to produce organics from the inorganics, in which the form had been achieved but no life bestowed. Amongst the few real thinkers on the subject was Leonardo da Vinci, whose views were remarkably clear and sound. He said that fossils, which he found when excavating canals, were the shells of sea-snails and mussels, and had been buried in the mud at the bottom of ancient seas which afterwards became dry land.

When ecclesiastical circles could no longer deny the organic nature of fossils, the diluvial theory had their enthusiastic support. This stated that fossils were truly the remains of animals and plants, but had all been left in their present positions after the subsidence of the biblical flood. Other theories were more fantastic; fossils were illusions of nature, or mere imitations of real forms produced in the rocks by some malign influence.

The theories of the unfortunate Johannes Bartholomew Beringer, a professor at the university of Würzburg, were typical of his times. In 1726 he published a palaeontological work, *Lithographica Wurceburgensis*, in which he illustrated not only a number of true fossils but many fantastic forms, suns, moons, stars and even Hebraic letters. Actually these had been buried as a joke by his students, and the attention of the professor called to the spots where he himself dug them up. The hoax was revealed when he found his own name so buried. After trying in vain to buy up and destroy his own work, he died, it is said, of a broken heart.

In spite of such incidents, the sixteenth, seventeenth and eighteenth centuries saw a great progress in learning, and it was then that the foundations of the science of geology were laid. Step by step facts were discovered and, though many of the hypotheses seem absurd by modern standards, the truth was slowly beginning to emerge. Many great scientists appeared during this period, in England, in France and particularly in Germany; their discoveries revolutionized the science and put it on a secure basis. Among them is an Englishman, with the plebeian name of William Smith.

William Smith (1769–1839) was an engineer who observed and collected fossils when carrying out excavations in the course of his work. Most of his collections were made from what are now known as the middle periods or Mesozoic Era of geological history. The first important fact that he noted was that in any series of strata imposed one above the other, each bed was characterized by particular species of fossils found at no other level. Moreover, in different parts of the country the order was always the same, and when fossils were discovered in a new locality it could be anticipated what would be found in the beds both above and below. From this was enunciated the doctrine that the strata of the earth belong to many periods, and that these periods may be recognized by the nature of the plants and animals which lived at the time of the deposition of the strata. Smith worked out the order of succession of many of the English formations, and published geological maps which were remarkably accurate and were for long accepted as standard.

Once the clue was found it was not long before the work was extended. Between 1794 and 1831 Smith had dictated a table of British strata from the Permian to the Cretaceous Period. Rocks containing similar fossils were known on the European continent and in the same order of succession. Other formations were studied, then other continents—Africa, America, Asia and Australia. Gradually, results of investigations into the sequence of events in different parts of the world were correlated, and the chronological table known as the geological record was built up. Much has been done, but the story is still far from complete.

Details of the geological record are not simple, but the broad principles can now be outlined. It is generally recognized that the known geological history of the world covers an enormous period of time, and that this may be subdivided into a number of natural eras, periods, or ages characterized by fauna and flora. Though a series of rocks, said to be of a certain age in, say, both England and Australia may not be exactly contemporaneous—in fact, the time of their beginnings and ends may be some millions of years apart—the duration of the main periods is so long as to make such differences negligible. Both the English and the Australian formations are then said to be of Silurian or Permain or Triassic or some other geological period.

There are great differences in the details. Owing to climatic and other conditions there has never been uniformity in the distribution of life throughout the world. Compare, for instance, the land faunas of Australia, Africa and America at the present day. Australia has its marsupials, Africa has its large carnivores, America has its tapirs and other peculiar forms. The same applies to the sea, though perhaps the difference is not so marked. In Australia there is much difference between the marine life of the colder waters in the south and the tropical corals, fish, shells, bêche-de-mer and other denizens of equatorial seas. Such conditions as apply today have also prevailed in the past, although in the very early periods there were fewer species and many of them may have had a greater geographical range. For instance, many species of fossils found in earlier Australian formations are either the same as or closely allied to species found in rocks in Europe and America.

Not only species but whole orders of extinct animals and plants have appeared and disappeared universally in the same succession. For instance, there is a curious order of marine Crustacea called trilobites, which were animals belonging to the same class as crabs, lobsters and prawns. These lived through several of the early main periods. They are divided into many families, which appeared one by one and then became extinct. Yet the sequence of the rise and fall of each of these families is everywhere the same, and, though species may differ, on the whole they alone enable formations in different parts of the world to be assigned to the same period. There are many other examples and some of these will be dealt with in more detail in later chapters, in which an attempt will be made to outline the past geographies of Australia.

In the meantime it is enough to give the general succession of the main periods into which the stratal record of the world is divided. They are simply natural and practical subdivisions indicated by the record; their duration varies greatly, and the positioning, in a series of strata, of the boundary between successive periods may often be impossible and remain a matter of opinion or convenience. Each country has in detail its own geological record and may choose to name subdivisions as it sees fit. However, apart from minor details, geologists the world over have long agreed to a broad understanding of the chronology of eras and periods as is outlined in Table (1).

MAJOR GEOCHRONOLOGICAL SUBDIVISIONS	Elapsed Time
CAINOZOIC ERA (65 M.Y.)*	
QUATERNARY PERIOD (2 M.Y.)	
Recent Epoch (10,000 Y.)	
Pleistocene Epoch (1,990,000 Y.)	2 M.Y. **
TERTIARY PERIOD (63 M.Y.)	
Pliocene Epoch (3 M.Y.)	
Neogene Subperiod (20 M.Y.)	
Miocene Epoch (17 M.Y.)	22 M.Y.
Oligocene Epoch (15 M.Y.)	
Eocene Epoch (17 M.Y.)	
Palaeogene Subperiod (43 M.Y.)	
Palaeocene Epoch (11 M.Y.)	65 M.Y.
MESOZOIC ERA (165 M.Y.)	
CRETACEOUS PERIOD (71 M.Y.)	
Neocretaceous Subperiod (35 M.Y.)	100 M.Y.
Palaeocretaceous Subperiod (36 M.Y.)	136 M.Y.
JURASSIC PERIOD (55 M.Y.)	191 M.Y.
TRIASSIC PERIOD (39 M.Y.)	230 M.Y.
PALAEOZOIC ERA (340 M.Y.)	
PERMIAN PERIOD (50 M.Y.)	280 M.Y.
CARBONIFEROUS PERIOD (74 M.Y.)	354 M.Y.
DEVONIAN PERIOD (40 M.Y.)	394 M.Y.
SILURIAN PERIOD (24 M.Y.)	418 M.Y.
ORDOVICIAN PERIOD (72 M.Y.) ***	490 M.Y.
CAMBRIAN PERIOD (80 M.Y.)	570 M.Y.
PROTEROZOIC EON (1830 plus/minus 100 M.Y.)	
ADELAIDEAN ERA (630 M.Y.) ****	1200 M.Y.
CARPENTARIAN ERA (600 M.Y.) ****	1800 M.Y.
NULLAGINIAN ERA (700 M.Y.) ****	2500 M.Y.
ARCHAEOZOIC EON (More than 2000 M.Y.)	
YILGARNIAN ERA (500 M.Y.) *****	3000 M.Y.
PILBARAN ERA *****	

    * Duration in Million Years
   ** Time elapsed since the beginning of the subdivision following above the line
  *** Includes the Tremadocian Epoch (**Gale et al., 1979**)
 **** Subdivision names used in Australia
***** New name introduced for the convenience of the readers

THE AGE OF ROCKS

A question often asked is: what is the length of these periods and can it be expressed in years? The use of the phrase "millions of years" might at one time have been somewhat figurative, but the fact is that we must indeed talk in millions.

With the compilation of the geological record geologists realized that a great period of time was involved, and were at first content to leave it at that. The first approximations in terms of years were made by estimating the rate of deposition of sediment in deltas and elsewhere and applying the estimated rate to the thickness of known strata, at the same time making a very uncertain allowance for the unconformities and intervals between the layers. In this way estimates as far apart as 150 million and 1000 million years were made for the interval which has elapsed from Cambrian times to the present day. It was realized that the results of these calculations, though interesting, were too vague an approximation and could be neither proved nor disproved.

The discovery of radium and other radio-active materials has thrown new light on the subject, and the great advance in the science of physics has disclosed a method which gives results that are satisfactory in many, though not all, cases. This method depends on the instability of radio-active elements such as uranium. They are constituents of certain minerals found in some—especially igneous—rocks belonging to various periods throughout the world. From the time of its crystallization within the cooling molten rock the radio-active mineral begins to decay and break up, an atom at a time, into different elements—uranium, for instance, into the noble gas helium and the metal lead. This goes on at a definite and known rate, so that by carefully analysing and determining the proportions of old and new materials in the minerals their age may be mathematically calculated. There are various pitfalls in this method but from all over the world much reliable data have by now been compiled, some of which are incorporated in the table of periods in the previous section.

HOW FOSSILS ARE PRESERVED

Though fossils are of the utmost importance in revealing the story of the past, our knowledge of them will always remain fragmentary. The chances of the remains of any individual organism being preserved from one age to another are very small. Not only individuals, but whole races, species and orders may have disappeared without leaving a trace. This applies particularly to soft-bodied organisms such as worms, sea-anemones, jelly-fish, slugs, and a host of others. Only animals which have a bony, calcareous skeleton, or secrete a calcareous or chitinous outside shell, are likely to leave some trace. But the great majority of these, too, disappear. The soft parts soon decay and the hard parts are eventually destroyed. It needs the odd chance and most favourable conditions for them to be preserved as fossils.

It is fortunate that this chance has on many occasions come off, and that even such insubstantial creatures as jelly-fish have been preserved in rocks of tremendous antiquity. Such finds were made in the Proterozoic of South Australia as well as elsewhere. A great boon was the discovery of a host of fossils in the Cambrian beds of British Columbia, including many soft-bodied creatures so beautifully preserved that even many anatomical details are visible.

Sea life has the best chance of being preserved, and the majority of known fossils are of marine origin. It was shown in the previous chapter how, simultaneously with the wear and tear of the land, there is the unceasing accumulation of sand and mud on the bed of the sea. The amount of life in the sea is prodigious, far greater than in corresponding areas of the land, and there is abundant evidence that this has always been so. As the creatures living either on the sea-bed or in the sea itself die, generation by generation, their soft parts decay, but their bones and shells fall to the bottom to add to the bulk of sediment in which they become buried.

Unless buried quickly, much of this material disappears. Carbonate of lime, of which bones and shells are largely composed, is slowly soluble in sea-water, and so is returned to the source whence it came. Even when marine organisms are buried, particularly in porous sediments such as coarse sand, corrosion may continue until nothing remains. Mud is a better preserving medium, or mud mixed with fine sand. During volcanic eruptions, showers of fine volcanic ash may fall into the sea and destroy much of the sea life. This is immediately buried, in material which quickly sets in a hard, compact mass. In this way fossils may be perfectly and indefinitely preserved.

In areas of the sea where no foreign sediment is deposited, life may be so prolific that its remains accumulate without any mixture of other material. In this way beds of organic limestone are formed, sometimes of great thickness. Ultimately the limestone may be composed entirely of fossils, or it may be partially re-dissolved and compacted with no trace of its organic origin.

Thus is preserved a small portion of the life of the sea, but the remains so preserved are still subject to change and possible destruction. If earth movements lead to the elevation of the new rocks into land or their folding into mountain ranges, these remains are not only subjected to enormous stresses and strains but are often shattered. Metamorphism, spoken of in the last chapter, may be on a vast scale, and it may effectively destroy all or most traces of organisms in the rock. Even when it occurs only on a small scale, the few fossils which remain are so distorted and shattered as to be unrecognizable.

Rocks belonging to the earlier formations, most of which have been subjected many times to the stresses of earth movements, have suffered more from metamorphism than later rocks. Thus, as we go back in geological time, fossils are apt to become more and more rare. However, even in the very oldest rocks known, traces of organisms such as bacteria, or algae, do occur.

So far marine fossils alone have been considered. What of the plants and animals which once lived upon the land? Their chance of preservation is actually much smaller, and is only possible under an exceptional combination of conditions. In flood time, tree-trunks, leaves and the fronds of ferns are washed into rivers, and thence into lakes or even to the sea, where they may sink to the bottom and be buried and preserved. The bones of drowned animals, even insects, are sometimes preserved in this way; the skeletons are generally disintegrated and scattered, but the finding of an odd bone—a tooth, or particularly a skull—has thrown some light on the life of its period.

Marshes sometimes act as traps for land animals, which become bogged, die, and are entombed. The muddy bottoms of water-holes in drought time are frequently the depository of great numbers of dying animals. Such areas may subsequently be submerged, new strata of sand and mud deposited above them, and their fossil contents come to light in later ages. Lakes of natural pitch have proved gigantic animal traps, and have yielded much information about the life of the past. Even volcanic eruptions which have filled the valleys with lava have covered river-beds and the soil, preserving leaves, roots, wood, and the bones of animals. Such deep leads, as they are termed in mining parlance, have yielded fossils in Australia, as has the alluvium which covers much of the interior plains. In caves or rock shelters, to which carnivorous animals have taken the carcasses of their victims, or where animals have crawled to die, or into which flood waters have carried the bodies of drowned animals, the remains are often buried beneath layers of stalagmite and cave earth and thus preserved.

The frozen tundras of Siberia, which are of more recent origin, have yielded not only the bones, but the flesh, skin and hair of many animals, which became bogged in the half-frozen swamps in the latter part of the Great Ice Age. Insects may be trapped in the resin of pine trees, and exquisitely preserved specimens have been found in amber, which is itself a fossil resin.

THE COLLECTION OF FOSSILS

Fossils may be found in all sedimentary rocks, but not all sedimentary rocks contain fossils. Even when they occur in a formation, they may be confined to tiny patches in a large area, or to one or two layers or horizons in a great thickness of strata. A trained geologist may consider the appearance of a formation suggestive, and search systematically until success crowns his efforts, but most of the striking discoveries have been found by chance. The curiosity of a well-digger has been aroused by strange markings in the rocks he uncovers, the interest of a quarry man by something which looks like the imprint of a fish or a bone or the wing of an insect—and science has been enriched by a new discovery. The wonderful series of fossil insects from the brick pit at Brookvale near Sydney were collected, one by one over a period, by an interested quarry worker.

It is within the reach of everyone to make such finds, as it is within the reach of every prospector to find a gold nugget. The odds are, however, rather better for the fossil collector, for fossils are a good deal commoner than nuggets, and the competition is not so keen. It is curious how finds are made, as one or two personal experiences will show.

While in Cooma—in southern New South Wales—searching for building stones on behalf of the Technological Museum, Sydney, I* was shown a piece of slate by a monumental mason who thought it might be useful for roofing. It was not, but one corner showed the barely perceptible trace of a fossil known as a graptolite, a member of a group characteristic of the Ordovician Period. As rocks from this period had not been recorded from this part of the State, I was naturally very interested. The locality where the specimen was found was about 18 kilometres from the town, and, after some search, a very fine deposit of these fossils was discovered and a good collection made, which is now in the Australian Museum, Sydney. This find was of considerable importance; it led to the modification of the local geological map, and added considerably to the area of Ordovician rocks in the country. A similar incident happened a little later at Cobargo, New South Wales, when examination of a piece of slate revealed the presence of graptolites and showed that Ordovician rocks extended to this area also.

On another occasion, on the railway line near Tarago, also in southern New South Wales, I picked up a piece of limestone which contained a broken fragment of a bivalve shell in a beautiful state of preservation. Discovering that there was a deposit of limestone near Lake Bathurst railway station, some seven kilometres away, I made a

*Laseron

cursory examination of this and found only a few fossil corals, not especially well preserved; it took much searching before a stratum similar to the original piece was found. It was not more than a few centimetres thick, and only on the weathered surface were the fossils extractable, but the stratum was crowded with small treasures, some quite new to science, and all throwing more light on the life of this remote period.

In the National Museum, Melbourne, are fossil shells which were found when the foundations for a new wing were being dug. The famous Jurassic fish beds near Gulgong, in New South Wales, were also discovered purely by chance. A specimen picked up by a local farmer was sent to the Mines Department, and the ensuing systematic search by officials of the Department yielded hundreds of specimens of fish and plants in a wonderful state of preservation. This find is world-famous and has added much to the knowledge of the anatomy of Mesozoic fish.

The best advice to the would-be collector of fossils is to look everywhere at every opportunity, and never to be disappointed at lack of results. He will enjoy his cross-country walks even more if he keeps an open eye on the rocks on either side. Every road or railway cutting, every quarry is a potential field of discovery; every rocky creek-bed, the cliffs by the seashore, outcrops on a hillside, the spoil thrown from the digging of a well may reveal the life of bygone ages. There will be many false alarms; the freak shape of a pebble, a stain of iron oxide, may deceive for a moment, but ultimately there must come the reward and the thrill of real discovery. And it is a thrill, comparable in a way with Balboa's as he gazed for the first time on the Pacific. Every find of a new fossil bed is exploration into a remote past, into bizarre landscapes populated with strange animals and plants, on which the finder is the first human being to set eye.

Of course, it is easier to collect in the known fields, to join excursions of geology students to localities which have been well and truly exploited and studied. This method of collecting may seem to savour of travel to a tourist resort, where there are guides to show the sights, but it is good fun, and the chances of making a discovery, though small, do exist; for the centre of every unbroken piece of stone is a mystery, to be solved only by the geological hammer.

It is well to give some hints on how fossils should be collected. Much depends on the type of fossil and the rock in which it is buried. The scientific value of a fossil increases with improved preservation and completeness. It is true that many organisms will never be found in an even moderately complete state, so that even a fragment may be of considerable importance, giving a clue to the geology of a district; but where it is possible to choose, it is better to spend a maximum of time in seeking the spot where the preservation is best. In nearly every locality where fossils occur there will be one bed or one spot where preservation is better than elsewhere, or where the fossils are more readily separated from the enclosing rock.

If the rock be shale and the fossils plants, it is of little use seeking in the weathered portions for good specimens. The search may be long, but invariably one or more laminae will be found in which the best specimens occur. Once this place has been located it may be necessary to dig, and sometimes deeply, to secure unweathered blocks that are as large as possible. These are then split along what has already been noted as the most promising layer. In shales of marine origin, when the fossils are shells, single corals, or other organisms with the original material preserved, the presence of lime in the matrix often makes it very hard and compact and the extraction of fossils is difficult. The weathered portion is then generally the best source, and, where conditions are very favourable, specimens drop out entirely free from the matrix as kernels do from a nut.

Most sandstones are porous and not a good preserving medium, though there are exceptions. Among the best are those in which there is a percentage of volcanic ash. They are compact and impervious, but apt to be rather hard, which makes extraction of the fossils difficult. Much depends then on skill with the hammer and chisel, and also on the fortunate chance fracture.

Do not discard good specimens because they are broken, as many will be. Fragments carefully cemented together do not lose their value as does a stamp which has been repaired, and they are much more desirable than a poorly preserved specimen which is entire.

Sandstones containing well-preserved fossils are to be found at Gerringong on the south coast of New South Wales, at Allandale in the Maitland district, and in numerous other localities.

Limestones are perhaps the rocks that yield most fossils; in fact, many limestones are composed wholly of the remains of marine organisms. Unfortunately the massive limestones of the older formations in Australia are often too compact to permit easy extraction of the fossils, and reliance must be placed on chance weathering. Many fossils in limestone, particularly corals, have been subject to chemical change, and their original carbonate of lime has been partially replaced by insoluble silica. As the limestone is slowly dissolved by weathering, the fossils are unaffected and stand out in relief. Specimens of great beauty showing every detail are, as a result, often obtained, and the only labour involved in collecting them is that of walking over limestone outcrop and searching among the debris.

Though broken surfaces of limestone seldom reveal fossils, polishing discloses much detail, and is particularly valuable in studying the structure of organisms such as corals and sponges. The great wealth of fossils in limestone may be realized by doing a little "window shopping". In all our cities are many buildings with vestibules and stairways lined with polished slabs of Australian "marble" from Buchan, Borenore, Rockley, Spring Hill, Fernbrook, or wherever. In any such place, the tedium of waiting for an appointment may well be relieved by tracing the outlines of many strange creatures of the past, cross sections of shells, corals, sponges, and particularly the arms and stems of the curious sea animals known as crinoids or sea-lilies.

There are some fossils which require expert treatment, during both extraction and subsequent handling and preservation. These are mainly the bones and skeletons of larger extinct animals, the great lizards of the Middle Periods, and the early birds and mammals of Tertiary times. Very few of the former have so far been found in Australia, but mammals of the Tertiary Period occur in a number of strata. Should an amateur come upon such a find, the event would be of such importance that any of the museums and universities of Australia would at once respond, and no doubt a properly equipped team of experts would be very soon on the spot.

Finally, a word about labelling. It is very important that the exact locality where each specimen is found should be noted. Merely to state the district or town is not enough. Where strata are tilted, the difference of a kilometre or less in location may mean a vertical difference of thousands of metres and millions of years in time. Even more important than the exact geographical locality of the find is the exact level in the rock series from which each specimen is collected. This is particularly so when one collects from a cliff face cut into more or less horizontally disposed strata. It is precisely because of imperfect labelling and the mixing of specimens from different levels or horizons that there is still some confusion as to the correct sequence of certain strata in Australia.

Do not neglect to pack your specimens well. Each should be wrapped in one or more thicknesses of newspaper, and packed tightly so that it cannot move. Fragile specimens should be wrapped in cotton wool and preferably placed in small individual boxes.

II
BIRTH AND GROWTH OF AUSTRALIA IN THE PRECAMBRIAN EONS

3 The PRECAMBRIAN EONS

BEHIND THE veil cast by almost unfathomable eons of time there looms the shadowy outline of a stupendous picture—the birth of continents and therewith of Australia. Of the titanic events which led to this there is no knowledge; we can only imagine the crustal convulsions which created the first lands and oceans on our planet, and it is doubtful whether even in the very oldest visible parts of the earth's crust any remnants of the very first lands are preserved. The extremely ancient gneiss and granite bastions, which we call shields and interpret as remnants of the primeval nuclei around which the continents grew, may contain evidence of the earliest geological features on earth, but we can probably never be certain of that.

Reading the early chapters of our planet's geological history is like trying to make sense of a book the first three-quarters of which has been not so much destroyed as rendered illegible by the ravages of time. The pages have stuck together, and the printing has faded or become so blurred as to be indecipherable. The odd sentence that has survived has great value, but this can be overestimated, and the ideas formulated about the context can be very varied. If we start to read the book in its last quarter, the contents of these later chapters may enable some deductions to be made about what has gone before, but all the details can never be known, and even the broad concept may be greatly distorted. So it is with the earth's history.

During the nineteenth century, when the geological record was compiled back to the Cambrian Period, geologists were satisfied that they were approaching the beginnings of the earth's geological history. The occurrence of extremely old, or "Archaean", sediments below the Cambrian formations was known, but the magnitude and complex nature of these more ancient systems were not at first realized. It was anticipated, though, that they would be found to contain the traces of the first, or "Archaeozoic", living things.

Such very old Precambrian rocks have since been located in many parts of the world. It is also known that in many places they are overlain by formations much younger than Cambrian, and must therefore have remained exposed as land surfaces for long ages before they were again submerged beneath the sea. It is also obvious that vast areas of such ancient rocks are still hidden, buried under accumulations of younger sediments, and will so remain until mountain-building events of the future expose them on the surface.

These days it is no longer necessary for such ancient formations to be overlain by fossiliferous Cambrian rocks for us to prove their Precambrian age. Geologists have learnt to make use of a fossilized radiation-physical condition in certain minerals to gain that information, which can be directly expressed in years. This is a kind of information which organic fossils cannot give us. On the other hand, the technique of radiometry will never provide such precise evidence as organic fossils do for correlating formations, even individual and perhaps very thin beds, in widely separated parts of the world.

However, Precambrian rocks contain few organic fossils, certainly far fewer than Cambrian and later formations, and the few there are happen to be of a type that, though moderately useful, is not very efficient for the purpose of exact correlation—to correlate, in geology, meaning to show that a rock or a sequence of rocks in one place has the same age (was formed at the same time) as a possibly quite different sequence somewhere else, even on the other side of the world. These days, the degree of accuracy obtained by radiometry is infrequently as good as, or better than, that from such organic fossils in the Precambrian. Provided one keeps in mind that the radiometric technique is based on a number of unproven, though reasonable, assumptions and therefore liable to change, its services to geology can be of immense value. Even in the current early stage of application its successes in unravelling the secrets of Precambrian sequences have been quite spectacular, for they have already helped geologists to avoid correlating, as they did until a few tens of years ago, formations which, in fact, were formed hundreds of millions of years apart.

The interval covering the whole of Precambrian times is immense; in fact, well over six times longer than the whole of the time elapsed since the beginning of the Cambrian. Modern geological nomenclature calls such extremely long intervals of time Eons (the then formed rocks are eonothems), and in this book we subdivide the whole of the Precambrian into two Eons, the Archaeozoic and the Proterozoic. We are part of and are living in the Phanerozoic Eon since the beginning of the Cambrian Period—our Eon having reached the age of about 570 million years, still only a third or less of the two preceding Eons. But who knows what the future holds?

From data now available it is possible to loosely link the succession of geographical changes which took place in different places in the world during these vast eons, and one can thus attempt to reconstruct approximately the types and extent of the events which affected the continental nuclei after their unknown earliest history. We can see widespread mountain-building, the appearance and disappearance of oceans and seaways or, vice versa, the emergence and submergence of land areas, and regions of intensive volcanic activity with large lava flows and ash rains spreading over vast tracts of the surface of the earth. Most importantly, it has become possible to give a fair estimate of the time intervals in which such events took place, and therefore roughly to correlate them all over the world. In other words, we can put them on a map with a lot more confidence than a few decades ago.

The greatest expanse of Precambrian rocks is in Canada, of which they cover practically the whole of the eastern half, but large and small exposures of these ancient formations occur elsewhere (see Fig. 2) and on all the continents. Everywhere one has come to recognize the two major subdivisions mentioned above, the Archaeozoic (beginning of life) and the Proterozoic (early life) Eonothem, that is, the rocks and rock sequences which were formed in the respective Eons (of time).

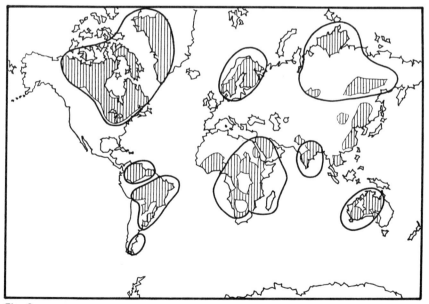

Fig. 2
The major exposure areas (shields) of Precambrian rocks. Almost all of them have also yielded organic fossils. After Murray (1965).

In recent years these two grand subdivisions of the earth's history have found a great deal more attention than before, and in various regions the complex systems of rocks, which represent them, have been very carefully studied. The results of these studies have changed previously held concepts quite considerably. For example, it has become evident that the Proterozoic Eon can, on the basis of certain widespread contemporaneous geological events, be subdivided further into three main eras, each of which is about as long as the whole interval of time since the beginning of the Cambrian. Moreover, simple organic fossils such as algae, fungi, and bacteria, and organisms consisting of a symbiotic (living-together) combination of two or all three of them, are now known to occur in rocks which are well over 2000 million years old. Thus more and more light is being shed upon these remote dark ages, during which the main foundations were laid of all that we see on the face of the earth—ages which hold the keys to many secrets shrouding our understanding of our planet's history. Let us now see what happened in these vast eons of Precambrian time.

THE MYSTERIOUS ARCHAEOZOIC EON

The ideas about what is, and what is not, Archaeozoic have changed considerably in the past century. Once everything older than Cambrian was simply called Archaean. Thereafter the term was applied only to strongly metamorphosed rocks underlying non-metamorphic, though Precambrian, formations. However, even if it was suspected that many of the metamorphosed rocks in one area could well be of the same age or even younger than non-metamorphic Precambrian rocks in another, there was no way of proving the point. Then came the development of the radiometric techniques—that is, the measuring of the time elapsed since the formation of certain minerals from the radioactivity characteristics of their component elements. This was a mighty step towards proving what had long been suspected. At present we still stand in the early phases of the large-scale application of these techniques, but there is already a large amount of new information.

What, then, is meant nowadays by the term Archaeozoic or Archaean? The first answer that one is tempted to give is still: "The Archaeozoic is so early in the earth's history that its rocks and records are all and everywhere found strongly metamorphosed and thus rendered unreadable." Unfortunately, this is no longer a good and practical definition, simply because we might perhaps never know with reasonable accuracy what, where, and how old—the oldest non-metamorphosed Precambrian rocks are. One need not be overly surprised to hear that geologists find it difficult to agree worldwide when Archaeozoic times ended, and the Proterozoic Eon began. Thus, as far as we know at present, the oldest still non-metamorphosed beds of Australia in the Pilbara were laid down 2600–2400 million years ago (*Plumb, 1979*). Yet in southern Africa's Swaziland we know of unaltered dolomitic beds which are older than 3000 million years. For us it is the much younger dateline which is meaningful; for geologists in southern Africa it is the older one, and theirs appears meaningless to us—at least as yet.

There are many geologists who consider the disagreement to be merely a matter of semantics. One might as well choose that kind of dateline simply for reasons of convenience—as is done in this book—because nature itself has apparently decided to be utterly unconcerned about our worries. Therefore, because in Australia the period between 2600 and 2400 million years ago seems to be geologically significant, I* will terminate the Archaeozoic Eon at a neat and round figure of 2500 million years before Present. The small matter of give-or-take a hundred million years need not bother anyone much, provided it is realized that the order of magnitude of misinterpretations geologists may make when trying to unravel secrets of such immensely long past ages, remains very large indeed.

On the Australian continent the oldest of what we have just decided shall be Archaeozoic rocks are over 3000 million years old. As would be expected, they have been intensively folded and metamorphosed not long before the end of the Pilbaran Era, and consist of tough gneisses and schists occurring widely in the Pilbara region of Western Australia as well as in a northward narrowing belt from the Albany-Esperance coastal area to the upper reaches of the Murchison River, that is, in a zone along the western margin of that large and ancient land bastion known as the Yilgarn, or West-Australian Shield.

The well-worn Yilgarn plateau stretches from the Stirling Range in the south to the Ashburton River in the north, from the Fraser Range in the east to the coastal plains in the west (*See Geological Map 25*). Most of its formations have been laid down in the earlier part of what we here call Yilgarnian times. Shortly thereafter, around 2700 million years ago, they were folded, metamorphosed and consolidated, and welded to the older, Pilbaran, zones in the west and north. Thus at least for a time towards the end of the Archaeozoic the two most ancient and obvious building blocks, or structural nuclei, of Australia, the Pilbara and the Yilgarn, probably formed a loosely connected but contiguous shield. Contiguity is indicated by a number of smaller exposures of Archaean rocks in the region between the Ashburton and Fortescue rivers, because these occurrences even now almost bridge

*Brunnschweiler

the gap between the two main blocks. The combined Yilgarn-Pilbara Shield—one of the largest truly Archaean bastions in the world—constitutes the principal structural nucleus of our continent.

No doubt within that ancient land there were once lofty mountain ranges, bare and stark, reaching into the clouds and covered with snow and ice, but they have long since been worn to their roots. Moreover, southern and central parts of this shield have scarcely at any time since been submerged beneath the sea, and belong therefore to the most ancient land surfaces in the world, having remained above sea-level almost continuously for 2500 million years or more.

These Australian areas are not the only Archaean lands on our globe, and the still popular notion that Australia is the oldest of the continents is therefore not true. Eastern Canada has already been mentioned as another, and so are parts of other ancient shields such as those in Africa, Scandinavia, Siberia, Brazil, and the Antarctic. In other words, all continents have their origins way back in the Archaeozoic Eon.

Are there other regions in Australia where Archaeozoic formations appear?

Considering the many and large areas of ancient and intensely metamorphosed rocks, it had, until a few decades ago, been quite reasonable to assume that many of them were as old as those of the Yilgarn-Pilbara Shield. Thus the high-grade metamorphic formations in the Kimberley district, the Katherine-Darwin region, the Tanami-Alice Springs area, the Harts Range, the Mann-Musgrave Ranges, Eyre and Yorke Peninsulas, Mount Lofty-Flinders Ranges, around Broken Hill, Mount Isa, and from there northward into Cape York Peninsula were commonly classified as Archaean which, at the time, meant at the very least 2000 million years old. But radiometry is a proper spoil-sport. These all too simple concepts have had their day.

As far as one has been able to ascertain to date, rocks of greater age than 2500 million years occur only in two of all these areas. One is in the north of the Northern Territory and comprises the Rum Jungle gneisses of uranium fame and those of the Nanambu Complex about 190 kilometres to the east. The other area includes Eyre Peninsula in South Australia and is known as the Gawler Block or Craton. Here granites, which have an age of about 2400 million years, have been intruded into an older series of rocks which is known as "Flinders Series". They were probably laid down in the latter part of the Archaeozoic, that is, during the Yilgarnian. The same applies to the occurrences in the Northern Territory (*Plumb, 1979*). Thus we do have some other structural nuclei by the beginning of Proterozoic times, but they were nowhere near as prominent and permanent as the West Australian one. They became repeatedly submerged and involved in crustal orogenies, especially during the Proterozoic, before they emerged as stable blocks of the continent's foundations in the north and south.

Concerning the other areas of ancient metamorphic rocks, none of them contains formations which are older than 2200 million years. That does not mean, of course, that older rocks could not eventually be found. The main problem with the identification of such ancient rocks is that the original setting of any radiometric "clock"-mineral is always changed by subsequent heat-and-pressure events of metamorphism. Thus, when a metamorphic rock's age is determined, one is never sure whether one has the age of the last metamorphic event, or a mixture of original rock age and one or more ages of events of metamorphism. Usually one can only say that the result obtained determines the minimum age of the rock. The oldest formations in the Carpentaria-Mount Isa region, for instance, are older than 1800 million years, and in the Arunta System of central Australia more than 1900 million years. How long before that time they were really formed we know not.

In the second edition I have shown some patches of Archaean in the latter areas on Map 25, simply to provide for the possibility that they might be there theoretically. This time I am omitting them and shall confine myself to what we really know. It is interesting enough.

There is an important point to be repeated and remembered from all this. The fact that a formation is strongly metamorphosed does not in itself mean, as once was believed, that it is very old geologically; or that an extremely strongly metamorphic series is even older. Indeed, we know of extremely altered rocks (not Australian ones) which are twenty times younger than some over 2000 million years old, quite unaltered, early Proterozoic formations of the iron-rich Hamersley district in northwestern Australia.

Metamorphism on a large, or regional, scale is the result of great heat and/or pressure created in the course of the severe upheavals in the earth's crust from which arise intercontinental and even circumglobal mountain systems. Evidently, formations of great antiquity, which are not metamorphosed, owe this to the fact that they have remained for some reason outside major orogenetic zones ever since they were deposited. There is nothing very unusual in the fact that such relatively stable regions exist; what is surprising is that there are more of them than one might reasonably expect, considering the rather turbulent history of the earth's thin outer crust.

Can there be anything said about the pattern of land and sea as it may have been toward the close of the Archaean Era? That is, do we have sufficient information to sketch a rough map of the geography of that part of the globe which was to become the Australian continent

2500 million years later? The answer is both yes and no. Naturally, the further back we step into the past the less reliable becomes our reconstruction of the then geography, because we cannot avoid risky deductions and much imagination entering the picture. Nothing, or almost nothing, can be proved; everything, or nearly everything, is disputable. However, for what it is worth, we shall try to create a little such imagery from the few scraps of information at our disposal.

Maps which show such reconstructed ancient landscapes are called palaeogeographical maps. This book contains twenty-four of them. They are not factual, as are maps of the present face of the earth, but more or less disputable interpretations of the fragmentary records contained in the rocks. One should therefore regard them merely as playful sketches of the general pattern through which the continent may have developed into its present shape. Incidentally, rather than presenting palaeogeographies for a number of subdivisions of the conventional geological time-scale (p. 23), the maps have been selected in such a way that they depict the state of things before, in some cases during, and after certain events which are of great significance in the continent's evolution.

In principle each map simply tries to give a picture of the distribution of land and water during the specified period. How much of the area shown as water is deep ocean, and how much is shallow sea covering underwater extensions, or shelves, of the continental blocks, is not indicated, but is discussed in the text. As a general rule, though, we assume that zones of strong mountain-building emerged out of deep-sea trenches along the margins of the continental nuclei. The Australian continent therefore grew from its nuclei outwards by adding one folded crustal zone after another. In places, though, the continent could also break apart, or newer folded belts would cross older ones, thus complicating or even obliterating the older structures. Fold belts can also split up into two or more individual zones with little or non-folded crust between them, they can be almost straight or arched or even sharply bent. The older the mountain-building is the more difficult it becomes to trace its ways and put the pieces together into a believable reconstruction. However, let us now see what kind of story we can make of it all. We will start with two maps from the latter part of the Archaeozoic, that is, the Yilgarnian.

Map (1) gives an idea of how the Australian region may have looked around mid-Yilgarnian times, say 2700 million years ago. There was no Australian continent proper in those times, and in the regions which we inhabit today much of the earth's crust was of the thin oceanic type. But there was a continental mass—with a crust thickness of 25–30 kilometres of which western Yilgarn and the Pilbara formed marginal parts, and which extended into the area of

Map 1:

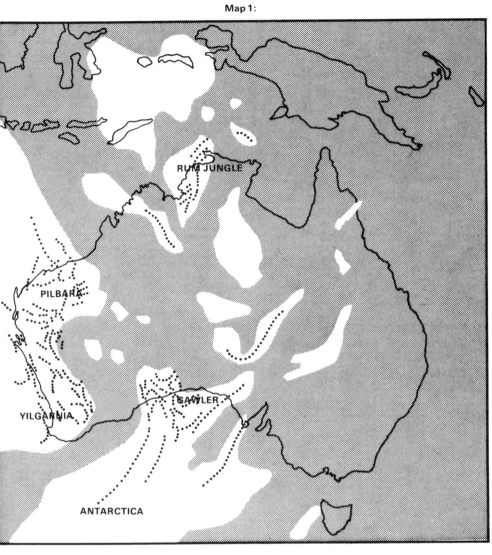

The Australian region during the late Archaeozoic, i.e. about the middle of the Yilgarnian Era, 2700 million years before present.

the present Indian Ocean. The fact that much of the ancient structural pattern, especially in the north of the Shield, seems today to be cut off by much younger geological features indicates such a former extension.

Separated from the western, Pilbaran, part of Yilgarn by a marine strait which may well have been a deep-sea trench similar to the present-day Timor Trough, we will notice another continental mass which sends peninsulas into the western Nullarbor Plain and the general region of Eyre Peninsula. The main part of this hypothetical land mass spread out into the Southern Ocean, and belongs now to Antarctica. A number of features in the region of the Great Australian Bight which, although not visible at the surface, are detectable with geophysical instruments, such as the gravitymeter and magnetometer, indicate this.

The various islands are shown on the map not only to allow for the possibility that smaller Archaean nuclei may have been in central and northern Australia, or because it is reasonable to assume that the ancient western land masses had some islands or even an island arc flanking them toward the aboriginal Pacific Ocean. The main reason is that the situation in Map (1) ought also to set the stage for a great mountain-building, or orogenetic, event that was soon to take place not only between visibly opposing land masses in the southwest, but also along their Pacific margins and what may be called the Rum Jungle Archipelago of the Yilgarnian Era. The East Kimberley and the "top end" of the Northern Territory have long been believed to contain continental nuclei of great antiquity. However, the very oldest rocks so far dated there are not older than late Yilgarnian. Formations of Pilbaran age do not seem to occur. Thus, on Map (1) the region is not shown as continental as are the western and southern nuclei.

It will have been noticed that throughout the past few paragraphs the position of a feature was not given in terms of geographical co-ordinates. The fact is that we do not know in the least where poles and equator lay in those times. For all we know, the Australian region may have been part of the Northern Hemisphere with the equator situated, say, along what is now longitude 140° east. We can turn the map through 90° clockwise and imagine Western Australia as part of some long vanished Asia; or turn it the other way and imagine it as anything we like. There is no way of proving or disproving our fancy. However, because it is easier referring to compass directions when describing features on a map, we shall do so from now on and trust the reader to remain aware of the pitfalls of such references whilst he is reading the next few chapters. Later, when we have waded through

Map 2:

 SEA

 LAND

 ZONE OF CONTEMPORARY
OR JUST COMPLETED
MOUNTAIN-BUILDING

 AXIS OF UPWARPING
—IN PLACES AS AN EFFECT OF
DISTANT MOUNTAIN-BUILDING FORCES

The Australian region at the end of the Archaeozoic, about 2500 million years before present, after the Kalgoorlie-Rum Jungle Orogeny. The very oldest rocks known in South Australia (southern tip of Eyre Peninsula) were also metamorphosed about that time.

much of the earth's early geological history, we shall see that records enabling us to determine the geographical location of the Australian region become more and more numerous in the rocks.

Map (2) illustrates how the geography may have changed after the mountain-building event towards the end of the Yilgarnian Era, the Rum Jungle Orogeny. A fair number of radiometric data from the central and eastern part of the Yilgarn Shield, the eastern Pilbara, the Tarcoola-Eyre Peninsula region, and Rum Jungle-Nanambu formations indicate that strong metamorphism and volcanic activity, both intrusive and extrusive, stemming from such an orogenetic event, affected the Yilgarnian-age rocks throughout these areas.

The Rum Jungle Orogenesis joined the Yilgarn-Pilbara block to the land mass in the south as well as to that in the north but, as will be seen, the weld was not very solid at this stage. Rather than effectively joining the various continental masses the orogenesis merely thickened and consolidated the eastern and northern additions (laid down in Yilgarnian times) to the primitive Pilbaran nuclei of the Antarctic, Yilgarn and Pilbara shield areas. In this way these ancient land masses grew and became stabilized eastward. On a lesser scale, similar features were probably added to the suspected nucleus in the north.

On the other hand, because mountain-building events not only create and consolidate, but also pull down and break up elements of the earth's crust, we suspect that a fracture and shear pattern developed to the north of the Pilbara Shield. This would mark the beginning of the eventual separation of the Yilgarn-Pilbara Shield from the continental mass to the north. Such fractures also prepared the reopening of a seaway through northern Australia, replacing a similar feature which had been closed by the latest orogenesis.

To the east the sea has swallowed most of the insular and peninsular elements of Map (1), as if to compensate for emergent lands in the south and west. The ancient Pacific coast now shows wide open gulfs and bays, probably with shallow or but moderately deep waters. These embayments are now the receptacles for colossal amounts of gravel, sand, and silt brought down by rapid erosion from the lately raised ranges in the coastal hinterland.

The nucleus in the Katherine-Darwin region has grown considerably and, although still poorly consolidated and partly submerged, its growing influence begins to be felt across the Carpentaria region into northern Queensland, Papua, and northward along the newly formed, largely submarine, orogenetic welt into West Irian and beyond. In the Indonesian area too it is likely that a fair part of the waters shown on Map (2) represent shallow sea, underlain by continental nuclei rather than by oceanic crust.

Thus then—with a large pinch of salt—may we imagine the

geography of the Australian region at the dawn of the Proterozoic Eon, some 2500–2400 million years ago.

GROWING PAINS—THE PROTEROZOIC EON

Unlike the Archaeozoic, of which one has so far only deciphered a few events which took place towards its end in the Yilgarnian Era, the story of the Proterozoic can be read fairly well from a beginning 2500–2400 million years back to its close around 570 million years ago. The picture is only a rough one, for much of the detail is lost or hidden under later rocks. Still, as we know now, there are various types of organic fossils which, however primitive and difficult to classify, help a great deal in correlating widely separated rock outcrops, especially if one can also apply radiometry to minerals in such fossiliferous outcrops.

No longer need we resign ourselves to say, as Andrews (*1938*) did: "It would need the close attention of an army of skilled field geologists for decades to provide a satisfactory broad classification or correlation of the now widely separated units". As so often happens in science, a different approach to a problem, the sudden emergence of new methods, or simply normal technological progress such as vastly improved means of transportation in difficult terrains—just think of the helicopter!—have changed the picture completely within a very short period of time. No longer does one need an army; a platoon equipped with better means and new skills will suffice, and a good job will be done, at least as much of it as Andrews wanted, in a few years or, with luck, even a few months. Our knowledge of the geology of Australia has, in fact, made tremendous progress in the second half of this twentieth century, and a great portion of this new knowledge concerns the Proterozoic systems.

It had been known for a long time, of course, that several major mountain-building events had occurred during the Proterozoic, for in many places one set of Proterozoic formations is resting at an angle—or, in geological language, unconformably—on the upturned and denuded edges of another. As a result of such events there must have been many radical changes in the distribution of land and sea, but it is only in the past two decades that both the sequence and extent of many of these events have begun to reveal themselves to geologists. Here again, though, the older the events, the less one knows about them.

The Nullaginian Era

In Map (3) we see a pattern of land and sea as it may have developed from that depicted in Map (2) through the first half of the

Map 3:

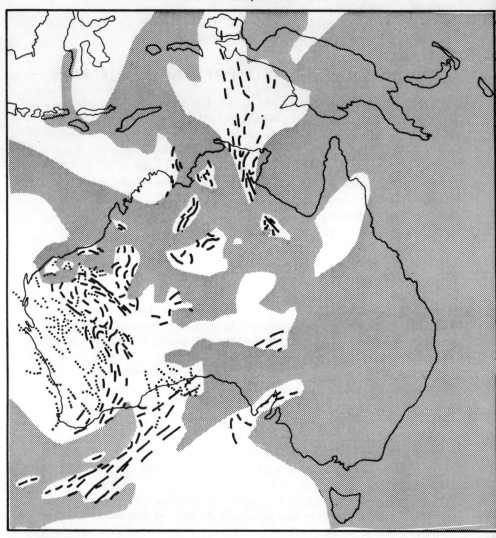

- SEA
- LAND
- STRUCTURAL TRENDS OF LAST COMPLETED FOLD MOUNTAINS—NOW UNDER INTENSE EROSION
- STRUCTURAL TRENDS OF ANCIENT FOLD MOUNTAINS—NOW MORE OR LESS PENEPLAINED

The Australian region in mid-Nullaginian time (early part of the Proterozoic), some 2300–2200 million years before present, when geosynclinal troughs developed through the Halls Creek, Pine Creek, and Eyre Peninsula areas, and the sea also inundated the ancient Pilbara Shield.

Nullaginian Era or, that is, the earliest part, from 2500 until about 2200 million years ago, of the Proterozoic Eon.

In the west lies the stabilized Yilgarn land, furrowed and scarred by the remains of ancient mountain systems. The youngest of these is winding its way from the southern coast through Kalgoorlie, Leonora, and Wiluna to the Ashburton River, then swings east and again northward through the region of the Great Western (or Canning) Desert. Its continuation into the Katherine-Darwin districts, however, is now interrupted. Through weakened or fractured zones created by the latest of the Archaean orogenies the sea is seen to have penetrated deeply into the land, and even the ancient Pilbara Shield is about to become inundated.

These rearrangements of parts of the earth's outer crust were in various regions accompanied by intense volcanism. Basaltic lavas, several thousand metres thick, were poured out over the western part of the Pilbara to form, together with many thousands of metres of sands and shales, what is called the Fortescue and the Hamersley Groups of formations. Extrusive and intrusive rocks of similar composition, associated with sediments, occur also farther north in the very metamorphosed Lamboo Complex of the Kimberley, and southward through central parts of the eastern half of Western Australia into the northwest of South Australia where they form part of the copper and nickel-bearing rock series known as the Giles Complex.

Great piles of sediments accumulated in the marine basins and channels of the Kimberley and in large subsiding geosynclines through the Pine Creek and Davenport Range areas, as well as farther south through central Australia. In the northwest of South Australia (Mann-Musgrave Range systems of formations) as well as in the southwest, from the Nullarbor to Yorke Peninsula, many thousands of metres of marine shales and sandstones, in many places interbedded with limestones and various kinds of lavas and tuffs, were deposited.

It is in the mighty series of this Era that the huge iron ore deposits of the Hamersley Ranges in the Pilbara as well as the smaller, but also important, deposits on Eyre Peninsula are found. The Hamersley Group, which is about 2500 metres thick, also contains the peculiar "blue" asbestos layers which have been commercially mined. The Pilbara-Hamersley region is still another of those cases of how the discovery of large ore deposits of one kind or another opens up and populates previously scarcely touched, tough and infertile country. The region south of Port Hedland is very rough and arid tableland intersected by the dry gorges and gullies of the Fortescue River and its tributaries. Through many areas the Proterozoic rocks have been very little disturbed and lie there almost as they were once deposited.

However, although they have been left untouched by the heat and pressure of orogenies, they are nevertheless altered chemically. Both the iron deposits and the asbestos are the result of such chemical solution, alteration, and reconcentration processes (*Knight, 1975*).

How much of the eastern Australian region was land, and how much sea, one does not know. Very likely, most of it was part of the ancient Pacific, dotted perhaps with groups and strings of small volcanic archipelagos which we will not bother to put on a map. On the other hand, considering that the pattern of Map (3) ought to develop into a framework for the next great mountain-building event which is shown on Map (4), there are good reasons to assume that peninsulas and islands of the kind shown on Map (3) have existed in those times in parts of South Australia and Queensland. The upwarping of the crust and therefore the emergence of island arches on the oceanic side of growing mountain systems along continental margins is, in fact, a characteristic and ever-recurring feature in the structural history of the earth. Out of similar reasoning a framework of large island masses is shown in the Arafura-West Irian sector, and a major continental feature is believed to have extended from northwestern Australia across to Borneo and beyond.

On Map (4) we have moved on through several hundred million years, and the Australian region is seen there as it may have looked toward the close of the Nullaginian Era, some 1900–1800 million years ago. Another great mountain-building event has just taken place, during which the accumulations of volcanic and marine sediments in the Kimberley (Lamboo and Halls Creek beds), Pine Creek (Agicondian system of formations), and in the Davenport (Warramunga Group) geosynclines became violently folded and metamorphosed. The same is just about to happen to the sediments in the central part of the southwestern embayment on Map (3) in South Australia (Middleback Group), and it is from a locality in this area that this great orogenic event is called the Kimban Orogeny (*Webb and Thomson, 1977*). It also affected the formations in what may be called the Argyllian Trough extending through the Mount Isa-Cloncurry area towards eastern Arnhem Land and across the Arafura Sea. In all these orogenetic zones we find also that the metamorphosed formations were eventually invaded from below by granitic magmas on a large scale. The volcanic activity on the surface, derived from the granitic magmas at depth, resulted in many places in the extrusion of large sheets of acidic, light-coloured lava and ash. Radiometric age determinations on rocks of this volcanic cycle almost always indicate 1900–1750 million years since their formation.

It will also be noticed on Map (4) that the Pilbara region remains almost unaffected by this great orogeny which, for a time, welds

Map 4:

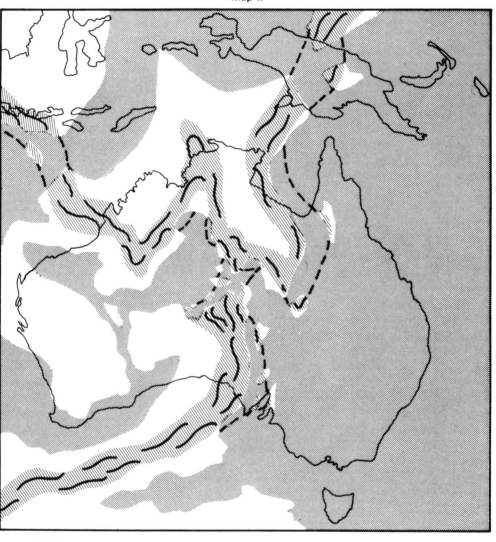

	SEA
	LAND
	ZONE OF CONTEMPORARY OR JUST COMPLETED MOUNTAIN-BUILDING

The Australian region after the late Nullaginian mountain-building events (1900–1700 million years before present) which folded the sediments laid down in the Halls Creek-Pine Creek-Hatches Creek geosynclines together with the oldest sequences known from the region of Mount Isa and Carpentaria. In South Australia (Eyre Peninsula) continued crustal unrest lasted into the early Carpentarian Era, i.e. to about 1600 million years before present (Kimban Orogeny).

together the western and northern Australian continental nuclei. On the other hand, a large crack lets the sea in around southern Yilgarnia to form a marine strait which connects with the Pilbara gulf and eastward through central Australia with the Pacific. Again thousands of metres of sediments are deposited in these seaways from the rapid erosion of the newly risen mountains as well as from the continued wearing-down of the more ancient land areas and mountain systems.

The Carpentarian Era

In Map (5), 200–300 million years later, the Australian "continent" is again seen broken up into larger and smaller islands. Yilgarnia is now completely surrounded by the sea. In the Pilbara sedimentation continued through most of the early Carpentarian Era (Wyloo Group). The late Nullaginian strait around southern Yilgarn has widened and the later metamorphosed Albany-Fraser sequence is accumulating there. The main region, however, of commonly very thick marine sedimentation extends from the East Kimberley across the "Top-End" (Katherine River Group) to the Gulf of Carpentaria (Tawallah Group, lower part of McArthur Group) and northern Queensland (Einasleigh-Georgetown Complex). This northern seaway connects through the area of the Davenport Range (Hatches Creek Group) with the Centralian straits.

A most significant feature, particularly with a view to the subsequent style of the growth of the continent, is the great geosynclinal seaway reaching from Cape York Peninsula via the Einasleigh-Georgetown area and Mount Isa to the Barrier Range, the northern Flinders Ranges and Eyre Peninsula, and westward across northern South Australia around the southern shores of Yilgarnia. In many parts of this seaway masses of marine sediments together with lavas and ash are piling up. In some of the layers there arise important accumulations of lead, zinc, silver, copper, and other metallic ores. Whether many of these famous metallic deposits originated simply as ordinary sediments, or were introduced through volcanic events after the sediments were laid down and consolidated into rocks, has in recent years been a much disputed question—one which has considerable bearing on the way future ore exploration is going to be conducted.

It seems remarkable that in all these thick series there is no evidence of truly deepwater sediments. Evidently, the rate of sediment deposition in this geosyncline must have kept pace with the rate of subsidence; as a result the waters remained generally shallow. After the mountain ranges in eastern North America, where the feature was first described, this is called the Appalachian type of geosyncline, and it has long been realized that most of the geosynclinal depressions

Map 5:

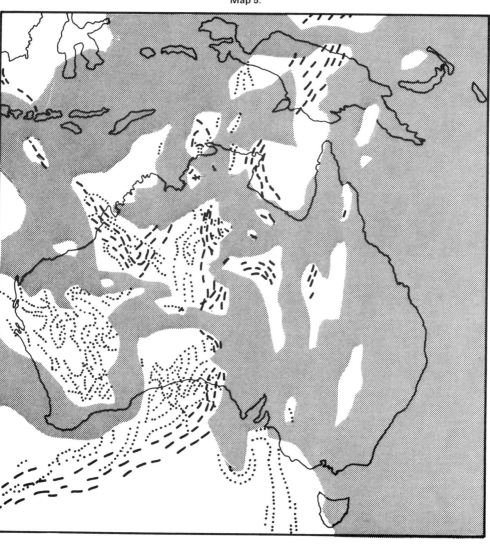

 SEA

 LAND

 STRUCTURAL TRENDS OF LAST COMPLETED FOLD MOUNTAINS—NOW UNDER INTENSE EROSION

 STRUCTURAL TRENDS OF ANCIENT FOLD MOUNTAINS—NOW MORE OR LESS PENEPLAINED

The Australian mid-Proterozoic archipelago, about 1600 million years before present towards the end of the early Carpentarian, when the main part of the Mount Isa and Broken Hill geosynclinal sequences accumulated.

which have at one time or another crossed the Australian region were of this Appalachian type. Of the other, or Alpine, type, which is characterized by very deep waters and little sediment deposition, there is as yet scarcely any evidence on this continent.

Similarly great thicknesses of shallow-water deposits occur also in a lesser marine trough which roughly parallels the Mount Isa-Broken Hill geosyncline and extends from eastern Arnhem Land southward through the area of the Barkly Tableland toward the Davenport Ranges. It is in the northern part of this crustal depression (Limmen Geosyncline) that the large McArthur River lead-zinc ore deposit is located.

For the areas east of the Mount Isa-Broken Hill geosyncline there is no information pertaining to the distribution of land and sea for this period. The pattern shown in Map (5) is therefore again based on the idea that it should appear to be a likely framework for the great orogenetic event which was to follow, and about the course of which we are reasonably well informed. From it one may surmise that oceanic deep sea with possibly one or two island arch zones extended over the eastern Australian region.

The course of this next mountain-building event, which characterizes mid-Carpentarian times and folded and metamorphosed the thick geosynclinal sediments from northern Queensland through to Eyre Peninsula, is shown in Map (6), which represents a period some 1500–1400 million years ago. As might be expected, this orogeny did not only affect the great geosyncline in the east but also the central Australian and southwestern seaways.

The degree of metamorphism varies quite considerably from one area to another. In the Georgetown-Einasleigh district of northern Queensland, as White (*1965*) has shown, one can see how a 9000-metre-thick sequence of mostly fine-grained sediments gradually passes from a non- or little altered into a more and more metamorphic state over a distance of fifteen to twenty kilometres. Shales turn first into slates, then micaceous slates, finally into micaschists and phyllites (soft, laminated, rocks consisting almost entirely of various kinds of mica); limestones become completely crystallized into marbles, and so on. In the Mount Isa district too one observes an overall increase in metamorphic grade from west to east, but small inliers of extremely altered rocks nevertheless do occur in places already among the low-grade metamorphics in the western areas (*Joplin, 1955*). The famous lead-zinc-silver-bearing Mount Isa shales themselves and the associated copper-bearing rocks belong to the sequence which is only slightly altered.

A regionally more intensive metamorphism is found to have affected the mountain-building segment which extends from Eyre Peninsula through the Flinders Ranges (Barossa Metamorphics, Mount Painter Complex) and Broken Hill (Willyama Complex) to the Lake Eyre area (Peake Metamorphics) and from there via the Mann-Musgrave Ranges to the southern margin of the Yilgarn Shield (Fraser Range Metamorphics). This stronger metamorphism is in part due to the fact that the colliding blocks of continental crust—squeezing the thinner crust between them—were much larger than in the east. On the other hand, in various areas it is the result of heat and pressure from one or two or more later orogenetic events. What we see now is a combined effect, and such rocks are called polymetamorphic.

In the metamorphics of the Willyama Complex at Broken Hill there is, as in Mount Isa, an increase in the intensity of metamorphism in a certain direction which, probably because the orogenesis changes course from south to west here, is from northwest to southeast (*Binns, 1964*). It is not possible to say whether the regional change in metamorphic grade is simply due to differences in heat, pressure, and intensity of movement in the orogenetic zone, or to a combination of heat and mineral-laden solutions and vapours pervading the rocks from an intrusive volcanic source at depth. Such a volcanic influence could have been contemporaneous with the mountain-building event, or have taken place at its very end or shortly thereafter. Even in much younger mountain ranges, of which much more can be seen, such problems of the timing of significant events often remain unresolved.

It will, of course, be noticed in Map (6) that the entire northwestern and northern regions of Australia remained well outside the mountain-building zone. The sea withdrew from the Pilbara region and for a long time remained restricted to a fairly narrow strait linking the west with central and northern Australia through the still unstable zone between the two ancient Western Australian shields. This is probably the time when the earliest sediments of the Badgeradda and the Moora Group in the west were laid down, together with an unknown portion of the metamorphosed sequences in the Precambrian Gascoyne Province. In central Australia there is the beginning of a depositional cycle which is to become the Pertaknurra Group. On the northeast of the Pilbara-West Kimberley block the slowly subsiding Kimberley Basin receives great masses of sands and shales, lavas and tuffs, occasionally also beds of carbonate rock partly constructed by marine algae. In fact, the beginning of this sedimentary story goes well back to the earliest times of the Carpentarian Era (Speewah and Kimberley Group). Where we are now, toward the close of the middle Carpentarian, this story approaches its final stages although shallow marine sedimentation still reaches across the Northern Territory (unnamed sedimentary sequences below the Victoria River Group and the Mount Rigg Group) to northwestern

Map 6:

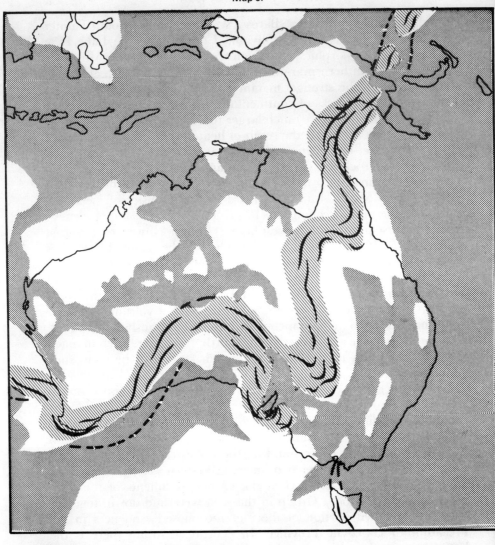

SEA

LAND

ZONE OF CONTEMPORARY
OR JUST COMPLETED
MOUNTAIN-BUILDING

The Australian region about 1500 million years ago at the end of the Olarian (Willyama) Orogeny which folded and metamorphosed early Carpentarian sequences of Mount Isa, Broken Hill, Eyre Peninsula, Musgraves, and the southern coastal zone of Western Australia. Note the continuation of marine sedimentation in the southern Pilbara, the East Kimberley and the Carpentaria provinces (compare Maps 5 and 7).

Map 7:

 SEA

 LAND

 STRUCTURAL TRENDS OF LAST COMPLETED FOLD MOUNTAINS—NOW UNDER INTENSE EROSION

 STRUCTURAL TRENDS OF ANCIENT FOLD MOUNTAINS—NOW MORE OR LESS PENEPLAINED

The increasingly complex structural pattern of the growing components of the Australian continent shown at the time of the deposition of the Bitter Springs Dolomite in central Australia during the later part of the Carpentarian Era, 1300–1200 million years before present.

Queensland and the Gulf of Carpentaria (upper part of McArthur River Group and Fickling beds).

The orogenetic event in the south and east—Thomson (*1969*) gave it the name Olarian Orogeny after the Olary district in the northeast of South Australia—did nevertheless affect the western and northern regions indirectly by continuously changing the configuration of land and sea. Because much of the northern waters were shallow even slight ups and downs resulting from repercussions of more or less distant mountain-building movements were able to effect vast shifting of shorelines. Areas of deposition of sediments became areas of erosion and vice versa. It is all a rather complicated story and, at present, only its general outlines are known.

All this helps to illuminate an important point concerning the growth and size—in fact, the whole concept—of what we call a continent. Evidently, to the geologist the outline of a continent is not only determined by what is seen above sea-level. Large shallowly submerged areas may in reality also belong to the continent because they are underlain by a crust which is of continental type and thickness. This means that the true size of a continent can be, and usually is, badly misrepresented by a picture which only shows a pattern of land and sea. We must always keep that in mind when looking at both palaeogeographic and present-day-geography maps.

By Map (7) we have arrived at a period about 1400–1200 million years ago, that is, in the late Carpentarian. The basic pattern still resembles that shown on Map (6) but the sea has again managed to breach a number of barriers which had for a time been raised by the Olarian Orogeny. Western and southern parts of Yilgarnia are further submerged, and the Badgeradda-Moora-Yandanooka type of shallow water sedimentation now extends southward and continues with the east trending Stirling-Mount Barren beds which, according to Prider (*1952*), are derived from a land to the south. These beds rest unconformably on the up-ended edges of the early Carpentarian Albany-Fraser sequences which had become metamorphosed in the mid-Carpentarian Olarian Orogeny.

While the Pilbara has emerged there remains a shallow seaway from the Gascoyne Province into central Australia in which the limestones, dolomites, salt and gypsum beds of the upper Pertaknurra Group (Bitter Springs Dolomite) are formed. From there this seaway opens through South Australia into the ancient Pacific past a large shallow Nullarbor embayment. Along the latter's eastern threshold much debris from the Olarian mountain-building event in that region is deposited in the form of boulder and pebble beds (Corunna Conglomerate). The centre of sedimentary accumulation seems to be in a trough extending from Spencer Gulf into the Lake Eyre area.

Although only about 2000 metres can be seen in outcrop the whole sequence, known as the Callanna beds or the "Willouran Series" in the descriptions of the late Sir Douglas Mawson, is probably up to 4000 metres thick. It consists of conglomerates, sandstones, shales and slates, limestones, dolomites, and is interbedded with basaltic lavas. Formations of similar composition and similar, or even greater thickness are also known farther north around the Peake-Denison Range and from there westward as far as the Warburton Ranges in Western Australia. To the east they occur in the Mount Painter region of the northern Flinders Ranges, and southward in places along the western foot of the southern Flinders and the Mount Lofty Ranges—for example at Depot Creek west of Quorn, then near Port Pirie and around Clare. It is noteworthy that between the top of the Callanna beds (of Willouran age) and the basal beds of the next younger sequence (Burra Group of the Adelaidean System) there is commonly evidence of an interruption in the progress of sedimentation, although only rarely of an unconformity.

Widespread volcanic activity is rather characteristic of the late Carpentarian of southern Australia. The basaltic eruptions have already been mentioned in the Willouran sequence, but they occur also towards the southwest (Roopena lavas), and far to the south in Tasmania there appear basalts and tuffs at this time. More acid magmas erupted in the Gawler Range west of Port Augusta. Even in the Pilbara and across in the Gulf of Carpentaria various types of volcanics occur.

In northern Australia the situation has otherwise remained rather similar to that in Map (6). In the west the youngest sequences of the Kimberley, the Victoria River and Mount Rigg Groups are laid down, and here too we frequently find lavas interbedded. In the Carpentaria region the sediments of the Roper Group are deposited over the McArthur sequence. The wide marine strait across the Carpentaria Gulf area shown in Map (7) was very shallow and may, in fact, not have opened into the Pacific all the time during the period under review. Tasmania, which had just been involved in a mountain-building event (known as the Frenchman Orogeny) that resulted in the Franklin and associated formations becoming metamorphosed, is again submerged but much of southeastern Australia appears as a large island extending southward from the Broken Hill (Willyama) region.

The Adelaidean Era

At the turn from the Carpentarian to the Adelaidean Era—about 1300-1100 million years ago—we again observe a rather unruly behaviour of the earth's crust in the central and western Australian regions. Strong folding accompanied by metamorphism, magmatic intrusions, and volcanism affected the southern parts of the Hamersley Basin sequence (of late Nullaginian age) in the labile zones around Yilgarnia, especially in the north. From the Gascoyne Province this activity spreads eastward towards Lake Eyre, first through the Paterson Ranges and the Nabberu Basin, and from there into the Mann-Musgrave Ranges and southwestward into the Albany-Fraser Province. The effects are particularly well expressed in the strong deformation and alteration of the Wyloo Group, especially its youngest part, the Ashburton Formation of several thousand metres of shales. The metamorphism in that area falls into the time interval 1200–1100 million years (*Wilson, 1961*). In the Musgrave Ranges it is around 1200 million years, and followed by granitic intrusions 1150–1100 million years old. Through the Albany-Fraser Province the time slot of these events, which are the last ones to affect this region, is also 1300–1100 million years.

Map (8) shows the changes which resulted from these crustal movements which are known as the Madiganian Orogeny (*Brunnschweiler, 1961*). It will be noticed that the central and southeast Australian seaways are only shifted, not closed. This is due to the movements having been considerably weaker in areas which are removed from the immediate neighbourhood of the principal orogenetic belts. Thus the boundary between the pre-Madiganian (late Carpentarian) Pertaknurra and the post-orogenetic (Adelaidean) Pertatataka Group in the Amadeus Basin region north of the Mann-Musgraves, as well as through South Australia into Tasmania (break between Willouran and Torrensian and their equivalents in Tasmania), shows good evidence of a lengthy interruption in the process of sedimentation and of a concomitant period of erosion. Yet there is rarely an angular unconformity as there is between the folded Wyloo sequence and the early Adelaidean formations of the Bresnahan and Bangemall Group in Western Australia. The latter, incidentally, represent the last manifestation of the sea transgressing across what emerges and remains thereafter as the great unified Pilbara-Yilgarn Shield.

Together with its younger appendages eastward into South Australia, this shield area now supplies large amounts of sediments into the winding seaway of shallow waters which crosses the Australian region from Tasmania northwestward through South Australia and around the Mann-Musgraves Peninsula of the Carpentaria land to the oceanic waters in the north.

In the south, where subsidence is most pronounced, this seaway is known as the Adelaide Geosyncline. Taking on varying shapes it existed not only during the latter part of the Proterozoic Eon but also

Map 8:

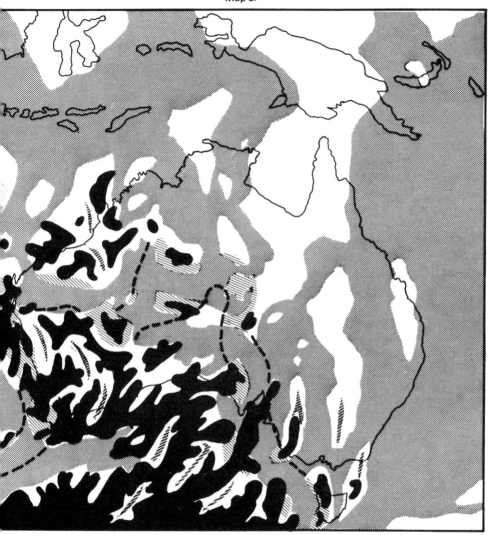

 SEA

 LAND

 GLACIATED AREA

 ZONE OF CONTEMPORARY
OR JUST COMPLETED
MOUNTAIN-BUILDING

AXIS OF UPWARPING
—IN PLACES AS AN EFFECT OF
DISTANT MOUNTAIN-BUILDING FORCES

The Australian region early in the Adelaidean Era, perhaps 1000 million years before present, after the Madiganian Orogeny which affected particularly the zone between Northwest Cape and central Australia. In Tasmania crustal movements at about the same time are known as the Penguin Orogeny.

through much of the earliest part of the Phanerozoic, that is, the Cambrian Period of the Palaeozoic Era which followed. There is no obvious break or unconformity in the mighty succession of strata, and there has long been uncertainty as to just where the Adelaidean system of the Proterozoic ends and the Cambrian sequence of the Phanerozoic begins. This question has in recent years been resolved with the help of the organic fossils which continue to be discovered in increasing variety in what were not all too long ago regarded as unfossiliferous Precambrian rocks. In the Adelaide geosynclinal sequence the Pound Sandstone (or Quartzite) used to be the main point of contention; for some it was the youngest formation of the Proterozoic, for others the oldest of the Cambrian Period. Named after the precipitous rim rocks of Wilpena Pound in the Flinders Ranges, it has since yielded a most interesting primitive assemblage of peculiar soft-bodied marine animals such as jellyfish, seapens, as well as flat and segmented worms. This fauna is known as the "Ediacara Fauna". It has no known continuation in Cambrian rocks in the world but has been found elsewhere also in rocks which are obviously older than Cambrian, and it is therefore now recognized as characteristic of the latest Proterozoic. In the Adelaide Geosyncline proper the youngest formation of Precambrian age is therefore the Pound Sandstone (*Glaessner and Daily 1959*).

Because of local variations in the rate of subsidence in the geosyncline the thickness of the rocks comprising the Adelaide system of formations varies considerably from one area to another. In some areas the sequence aggregates as much as 10 000 metres, in others it is barely 1500 metres thick. It is not surprising then that this system's composition and history is very complex. In spite of it having been the happy hunting ground of numerous geologists, some of them of great fame, there are still a great many important secrets to be uncovered there.

The oldest development of the Adelaide Erathem—the rocks formed in and therefore representing, as far as they are preserved, the geological events of the Adelaidean Era—is known as the Burra Group ("Torrensian Series"). It consists mainly of sandstones in its lower part, whereas its upper formations are predominantly shales and dolomites in which layers and concretions of magnesite are found. The next younger series is called the Umberatana Group, or "Sturtian Series". It attains in places as much as 4500 metres in thickness and is especially remarkable for locally colossal accumulations of tillite. Tillite is an interesting rock, consisting of clay and silts and iceworn fragments of other rocks brought down by glaciers. In its strict sense, derived from "till" (Scottish: boulder clay), it means consolidated moraine and should thus have been deposited on land, but for a long time now one has also included marine and freshwater, even riverbed "boulder clays" which, in fact, are glacial tills redistributed by the action of water. The point is, of course, that in spite of redistribution such rocks retain a number of features which allow us to recognize that the action of glaciers was at least in part responsible for their formation.

The presence of glacial deposits on such a scale—they are also known from some other Australian regions—throws some light on the climate of the time. If the glaciers were simply due to the existence of some high mountain ranges in the area, tillites would not be so widely spread over far apart regions of the Australian continent. Evidently, this is the first of several great glacial periods in Australia of which we have knowledge ever since one of the great pioneers of geology in this country, Professor W. Howchin of Adelaide, recognized it seven decades ago in 1908. Since then it has become clear that there were, in fact, two such periods during the Adelaidean, both of great intensity and probably considerable duration. Professor D. Mawson (*1949*) even thought he had evidence of three glaciations but this has been shown to be unlikely. During the first advance of the glaciers over 1200 metres of tillitic and other erosive material was laid down, then in a long warm interglacial interlude 600–800 metres of non-glacial marine sediments and, with the return of the cold climate, another 1200–1500 metres of glacial debris rocks.

Some of the South Australian tillitic rocks may have been deposited from floating icebergs, but most of them were left by the melting of glaciers on land nearby, or from great floating ice-sheet aprons similar to those of Antarctica at the present day. One wonders, of course, just where the lands were from which the glaciers and ice aprons came. There has been a great deal of research on the subject, and the predominant opinion now is that while much of the tillite material comes from the immediate vicinity of the Adelaide Geosyncline there is a lot more that came from farther away in the south and west.

This is, among other things, what Map (8) tries to show: in mid-Adelaidean times ice flows off a great land in the south and southwest into the Adelaide Geosyncline and towards Tasmania on the one hand, and on the other into a shallow embayment south and southwest of the West Australian Shield. In the Kimberley and through central Australia it seems preferable to think of isolated glaciers coming from mountainous terrains which, incidentally, need not have been of great elevation because the nearness of the great glaciation in the south indicates that they were anyway in rather high latitudes.

The Great Land in the South which has figured in various shapes on all our palaeogeographic sketches so far, but for the nearness of which there was never any proof, makes itself strangely apparent for the first time through the exotic materials in the tillites of the Adelaidean. This is where we first encounter the concept of "Gondwanaland"—the Great South Land—a very large and hypothetical continent which once took the place of much of the oceanic waters in the Southern Hemisphere—only to disappear, or break apart and float away in pieces, at the latest towards the end of the Mesozoic Era. Ages after this late Proterozoic period, in another great glaciation at the beginning of Permian times, glaciers and ice aprons again flowed over the west and south of the Australian region from that land in the south.

There is considerable evidence that at least some of these earlier ice ages, as well as some others in later periods, represent times of colder climates throughout the world. Late Precambrian glacial deposits are, in any case, known to occur in many places, in China, India, central Asia, Norway, South Africa, Greenland, Canada and Alaska, and the United States, and it is quite likely that a number of them were contemporaneous with our Adelaidean ones. Others, however, are definitely not, and are simply evidence of quite local glaciations on high mountain ranges which have long since been worn down.

There is an interesting point about ice ages which is of particular significance to scientists who try to unravel the evolution of the planet earth; geologists, geophysicists, and others. It was once thought that the earth is a slowly cooling body, and that evidence of generally much warmer climates would be found in the rocks of earlier eras. If so, as we know now, these eras must have been well before the later part of the Protcrozoic Eon, where we have proof of ice ages. Since that time the range of climates on Earth has certainly not changed appreciably. No one really knows what has caused the repeated fluctuation in temperature. It has been compared to the alternation of warm with chilly days, but on a far larger and more protracted scale. Yet that would exaggerate its importance, for, compared to the temperature range of the universe, from the millions of degrees in the sun (and much more in some other celestial bodies) to the immense coldness of space, the fluctuation of a few degrees in world climate, though sufficient to produce an ice age, is infinitesimal.

After the retreat of the glaciers the climate in the region of the Adelaide Geosyncline became warm and arid. This is indicated in places by very thick sequences of desertic "red beds" of shales and sandstones which are interbedded with large and thick sheets of clean sandstones, grey and brown shales, and dolomitic limestone layers. This post-glacial series, the youngest formation of which is the previously mentioned Pound Sandstone, is known as the Wilpena Group or "Marinoan series". It is the most widely spread series of the local Adelaidean system and occurs from the Western Australian border, between the Bight and the Musgrave Ranges, across the Lake Eyre-Lake Torrens region to Broken Hill—and even farther eastward where, in the far west of New South Wales, it is known as the Torrowangee beds.

In central Australia rocks of similar age as those of the Adelaide system are of almost the same composition, but the sequence is less varied and much thinner, in fact, on average no more than 2000 metres. The reason for this is probably that after the Madiganian orogenetic movements there was for a long time no sedimentation in the area. Not until the advent of the first glacial period during "Sturtian" times did the sea invade the centre again. In other words, when the bulk of the Burra Group was laid down in the south, central Australia was still an area of erosion. With the renewed transgression of the sea the first sediments are the conglomerates and sandstones of the Areyonga Formation which, in some places at least, show influences of a glacial environment. Thereafter, as in the Adelaide Geosyncline, "red beds" follow, and chocolate shales, ferruginous sandstones, and dolomites, all together known as the Pertataka Group. Here too the last of the Proterozoic formations consists of a large sheet of sandstone.

A similarly reduced sequence, beginning also with glacials corresponding to one or other of the two "Sturtian" tillites in the south, was laid down in the Kimberley region (Moonlight Valley, Egan, Walsh tillites) and across the Ord-Victoria River districts (Tolmer Group) as far east as about the meridian of Darwin.

In the equivalent sequences of the Carpentaria province, e.g. the South Nicholson Group, and those farther south near Mount Isa (Pilpah Sandstone) tillitic formations are absent. Here, therefore, we are well out of reach of the advance of the "Sturtian" glaciers—or, at least, the glacial debris was unable to move so far north. It will also be noticed that the marine province of Carpentaria is being steadily reduced in size.

It is also in the late Proterozoic, the Adelaidean, that the southeast Australian story begins to become legible. In Tasmania, which lay in the path of the southern extension of the Adelaide Geosyncline, up to 4500 metres of quartzitic sandstones, shales, and dolomites, together with some basaltic volcanics are deposited. However, because on King Island the youngest member of this sequence is a tillite (*Carey, 1947*), the beds below that tillite correlate only with the lower part of the Adelaidean succession in its type area, that is, with the "Torrensian" Burra Group. During the later "Sturtian" and the "Marinoan" Tasmania had emerged as a land area. It is interesting to note that this is precisely the opposite of what happened to central and northwestern Australia.

In Victoria no Precambrian formations are known, but it is likely that the emergence of Tasmania was part of a bigger southeast Australian story which raised, probably in the form of rather large islands in Victoria, New South Wales, and eastern Queensland, portions of the floor of the ocean in the east above sea-level. Map (8), which represents a time at about the turn from "Sturtian" to "Marinoan", say 850-750 million years ago, shows an early stage in this emergence of the southeastern region.

It must always be remembered that on our palaeogeographic sketches features appear together which may, in fact, have been separated in time by several tens of millions of years. Such considerable telescoping of events and features is unavoidable and comparable to what in other fields is called artistic licence.

The last of the seven sketches of Proterozoic palaeogeographies, Map (9), combines the picture of late Adelaidean times with that of the earliest Cambrian, say an interval some 650-550 million years ago.

Things have again changed very considerably. Instead of a complicated array of large and small lands, with many peninsulas and islands, we get for the first time a glimpse of the shape the Australian continent may ultimately come to assume. Two large slabs of subcontinental size, separated by a shallow marine strait, fill the central part of the picture. In the south there still looms the Great Southland. On the whole, the geography of Map (9) develops naturally from that of (8) through a narrowing of the central straits and concomitant emergence especially of the eastern flanking land, the latent presence of which was already evident in the mid-Adelaidean, or "Sturtian", pattern on Map (8).

There is an important climatic implication in this large-scale land emergence—it must have brought considerable aridity to interior parts of the western subcontinent, and this remains the case even if we turn the map through 90 or 180 degrees. It is therefore not at all surprising that an aridic "red bed" sequence characterizes the late Adelaidean, or "Marinoan", in central and southern Australia.

The changes in the pattern of land and sea at the close of the Proterozoic could, of course, have been due to simple up and down movements in the earth's crust. However, yet again things were not quite as uncomplicated. There is clear evidence of strong mountain-building events in at least three separate regions on the continent—along the King Leopold Ranges in the West Kimberley, from Cape Naturaliste to Cape Leeuwin in the deep southwest and, most prominently, in the persisting labile zone from the Gascoyne Province through the Bangemall Basin towards the northern foreland of the Musgrave Ranges. All radiometric age determinations on rocks from these deformed zones indicate that their folding and metamorphism

Map 9:

 SEA

LAND

 ZONE OF CONTEMPORARY
OR JUST COMPLETED
MOUNTAIN-BUILDING

STRUCTURAL TRENDS OF LAST COMPLETED
FOLD MOUNTAINS—NOW UNDER
INTENSE EROSION

 STRUCTURAL TRENDS OF ANCIENT FOLD
MOUNTAINS—NOW MORE OR LESS
PENEPLAINED

The meridional straits through the Australian region at the end of the Proterozoic Eon, about 650–550 million years ago. The mountain-building events, especially strong in the southern part of the Northern Territory and from there across into the southern Pilbara, as well as southward from there into the Cape Naturaliste-Cape Leeuwin area are associated with the Peterman Ranges Orogeny.

took place between 800 and 550 million years ago, that is, during the late Adelaidean. Whether these orogenetic zones were actually interconnected—for instance in the manner proposed on Map (9)—is not certain.

From the ranges across the Northern Territory-Western Australia border west of Ayers Rock, the central Australian segment of this mountain-building event is called the Peterman Ranges Orogeny (*Forman, 1966*). It is of particular interest because it is believed to be one of the rare examples of violent deformation to have taken place in the interior of a relatively well consolidated, thick-crusted (cratonic), region of a continent (*Forman and Shaw, 1973*). Normally, that style of mountain-building, which includes widespread overfolding and overthrusting of older over younger rocks, happens only along the thin-crusted margins of the continents when they collide with each other or with smaller fragments of the crust. Central Australia is unique in this regard because the same thing happened there again some 250 million years later at the end of the Devonian Period, when the MacDonnell Ranges were folded and pushed southward over the northern margin of the Amadeus Basin.

The Great Southland, in fact the Antarctic continent, remains attached to Australia. Tasmania in the far southeast remains largely above sea-level, but the oncoming early Cambrian transgression of the sea is indicated. Other places in the east and southeast, where marine sedimentation may have taken place at the close of the Proterozoic are the Snowy Mountains area and southeastern Queensland (part of the "Brisbane Schists"). The Torrowangee embayment east of Broken Hill is about to dry up. Within the Adelaide Geosyncline the deposition of the Wilpena Group comes to a close with the Pound Quartzite, as does that of the Pertataka Group in central Australia with the Arumbera Sandstone.

The Kimberley and Ord-Victoria country, although still shown as inundated, is about to emerge until in early Cambrian time the whole region is land. Arnhem Land and Carpentaria are now all land and will remain so until mid-Cambrian time.

Finally, the reader may ask what information there is on Precambrian events in Papua-New Guinea, Irian Jaya, and other Southeast Asia regions appearing on our maps. Frankly, nil—and it is not very likely that such will be forthcoming because it is doubtful that formations of such antiquity are exposed and preserved on the islands to the north.

The matter is rather different in the south. On the Antarctic Continent quite a number of Precambrian formations are known, and some of them are very similar and obviously closely related to rocks in Western and South Australia. There are therefore very good reasons for showing such close relationship to the Southland as is pictured on our palaeogeographic maps.

Here then, at last, we have arrived at the dawn of the Palaeozoic Era about 570 million years ago. It has been a long and complex journey, much of the story of which we were able to reconstruct from disconnected words and phrases in faded chapters. To fill in the context was a matter of both deduction and imagination, and nobody need be surprised when another ten years hence, perhaps in still another revised edition of this book, many of the map sketches look different again.

4 LIFE and EVOLUTION

UNTIL A FEW decades ago most geologists were doubtful about finding organic fossils in rocks older than Cambrian. True, one had found in various places such things as the trails and burrows of mysterious animals, as well as structures interpreted as algal growths. A number of round impressions looked like fossils of jellyfish, and there were even some specimens ascribed to the primitive hingeless lamp-shells. Yet on the whole the evidence was regarded as not very convincing.

Today, things are rather different. Not only in Australia, which has contributed finds of major importance, but in many other corners of the world there have been numerous discoveries of various types of fossils of which we shall hear later on. No longer can it be doubted that fairly highly developed organisms populated the Proterozoic seas in great numbers. Recently organic fossils of microscopic size have even been found in Archaeozoic formations, and it seems now obvious that life on earth created its first and simplest forms in the molecular and microscopic realm already well over 3000 million years ago.

What life really is, or means, we do not know, and philosophical discussion of the abstract concepts involved is beyond the scope of this book. How life originated is another matter, and modern science has some fairly clear ideas on that. Furthermore, it has always been possible to study how life functions. From such studies evolved certain broad principles by which living things can be classified.

The forms of life, as known on this earth, may be divided into two groups: the vegetable kingdom and the animal kingdom. These are separated according to the way their members derive the materials of which they are made. Plants alone can derive their substance from purely inorganic material, and under the influence of light can convert such elements as carbon, nitrogen, and oxygen into the complex chemical compounds which constitute living tissues. Animals cannot do this, and though the cycle of their life history may be complex, it can always be traced back to the point where it is dependent on plant life. Even though purely flesh-eating animals may live on other flesh-eaters, and these in turn on others, sooner or later in the chain comes the animal which feeds directly on plants or on the organic matter derived from their decomposition. In other words, plants can live without animals—at least it seems so—but animals cannot live without plants. From this it can be inferred that either plants and animals originated simultaneously on the earth, or, and this is more likely, that plants came first and animals later. It is also possible that both were derived from some common source.

Animals and plants are alike in one way—both are composed of cells, and in their simplest forms can exist as single-celled organisms. In fact, every plant and animal commences its existence as a single-cell which, by continual multiplication, builds up the most complex organs and structure. Amongst the unicellular organisms are some which seem to combine both animal and plant characters, and ancestral forms of these might once well have been the connecting links between the two kingdoms.

Let us at this stage consider the major divisions into which the plants and animals are classed. In later chapters many references will be made to organic fossils, and without some idea of their general classification the discussion would be unintelligible. Information given here will be elaborated as occasion arises, but for more advanced and detailed study it will be necessary to consult text-books on botany, zoology and palaeontology—the last named being the specialized study of fossil organisms.

THE PLANT KINGDOM

Plants may be put into two primary divisions, the flowering and non-flowering plants, but a more convenient division is into a number of major sub-kingdoms or phyla. Their classification is then similar to that of the animals. The groups that constitute the phyla are given below.

THALLOPHYTES. These are the most primitive of plants, and they include the unicellular forms such as bacteria, certain flagellates, fungi, algae, charophytes, and lichens (a marriage, or symbiosis, of a fungus with an alga). Among the algae the diatoms, though minute, are of great importance in the economy of aquatic life, and they have also a long geological history. They secrete a complexly shaped and beautiful skeleton consisting of silica (chert), and swarm both in the sea and in fresh water. So abundant are they that, together with their lime-secreting brothers, they form the staple food of much marine life.

Accumulations of fossil siliceous algal skeletons are known as diatomaceous earth or diatomite, a material which has many applications in industry. The tiny platelets which are secreted by the calcareous unicellular algae to form a sort of armour are found in many marine fine-grained rocks. They are known as nannofossils and have become fairly important recently as palaeontological tools, especially in petroleum exploration. The large algae or seaweeds, though more complex, are still lowly plants, and as such they belong among the earliest plant fossils known and still abundant on earth.

BRYOPHYTES. The mosses, living in damp places without possessing roots, are non-vascular plants the chief geological interest of which is that they were among the first to come out of the sea and gain a footing on the land. They are very rare as fossils.

PTERIDOPHYTES. These are vascular plants provided with roots, but there are no flowers and therefore no seeds. Reproduction is by means of spores. Five phyla are included here.

The psilophytes are extinct but were the very first plants on land over 400 million years ago. The lycopods or club-mosses are also almost extinct. A representative known as mountain moss can still be found on the Blue Mountains near Katoomba. In the past club-mosses grew into large trees and for a time dominated the first forests. The leaves grew on the trunks and branches, and when shed left scars, the shape of which was different for each genus. The fruits were cones borne at the end of the branches. The equisetales or horse-tails are also nearly extinct, but were very important in the past. The stems of the plants which also grew to large size were vertically ribbed and they had nodes like a bamboo, from which grew circlets of narrow leaves. The filicales or ferns carried spores on the back of the fronds and have remained an ubiquitous plant through the ages. The cladoxyles were bushes or small trees and have long been extinct.

SPERMATOPHYTES. These are vascular plants with flowers, seeds, and roots—in short, complete. There are two main groups: the gymnosperms with exposed naked seeds, and the angiosperms whose seeds are protected by some form of capsule.

The more important among the gymnosperms are the seed-ferns, the cordaitales (at times grouped with the pines), the cycads, the bennettitales, the gingkos, and the pines or conifers. Seed-ferns are only known as fossils and are often difficult to distinguish from the more lowly organized ordinary ferns; often it is only the absence of spores on fossil specimens which suggests their true classification. Some of the cordaitales were tall trees up to 30 metres in height and surmounted with a crown of narrow, sword-like leaves. They too are extinct. The cycads in form resemble palms and are exemplified by the sago palms in the tropics. Other living examples are the burrawang (*Macrozamia*) common in Australian forests, and the cycads in the rain forest of Mount Tamborine, southern Queensland. The bennettitales, which partly resembled the cycads and partly the pines were a prominent group of which one form, the gingles, was a common plant all over the world, but is now a living fossil. Only one species remains, a very large tree whose sole survival in the Far East was largely due to the Buddhists for whom it has a special significance. The pines are familiar to everybody and were abundant in past ages too, as were related groups such as the *Araucaria* and the Tasmanian celery-top pine.

The angiosperms constitute the great bulk of living plants but, like the higher animals, they do not go far back in the geological record. As fossils they are found only in the later formations. Even so, they are difficult to classify. There are two main groups according to the number of seed-leaves: the monocotyledones with one (e.g. wheat, palms) and the dicotyledones with more than one (e.g. laurel, elms, heather).

THE ANIMAL KINGDOM

Like the plants, animals are divided into a number of sub-kingdoms or phyla, which may be summarized as follows.

PROTOZOA. These are minute unicellular animals of great variety. The most important geologically are the foraminifera, which secrete a minute calcareous shell of diverse form, and the radiolaria, which have a minute but beautiful hard shell composed of silica. Both groups are common as fossils and are abundant at the present day.

SPONGES. These are multicellular animals, but still of lowly organization. The body wall is composed of minute spicules, commonly of silica, and these when detached are often seen to be of intricate geometrical form. Sponges are among the very earliest known fossils.

COELENTERATES. These are still rather simple animals, though more highly organized than sponges. There are two main classes, the first including the Medusae or jellyfish, the second the sea-anemones and the corals. Jellyfish are not often found as fossils, but an allied group, the graptolites, are of great importance geologically. Corals also go far back in the geological record, and the remains of many extinct families are found in the early periods of the Palaeozoic.

WORMS. These form a large and very diverse group, split by most zoologists into a number of separate phyla. They range from the common earthworm to worms which are parasites in other animals and to the beautiful tube-secreting serpulids, which live in the sea. Though rare as fossils, they are of great importance not only because among them are possibly the ancestral types of many other animals, but also because they are among the earliest known fossils.

ECHINODERMS. There are several classes of these marine animals. The most familiar echinoderms are the starfish, the sea-urchin, and the sea-cucumber or bêche-de-mer of commerce. Not so familiar are the crinoids or sea-lilies, now nearly extinct, and the quite extinct cystoids and blastoids. The sea-lilies have the body enclosed in a calyx or cup composed of separate plates, surmounted by a crown of stony, jointed arms, and attached to the bottom by a similar stem. The cystoids also had a calyx composed of plates but were generally without arms or tentacles, and were attached by the base of the cup itself. Blastoids also had a cup and stem, but no arms, and the mouth was on top.

BRYOZOA. These are the so-called lace-corals, which were abundant in days gone by and are still equally abundant in the sea. Like the true corals they are compound animals, but much more highly organized. They may often be found as beautiful lace-like encrustations on reefs, on floating pumice, or even on driftwood. Some are foliaceous and simulate seaweed in appearance.

BRACHIOPODS. Brachiopods or lamp-shells are extremely important geologically, since they have existed already in the Latest Proterozoic. They were particularly abundant and varied in the earlier periods and many forms still live in the sea. They secrete an external shell consisting of two valves of unequal size, one on the back, the other on the front of the animal (in bivalves such as the mussel, which belongs to the next phylum, the valves are on either side of the animal).

MOLLUSCS. These may be marine, freshwater, or land animals, most of which secrete an external shell. These shells are familiar to all who frequent the beaches. Mollusca are divided into several classes. The first comprises the pelecypods or bivalves, such as the oyster and the cockle. The second contains the gastropods; among these are the snails and slugs, both marine or terrestrial, which secrete a shell in the form of a tube that is typically spiral; the pteropods or winged snails of the open ocean also belong here. Closely related to the gastropods are the chitons or coat-of-mail shells, which have eight overlapping valves and crawl under stones in the sea. The fourth class consists of the scaphopods, or tusk-shells, which are rarely found, being buried in sand on the sea bottom. The last class comprises the cephalopods, which include the octopus, the squid, the nautilus, the cuttlefish, and many large and diverse forms which have only been found as fossils.

ARTHROPODS. These are animals with jointed legs and bodies, and they belong to many different classes, orders, and families. The phylum contains many early, extinct groups, as well as marine creatures such as crabs, lobsters, sea-lice, shrimps, and the familiar barnacle, and all insects, spiders, scorpions, and many others.

ASCIDIANS. Here belong the sea-squirts and the cunjevoi, which is found attached to rocks and is used as bait. Though at first sight rather simple animals, ascidians are in fact very highly organized. It is interesting to note that in their larval state they are free-swimming and bear a strong resemblance to certain primitive fish. At this stage they have a notochord, which is practically a rudimentary backbone, and this would seem to bring them close to the next group, the vertebrates, which is the highest organized of all.

VERTEBRATES. These are characterized by an enlarged brain and the possession of a backbone. Included here are the fish, the amphibia (newts and frogs), the reptiles (turtles, lizards, snakes, and many extinct forms), the birds, and lastly, the mammals, which suckle their young and which culminate in man himself.

EVOLUTION

Members of all these main groups of animals and plants will be found as fossils. However, most of the fossils, although their relationships are recognizable, are very different from representatives of their groups living at the present day, and as we go further back in time this difference becomes more pronounced. There are exceptions, and a few forms have persisted with very little alteration right through the geological record we have of them. On the other hand, countless species, families and even orders have become extinct. The continual appearance of new species on the earth, their gradual rise and ultimate disappearance, naturally raises the question of their origin and development—in fact, the whole subject of evolution.

Evolution means the development of one race or species from another, and in its complete form visualizes the derivation of all the diverse forms of life, both present and past, from a common ancestor. Practically all geologists now accept evolution as the only rational explanation of what they have observed in nature. It is firmly established as a law of nature, though opinions may differ on the ways and methods by which it has been and is still being brought about. The truth, as with many things, probably lies in between the various theories and hypotheses advanced, because no single one of them can satisfactorily account for every problem raised. Darwin's theory of natural selection can be demonstrated and proved up to a point, but even in its very much improved modern form there are many things for which it cannot provide really satisfactory answers.

The same is true for the theories concerning the genetic mechanisms which provide the material on which natural selection goes to work. It is impossible to enter into a discussion of the merits and demerits of all these theories; but here are a few salient facts and arguments which it is well to keep in mind when following the progress of life through the ages:

1 In both the animal and plant kingdoms there is a method of functioning common to all the innumerable forms, from the simplest to the most complex. The method of the multiplication of cells and their growth into organs of digestion, propagation, feeling, hearing and seeing, though differing in detail, is everywhere the same in principle, and this suggests that there is a common origin.

2 The embryos of the more complex types of animals, from the time of their conception, go through all the phases that the evolutionist suggests their ancestors went through during bygone ages. Thus the frog in its tadpole stage shows a remarkable resemblance to certain types of primitive fish. More remarkable still, the foetus of a human being at various stages is not only practically identical with the foetus of other mammalian animals, but has many points in common with much lower forms of life. At one growth stage it even possesses gill slits similar to those of fishes, suggesting that long ago some of our ancestral forms swam in the sea. Reference to those parts in text-books of zoology which deal with ontogeny and comparative anatomy will multiply these examples a hundredfold.

3 Many animals retain traces of organs or appendages, which are apparently useless in their present environment, but which seem to be inheritances from some remote ancestor living in conditions where these organs were perfectly useful. Examples are rudimentary bones in the human ear which would, if more developed, allow the ear to be freely moved as in many other animals. Rudimentary legs in the whale point to an ancestor once living on the land. Once again these examples may be multiplied many times.

4 One of the major aspects of evolution visualizes the gradual development of simpler forms into those which are more complex.

Within limits, the geological record bears this out, though there are apparent anomalies (an important one will be dealt with in the next paragraph). The development of fish into amphibians and amphibians into reptiles is reasonably well demonstrable, as is the development of the reptiles into birds on the one hand and into mammals on the other. Within narrower limits the ancestors of the horse have been continuously traced backwards to a small free-toed creature about the size of a dog. Another interesting aspect of evolution is that it can also lead to apparently less complex forms. Under the pressure of adaptation and natural selection a genus, or family, or even a single species may in fact radically reduce its complexity, but more than compensate its "wrong-doing" by propagating the simpler form x-million times more often than it did the complex form, thus assuring its survival. This kind of evolution has been called the "survival of the mediocre"; but then, who are we to pass judgement on nature?

5 Another very important apparent anomaly is the fact that as far back as the rocks of the Cambrian Period the main subdivisions of the animal kingdom were as definite as they are today. But while it is possible to trace the ancestry of the horse, and even that of birds from reptiles, no links have been discovered connecting, say a starfish with a snail. This applies not only to the main sub-kingdoms but also to many of the classes within them. Bivalve shells, or pelecypods, and sea-snails or gastropods are found right back to the Cambrian, and though the species were different from those living today, they were undoubtedly pelecypods and gastropods; yet no links have been found to bridge the gap between the two classes. The curious group known as trilobites, predominant in the early part of the Palaeozoic Era, are very different from anything living now, though structurally they were undoubted crustaceans. However, all these, the snails, trilobites and others, though found as early as the early Cambrian, are animals of high organization, and not the rather simple organisms one would still expect at such an early period. It is a simple fact of nature—these highly organized animals appear suddenly in the geological record, without forerunners, as if thrown in from nowhere.

6 A partial explanation of this apparent contradiction of the principles of evolution is that our knowledge of life on earth before the Cambrian Period is still very limited—and, after all, the period lies only about one-sixth of the way back in the geological record. Before it occurred eons and eons, during which innumerable formations of the Archaeozoic and Proterozoic were laid down and during which, as we can fully prove now, various forms of life were in existence. It is among these that we must find the generalized species, the missing links, from which developed the later, more sharply defined groups.

THE ORIGIN OF LIFE

On current knowledge and theory our solar system is between 5000 and 6000 million years old, and the cooling of our planet earth began about 5000 million years ago. When it had developed into a solid (relatively!) body it would have had a first atmosphere consisting of straight, non-molecular, oxygen, hydrogen, carbon dioxide, carbon, and nitrogen. By various physico-chemical processes this was eventually turned into a mixture of molecular nitrogen, oxygen, hydrogen, water, ammonia, and methane.

One can say that this physico-chemical change was the first of the countless steps in the evolution of life on earth because, whatever special qualities living matter may have, it must also follow the molecular laws of chemistry and physics, just as do ordinary chemicals in jars and bottles. However, for a while there was one troubling question, raised at about the time when the first edition of this book appeared. The physicist and Nobel-Prize winner Erwin Schrödinger suggested in a little philosophical essay (*Schrödinger, 1951*) that the principles of probability calculus, on which all known laws of nature are based, may not be applicable at the level of the molecular structure of chromosomes, when it comes to explaining why such comparatively simple structures are capable of storing, computer-like, an extremely complex building plan for, say, a fish, a spider, or a human being. Schrödinger therefore posed the question: "Are there laws of nature we do not yet know of?"

Today we know that the answer to this question is: No. First, Englishman Francis Crick and his American colleague James Watson managed to solve the problem of just how nature is able to transmit a complex genetic code. This is done by means of a clever three-dimensional arrangement—a double helix—of what may be called the letters of the genetic alphabet, i.e. the nucleotic acids. Second, scientists like the Americans Nirenberg and Khorane and the German Matthaei have since deciphered in all detail what the individual parts of the genetic message (code) consist of chemically. So we know now how the boardroom language of the nucleotic management is translated into orders which can be understood on the shopfloor of the protein manufacturing plants. The veil has been lifted—how life operates and propagates is no longer a secret (*Crick, 1966, Ponnamperuma 1972*).

But we still haven't answered the question as to how life became established in the first place. What precisely, and in what sequence, happened in those faraway times will never be known, but as Professor and Nobel-Prize winner Manfred Eigen of Göttingen says:

"We are now definitely in a position where we can accurately describe what the absolute minimum conditions are, which permit that molecules will link up at all to form living structures. We believe we know the principle which governs the particular kind of natural self-organization—and self-organization it must be!—nature must have applied to create life on the basis of purely accidental events, and within the known laws of physics and chemistry."

Very early in the earth's history all the necessary materials for the construction of organic macromolecules were freely available, and as Urey (*1952*) and others have shown, atoms can be combined into simple as well as complex molecules in test tubes today in "much the same way as was possible under the conditions when earth was new". All that was needed thereafter was a combination of time and the effects of various forms of energy to compound these basic chemicals into such molecules which we call "living". A molecular system is called "living" when it is capable of metabolism, reproduction, mutation, and therefore of evolution through natural selection and survival of the fittest. As Oparin (*1961*) suggested, already at the very earliest stages of the formation of living molecular systems, the process of adaptation and natural selection must have been operating. After all, as the molecules, even non-living ones, unite to form aggregates, the aggregates must compete with each other for building materials. There must be some reasons—which need not have anything to do with complexity as such—why some aggregates will be better at this job than others and therefore not only become stabilized in that efficient form but assure their own survival. In other words, they become dominant and self-propagating species, at least in those environments which suit them best.

Even at this point of understanding we have, however, still not answered how the very first molecular living systems originated. We only know how the process carried on from that point. In other words, the principle of natural selection of the fittest is indispensable in the story—let us say it is 50 per cent of it. The other 50 per cent of the story, the other indispensable principle, is that the systems had to organize themselves in a special manner, to which Professor Eigen and his collaborator Professor Schuster (Vienna)—the discoverers of the principle—gave the name "Hypercycle". A hypercycle is a molecular ring system consisting of self-catalytic cycles, each of which not only propagates itself but also produces enzymes, which assist the neighbouring cycle by acting as catalysts in the latter's reproduction (*Eigen and Winkler, 1975*).

Another important point is that any build-up of organic molecules must have taken place in the primeval oceans, not on land. Before there was enough oxygen in the atmosphere in general, and a significant amount of ozone in its outer layers, only the deeps of the oceans would afford sufficient protection for the tender beginnings of life against deadly solar and cosmic radiation.

This gives us another clue concerning the time when the first life molecules might have formed in the seas, that is, life as we know it under today's physico-chemical conditions. Far back in the Archaean we know of extensive limestone formations. This means that enough oxygen must by then have entered the atmosphere to permit some or all seas to become saturated with lime ($CaCO_3$). These carbonates, as well as other rock types, are so similar to younger sedimentary rocks that one cannot but assume that the oxygen-carbon dioxide balance in such early Precambrian times was not markedly different from that of today. Another important aspect of this observation is that probably the strongest elements among these earliest living things were organisms which produced oxygen—that is why plant life in the form of marine algae, e.g. stromatolites, is predominant from the start. Organisms which consume oxygen obviously had to wait until such time as there was enough of it around.

This also explains to some extent why there is such a scarcity of fossils with hard skeletal parts in these very ancient rocks. Another part of that explanation would say evolution had not progressed so far as to produce organisms which knew how to use lime for skeletal structures. In addition, all animals and many plants need to get rid of excess lime in their bodies. This need, because the most active animals solve that problem best, produces mobile organisms which are not hindered by internal or external hard parts. In other words, until the organisms had learnt to build lime into their structure without it becoming a hindrance—or an evolutionary disadvantage—there could be no hard parts. This makes it difficult for us. There can be little doubt that many kinds of these ancient bone- and shell-less creatures populated the Precambrian seas, but to find them as fossils is a rare event indeed.

THE OLDEST ORGANIC FOSSILS

By now we can fairly well imagine what the earliest organisms looked like. First there were various types of hypercycled macromolecules. One—and only one—of these survived the selection process and became the ancestor of all earth-type living organisms. Macromolecules combined and recombined to become bacteria, algae, simple protozoans, then free-swimming, soft-bodied, animals, becoming more and more diversified with time. Some animals continued to feed on plants, some learnt the method of preying on others.

These others in turn sought escape by rapid swimming, or by sinking to the bottom and hiding, or by camouflaging themselves to look like plants, or eventually by developing a protective armour or shell.

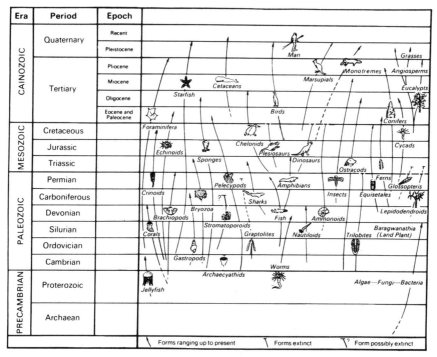

Fig. 3
RECORD OF LIFE IN AUSTRALIAN ROCKS
(Modified from Bur. Min. Aust., Canberra)

Many such Precambrian fossils have been found in Australia. The oldest are bacterial or single-celled algal organisms found in the Pilbaran of Western Australia *(Buick, 1981)*. From the earliest Proterozoic onward, algae evolve very strongly all over the oceans of the earth. Their best known types are called stromatolites, which are layered structures mostly in limestones and dolomites due to the activity of algaes and bacteria on the floor of shallow seas—successive algal lawns, fossilized. They have many shapes and sizes, from small, densely crenulated "flowers" to man-sized columnar growths, and carry names such as *Cryptozoon, Archeozoon, Gymnosolen, Collenia, Conophyton, Jurusania, Tungussia, Osagia, Newlandia (Walter, 1972)*.

First among the animals were probably the Medusae or jellyfish and the worms, representatives of the former occurring apparently as early as Nullaginian times (Hamersley Group). There is a sharp increase in the number and in the diversity of animal remains during the latest part of the Adelaidean Era, that is, shortly before the beginning of the Phanerozoic. The Ediacara fauna from the Pound Sandstone in South Australia (*Sprigg, 1947, 1949*), for instance, comprises well preserved forms of worms, jellyfish, sea-pens, as well as some other fossils the nature of which is still uncertain. Similar animals are known from the Arumbera Sandstone in central Australia and from many localities in other parts of the world. It has even been claimed that such higher organisms as lamp-shells (*Lingula*) and echinoderms occur first in the latest Proterozoic (*Murray, 1965*).

There is a fairly voluminous literature in which these Precambrian fossils are described. The drawings and photos on Plates 1 and 2 give a general idea of what the forms look like. For detailed descriptions and analysis the keen fossil hunter must consult palaeontological papers such as Glaessner (*1960, 1961, 1962, 1971*), Edgell (*1964*), and Muir (*1976, 1978*). In these, as well as in the previously mentioned publications, the reader will also find references to literature from other parts of the world. How fortunate the naturalist, amateur or professional, who makes another discovery in Australia as wonderful as that at Ediacara—especially if it were to be in even older rocks!

III
ADOLESCENCE AND ADJUSTMENT IN THE PALAEOZOIC ERA

III

5 The CAMBRIAN PERIOD

ALTHOUGH, as we have just seen, Precambrian rocks do contain organic fossils, one can nevertheless say that the close of the Proterozoic ushered in what is known as "Historical Geology", for at this stage the writing in the Book of Time becomes for the first time readable in considerable detail. Naturally, many of its pages are lost for ever and others are blurred, but it is possible to gain a much better idea of the geographical scenes in the Cambrian Period, 500–600 million years ago, than we were able to get of the Precambrian eras. This is due to the fact that a significant portion of the organisms which swarmed in the Cambrian seas is preserved in fossilized form.

It has been shown how the great orogenetic movements of Proterozoic times formed high mountain ranges in various parts of the continent. Erosion of, especially, the youngest of these, Map (8), provided the sediments deposited in the central sea, Map (9), during the latest Proterozoic and the early Cambrian. Moreover, on comparing Maps (8) and (9), we have seen that toward the end of the Proterozoic the sea gradually withdrew from wide areas in Western Australia and the Northern Territory, thereby narrowing the northern waters to a strait connecting the Timor Sea with central and southern Australia. The crustal movements which brought about this regression were only mild. In spite of that they produced many cracks and fissures through which in earliest Cambrian times colossal streams of basaltic lava escaped to form huge, flat sheets of these dark grey volcanic rocks over much of the newly emerged land. These are known as the plateau basalts of the Kimberley and Victoria River districts. Similar volcanics were extruded also in the Katherine-Darwin region and in Arnhem Land, but these are not as extensive as the plateau basalts.

These volcanic events notwithstanding, at the beginning of the Palaeozoic the process of consolidation of the Precambrian Shield underlying the western half of Australia was almost completed. This entire region emerged as a small continent the continuation of which across the Northern Territory into Queensland and Carpentaria was, if at all (early Cambrian fossils are not known here), only very shallowly submerged. Even in later Cambrian times the rate of subsidence throughout this northern region was very small, and the same goes for the central Australian gulf and the embayment just east of Broken Hill, Map (9).

Considerably more pronounced subsidence took place in the narrows of the Adelaide Geosyncline although, here too, the waters remained shallow or of only moderate depth because sedimentation filled the trough as fast, or almost as fast, as it subsided.

Because of the scarcity of exposures of Cambrian, especially early Cambrian, rocks there is very little information on the distribution of land and sea in this period and in the region east of the Adelaide Geosyncline. What is shown on Map (9) is purely speculative. The eastern half of the Australian region was in any case a zone of increasing crustal unrest through the Palaeozoic Era—from the mid-Cambrian onward lands came, and went again, seaways changed their course, mountainous island ranges rose from and sank back into the ocean, and so on. Many of the details of these complex and protracted happenings will unfold as our story develops in this and the following chapters.

The instability of the central and eastern regions, including Tasmania, manifests itself quite suddenly at the beginning of the Middle Cambrian period. During the unknown number of million years when the trilobite genus *Redlichia* populated the sea—that is, as the geologists say, "in Redlichia time"—the sea invaded almost the whole of eastern Australia, reaching westward almost as far as the eastern shoreline of the central sea (*Opik, 1957*). Only a narrow and probably mountainous land bridge—as Dr. A. A. Opik calls it, a Meridional Divide—running from Carpentaria through Mount Isa and Broken Hill to the Southern Ocean remained between the central and the eastern sea.

This was also the time when the previously small embayments in Victoria, New South Wales, and Queensland, Map (9), expanded rapidly to break up and submerge the eastern land of "Willyama". Great deep-sea basins and trenches developed here in which thousands of metres of fine-grained sediments and submarine basalt lavas accumulated. This is the age when the Brisbane Greenstones and Schists and much of the oldest sandstone and slate series in the Australian Alps south of Canberra were laid down (*Opik, 1957*). All these later became intensely folded, shattered, and more or less metamorphosed so that, unfortunately, no recognizable fossils are left in them by which we could prove their Cambrian age.

In northern Australia conditions were entirely different. In the shallowly submerged and only gently subsiding intracontinental basin, extending from the Queensland border to the Joseph Bonaparte Gulf, the Middle as well as later the Upper Cambrian sediments accumulated very slowly, and the varied and ever changing marine organisms which swarmed in the sea were gently buried, generation by generation, and preserved. The sequence is remarkably clear, and we have here one early chapter of geological history which has miraculously retained its legibility. The reason for this remarkable preservation of the geological record is that after deposition the strata remained not only undisturbed in the position in which they were originally laid down, but also at such a low level that they have never since been subjected to extreme erosion.

This land surface, though only a quarter of the age of the Yilgarn surface in Western Australia, is yet one of the oldest land surfaces in the world. There is something which makes an extraordinary appeal to the imagination in this part of Australia. It is a backwater, not only in geological history, but in human history as well. It is so old, so quiescent, so unshaken by the titanic events which have elsewhere convulsed the earth. Such events have been the prerogative of youth, and here maturity was reached some 500 million years ago. There are no towering mountains, no mighty rivers, no great gorges, no rushing waterfalls, no lakes, no spreading forests, no spectacular scenery. Instead, the hills are rounded and low; there are great expanses of monotonous plain; in the river beds we see nothing but a few placid waterholes overshadowed by trees. Everywhere is the imprint of old age; even the wild animals which come down to drink at the waterholes are antiques, survivors of ages long since gone.

To the east of the Meridional Divide the Middle Cambrian sea deepened eastward across the largely, though not entirely, submerged "Willyama" region. There was, for instance, a large island—"Duchess Land"—east of Mount Isa (*Opik, 1957*), and between it and the Meridional Divide the crust subsided to form the Undilla Basin. The phosphatic sequences in it contain a magnificent record of faunal evolution in the Middle Cambrian. The formations and the distribution of the guide-fossils in them is shown in Fig. 4, which is taken from Opik (*1957*). Naturally, these rocks contain many other fossils, so that where one of the guide-fossils is absent in a formation one can still determine its place in the time scale from the associated forms. Fig. 4 illustrates the time-scale developed by Dr. Opik from his studies in the northwest of Queensland and the Northern Territory. Remarkably enough, the sequence of guide-fossils is precisely the same as in the so-called Acado-Baltic Province of the United States and Scandinavia, indicating that the marine organisms of those times

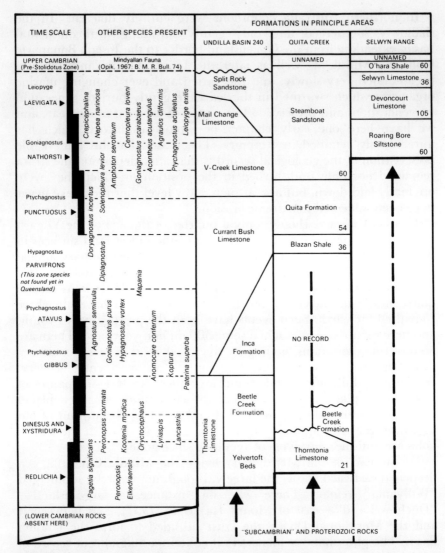

Fig. 4
CORRELATION OF MIDDLE CAMBRIAN FORMATIONS IN NORTHERN TERRITORY AND WESTERN QUEENSLAND.
Modified after Opik 1957

Figures on chart refer to thickness in metres; for Undilla Basin only total thickness of sequence is shown

could freely travel across such enormous distances and thus intermingle.

We can also see that this time-scale does not give us subdivisions in terms of years; nobody could state what the duration of, say, the "Atavus"- or the "Laevigata"-time was. What one can definitely state, however, is that should we find an isolated outcrop of sandstone with *Leiopyge laevigata* and, 30 kilometres away, another sandstone which contains *Ptychagnostus atavus*, then the latter sandstone was laid down long before the former, even though both of them are, broadly speaking, of Middle Cambrian age. Fig. 3 clearly shows how the formations exposed at Quita Creek and in the Selwyn Range, both some 80 kilometres south of the Undilla-Thorntonia area, fit into the much more complete rock sequence found in the Undilla Basin. We realize, for example, that after a first inundation in Redlichia-Xystridura Time the Middle Cambrian sea re-invaded the Quita Creek area in Parvifrons Time, whereas it never reached the Selwyn Range until the advent of Nathorsti Time.

This is a most important part of the method by which geologists reconstruct the geographies of long bygone periods. Evidently, the finer the fossil time-scale, the more accurately can palaeogeographic maps be drawn. The text has dwelt a little on the details of this case because the Cambrian of northern Australia represents a very fine example indeed of what can be done under favourable circumstances. By the same token it makes one realize why it is so terribly difficult to unravel the history of formations which contain few and poor fossils—as do the Precambrian rocks—or contain none at all.

As mentioned before, the southern portion of the central straits was a rather narrow trough, the Adelaide Geosyncline. The Flinders Range, with its southern extension, the Mount Lofty Ranges, is composed of rocks laid down in this seaway. The city of Adelaide itself is over its original bed, and also much of Yorke Peninsula and a considerable area to the east. Now, these strata have not lain undisturbed through the ages, but have been squeezed up, folded, and again worn down. Their remnants now lie at extreme angles to the horizon, and along the southeastern flank of the Mount Lofty Ranges they have been strongly metamorphosed. Their sequence is nevertheless quite clear and has been extensively studied and worked out. The sequence is up to 5000 metres thick in parts of the northern Flinders Ranges (Lake Frome region), up to 2000 metres elsewhere in the ranges southward to Kangaroo Island; but as much as 9000 metres were laid down during the early part of the Middle Cambrian interval alone in the form of the now metamorphosed Kanmantoo Group (*Daily, 1957; Daily and Milne, 1972*).

Many types of rocks make up the sequence, but there is a predominance of shales or slates, sandstones, and quartzites in many areas. In others there are many limestone beds, and in the north the series ends with a thick "red bed" sequence, thus indicating arid, desert conditions at that time and in that area. There is, incidentally, a complete absence of volcanic formations. Some of the limestones had magnesia added and have become dolomites, others are of organic origin and largely composed of layers of primitive coral-like animals known as *Archaeocyathus*. These, though superficially similar to corals in method of growth, are no longer regarded, as was previously believed, as the first builders of coral reefs of the world. Yet, they conjure up a picture of clear, placid seas under blue skies, and of a warm if not tropical climate with temperatures sufficiently high to support prolific growth. There is independent, geophysical evidence of Australia having been in tropical latitudes in Cambrian times. Magnetic sediment particles, such as flakes and grains of iron oxides, when laid down on the sea floor, come to rest with their magnetic poles orientated according to the direction of the earth's magnetic field. According to Irving (*1964*) studies on Australian Cambrian sediments indicate a position of the magnetic poles at that time which would have the equator running from Victoria through central Australia to the Kimberley District. The picture of tropical seas full of *Archaeocyatha* meadows is therefore substantiated, providing it is correct to assume that the spatial relations between magnetic poles, rotational poles, and climatic belts of the earth are for Cambrian times roughly similar to those of today. The assumption is "reasonable", or "probable"; we cannot, however, prove it.

From the late mid-Cambrian onward strong mountain-building movements in the Adelaide Geosyncline again changed the face of the earth in the Australian region considerably. The movements began in the south, so that by the beginning of Gibbus Time (see Fig. 4) the geosynclinal rock pile was definitely involved in the folding. Much of the area subsequently emerged from the sea, and a prominent mountain range arose during the Late Cambrian through these movements which, in South Australia, are known under the name of Delamerian Orogeny (*Thomson, B.P., in Parkin, 1969*), whereas in Tasmania the name Jukesian Orogeny is used (*Banks, 1962*). Northwestward to central Australia and north into Queensland the folding forces seem to have been considerably weaker and, because of a large blanket of younger rocks in the Great Artesian Basin, it is not even certain whether the movements in the north (Ravenswood-Anakie areas of central Queensland) are contemporaneous and tectonically connected with those in the south.

Map (10) shows that the "Meridional Divide" remains a prom-

Map 10:

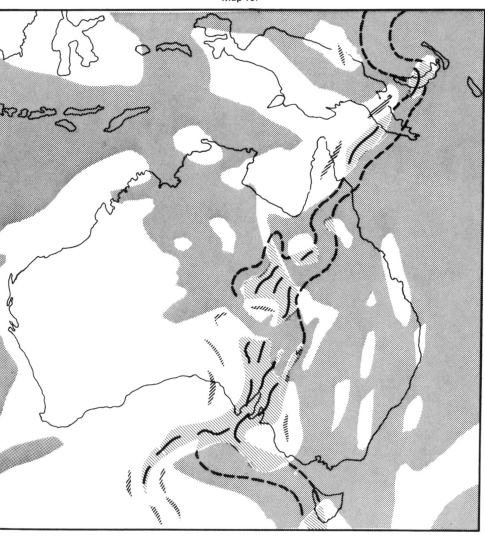

 SEA

LAND

 ZONE OF CONTEMPORARY
OR JUST COMPLETED
MOUNTAIN-BUILDING

 AXIS OF UPWARPING
—IN PLACES AS AN EFFECT OF
DISTANT MOUNTAIN-BUILDING FORCES

The north-south Meridional Divide which existed after the Middle Cambrian phase of the Delamerian Orogeny. It separated a provincial Trilobite fauna, living in the Northern Territory, from another contemporary fauna on the Queensland side of the Divide which was spread over a much larger part of the world. The events of the early Delamerian mountain-building are known in Tasmania as the Jukesian Orogeny.

inent feature, but also that it is beginning to break up both in the north and the south. The fairly extensive Willyama land areas are now mostly submerged, and so is much of Carpentaria. Yet for some time during the Middle Cambrian—in fact, from Atavus to Laevigata Time (see Fig. 4)—the sea withdrew from central Australia so that southern Carpentaria linked up with the Western Australian land mass. Thus during that particular slice of time the whole of Western Australia and South Australia, and a large part of the Northern Territory were above sea-level. In the New England region island arcs emerged late in mid-Cambrian times and supplied very coarse sediments (e.g. pebbles and cobbles with tell-tale fossils in them) into the seaways between the islands (*Cawood, 1976*).

With the beginning of the Upper Cambrian the seaway from the Kimberleys across the Northern Territory into Queensland was established again, although it remained at times rather narrow as is shown on Map (11). Towards the end of the Cambrian, however, and through the Ordovician this seaway widened very considerably, and much of the land gained through Cambrian times was lost again to the sea even though these waters were quite shallow—in places so shallow as to produce salt beds by evaporation.

It is only in eastern Australia that we find Cambrian sediments which were laid down in seas of greater depth. When it became submerged rather suddenly at the beginning of Middle Cambrian times. Willyama was broken up into deep-sea troughs and island arch—that is, a pattern such as is shown on Map (11). Among others, the island of "Duchess Land" (*Opik, 1957*) is submerged, but another island about, and west of, Mount Isa has emerged. A characteristic feature of the Middle and Upper Cambrian marine series in southeastern Australia is their content of volcanic rocks, a result of the break-up of the Willyama part of the earth's crust. No such volcanics are known to have been extruded over the more consolidated regions of central and northern, or western and southern Australia after the Late Proterozoic—at least not until very much later in geological history, and even then they are of very minor extent.

In Victoria Cambrian rocks are mostly buried beneath a vast thickness of later Palaeozoic rocks, and are strongly folded with them. Only in the centres of some of these folds, and where the overlying rocks have been removed by erosion, are they revealed. The land surfaces in this part of Australia are much younger than those in the Northern Territory, or the Flinders Ranges, and the sea transgressed not only once or twice, but many times in subsequent ages.

None of the Victorian series is older than Middle Cambrian. They are found in several north-south trending zones from as far west as the

Map 11:

	SEA
	LAND
	SALT LAKE
	STRUCTURAL TRENDS OF LAST COMPLETED FOLD MOUNTAINS—NOW UNDER INTENSE EROSION

The Australian region during the middle part of the Late Cambrian, ca. 500 million years ago. Note that the large Precambrian shield areas in the west and north of the continent are now fully emerged. It was also about this time that the oldest beds of the mighty Owen Conglomerate in northwestern Tasmania were deposited.

Glenelg River to Mount Wellington in northern Gippsland. One of the best exposures is near Heathcote, a town about 90 kilometres north of Melbourne. The locality is nine kilometres north of the town, and the fossil beds with trilobites are underlain by shales, hard black siliceous rock (called chert) with radiolarians and hydrozoans, flows of basaltic and similar lavas, and beds of volcanic ash or tuff, all together more than 1500 metres thick. A similar sequence is found at Lancefield, about 32 kilometres farther south, and along the Dolodrook River near Mount Wellington.

The story in Tasmania resembles that in Victoria except that the sea was probably much shallower in most parts. Volcanic series are common. As in the southeastern sector of the Adelaide Geosyncline, the prolonged crustal unrest brought granite magmas close to the surface toward the end of Cambrian times, and several of these granite intrusions in central northwestern Tasmania have since become exposed by erosion. The Cambrian rocks of Tasmania, known as the Dundas Group, are in many places over 3000 metres thick (*Banks, 1957*).

This concludes a general survey of the Australian scene in Cambrian times. In view of the earliness of the Period the extent of our knowledge seems astonishing, but what is known is only a fraction of the whole. There remains a vast field for research, by the enthusiastic amateur as well as the trained geologist. The occurrence of fossils in unexpected places should be particularly noted, and specimens, however poor, sent to one of the various museums or mines departments for identification.

LIFE OF THE CAMBRIAN PERIOD

The Age of the Trilobites

The life of the Cambrian Period is of extraordinary interest, for it is the earliest in which every main branch of the animal kingdom except land animals and vertebrates is represented. The fauna, although in many aspects still primitive, shows an astonishing diversity by comparison with what we know of life at the end of the Proterozoic Eras, and there can be little doubt that it culminated from long lines of ancestral types in the Precambrian of which we know little or nothing as yet.

Amongst the most interesting of the Cambrian fossils is the large and curious group of the Archaeocyathinae (the typical genus of which is *Archaeocyathus*). These are not only amongst the earliest animal fossils found in the world (they did not survive into later

Cambrian times), but their structure (cup-shaped—*cyathus*: cup) is such that they cannot be definitely assigned to any of the ordinary main groups of the animal kingdom. In some text-books they have been placed with the sponges; other authors consider them to be calcareous seaweeds. For still others they represent forerunners of the corals although much more coral-like forms such as *Cothonion* occur already as early as the beginning of Middle Cambrian times (*Jell and Jell, 1976*).

The classic work on the Australian forms is a monograph by Taylor (*1910*). After careful consideration of their structure and other evidence he came to the conclusion that, while some writers link them with the corals, on the whole they are more closely related to the calcareous sponges, and that they form a class by themselves. This view is now generally accepted and the class term Pleospongia is used for them. They are possibly one of those early and generalized classes anticipated by students of evolution, supposedly containing the ancestral stock from which both sponges and corals have been derived, and therefore one of the "missing links" in the evolution of life.

The Archaeocyathinae were conical or cylindrical cup-shaped organisms varying in length from less than 3 to 50 centimetres or more, with a diameter of up to 12 centimetres (Pl. 3). They had a central cavity (which, however, was not a body cavity) and a double wall. The animal tissue was between the two walls, both of which were perforated with numerous pores. This space was also divided by numerous vertical partitions or septa, which were generally plain, but in some forms were replaced by numerous rods, or in others bent and fused so that the living space was filled with vertical hexagonal chambers. In others again, cross plates or floors (tabulae), which divided the organisms horizontally, were present. The whole was attached at the base to the rock or to fragments of other dead individuals.

This group has been found in Cambrian, especially early Cambrian, rocks in many parts of the world—from Alaska to Nevada, from Siberia to Spain and Sardinia, from China to Antarctica—and hundreds of species have been described. The very fact of this well-nigh worldwide occurrence does, incidentally, shed some doubt on the tropical habitat of the Archaeocyathinae, and it seems not unlikely, as Opik (*1957*) suggested, that the general global climate in the Cambrian was different from that of today in such a way as to widen the latitudinal belts north and south of the equator in which Archaeocyathinae were able to live.

In South Australia and the Northern Territory specimens are very well preserved, and in places so abundant that whole masses of limestone are largely composed of their remains. Two of the handiest collecting localities in South Australia are south of Adelaide at Normanville, near Cape Jervis, and at Sellick's Hill a few kilometres to the north. Good specimens are also found at Ardrossan on Yorke Peninsula; 300 kilometres to the north on the railway at Wilson; at Wirrealpa, about 150 kilometres farther north again; and finally, one of the richest deposits of all, on a small limestone hill behind the Ajax Mine, 15 kilometres south of Copley in the Flinders Ranges.

Before other Australian Cambrian fossils are discussed, a remarkable Canadian find should be mentioned. It was made by Charles Doolittle Walcott in 1910 on the southwestern slope of Mount Wapta, near Field, in British Columbia. Here, not merely the hard parts of animals but many soft-bodied creatures have miraculously been preserved. They lie as thin carbonaceous films between the laminae of a fine black shale, and are so exquisitely preserved that fine, hair-like appendages, scales, and even the intestinal tracts are clearly discernible. Amongst these are jellyfish, annelid worms, holothurians (to which belongs the bêche-de-mer), small shrimp-like creatures, and curious many-legged little animals, belonging to a group known as the onchophoroids. These were members of the arthropoda—the phylum including crabs, insects, and spiders among others—and they looked something like caterpillars. There is still one living representative; it is called *Peripatus*, and it now lives in damp places on the land instead of in the sea. It has been intensively studied and, since some of its characters resemble closely those of the annelid worms, some zoologists see in it another "missing link", this time connecting the worms with the arthropods.

Many places in the world have yielded a host of Cambrian fossils— Great Britain, northern Europe, the Mediterranean, the United States, Asia from Russia across to China, South America and even Antarctica. Australia has made major contributions to the knowledge of Cambrian faunas, from their general aspects right through to the detailed life history and the intricacies of the intestinal systems of certain species of trilobites. The earlier work was mostly done in Victoria and South Australia. In Victoria the rocks are strongly folded and altered, the material is limited and on the whole rather poorly preserved. Until recently much of the material described from South Australia was in the same category, but the work of Daily (*1957*) has revealed much fine material which consists of at least ten distinct successive faunas. Very good finds have also been made in Tasmania (*Banks, 1962*).

By far the richest finds are being made in the Northern Territory and in western Queensland, where the rocks have remained undisturbed and fossils are not only prolific in many localities, but also of great variety and beautifully preserved. Of course, collecting there is not easy. The sparsely inhabited country, the lack of good roads, the scarcity of good rock sections such as are found in the gorges of higher country, and the covering of alluvium on the many plains, all add to the difficulties of the fossil hunter and geologist.

A highly interesting and important discovery was made in 1939 at Thorntonia Station in the far northwest of Queensland by Riek and Whitehouse. The discovery was of two new classes of animals belonging to the echinoderm phylum—the phylum which contains such classes as starfish, sea-urchins, and sea-lilies.

The echinoderms described by Whitehouse (*1941*) are of a very primitive type and, for some time, it was believed that each of the two forms corresponds very nearly to a distinct stage in the growth of the larva or embryo of living echinoderms, thus forming a link connecting very ancient fossils with present-day forms. However, both *Cymbionites*—a cup formed by five rounded plates of equal size, about one centimetre in diameter—and *Peridionites*—oval in shape, about half a centimetre in diameter, and formed by two unequal pairs of plates, surmounted by a fifth—are now recognized simply as very early sea-lilies, or eocrinoidea.

The bed of limestone which contains *Cymbionites* is so packed with specimens which have partially weathered out, that walking in the hills on the left bank of the Thornton River about a kilometre south of the station homestead is, to quote Whitehouse, "like walking over thickly strewn, embedded marbles". The other form, *Peridionites*, occurs on a higher level than *Cymbionites*, and a 1.5-metre bed of solid limestone is composed of almost nothing else but its remains.

Brachiopods or lamp-shells form another interesting group with a long geological history. They first appear in the Late Proterozoic, are fairly common though not much diversified through the Cambrian, and are still living throughout the world, quite abundantly also in Australian seas. They will not perhaps be familiar to most people, for they generally live in deeper waters and are rarely washed up on the shore. Though secreting a two-valved shell they are, as already pointed out in chapter 4, not related to the cockle or mussel. Within the shell, loops or spiral processes for supporting the brachiae or breathing organs are generally present. These internal structures are sometimes of great complexity, and are important for purposes of classification, though in fossil specimens they are not easy to see, except when revealed by a lucky fracture or by weathering.

Cambrian brachiopods were of a very simple type; in most of them the valves were not hinged, the shells were thin and of a horny texture, and internal processes were absent or little developed, such as in the form of two hooks. Two genera of these simple types still survive, *Crania* and *Lingula*. Living specimens of *Lingula* from the

Queensland coast grow attached to a stalk, and look not unlike the common "goose" barnacles. It is remarkable that simple, generalized organisms such as these should have survived under changing conditions, age after age, while more complex and specialized groups have developed one after another, became dominant for a period, then died out.

True bivalve shells or pelecypods were not at all common until latest Cambrian times although their immediate forerunners, the rostroconchs, are not rare. The more or less spiral shells of the gastropods, however, occur quite frequently already in the early Cambrian, whereas the first cephalopods, the class which contains the pearly nautilus and the octopus, do not appear before the Late Cambrian epoch (*Runnegar and Jell, 1976*). Two other curious organisms have also been found, *Hyolithes* and *Tentaculites*. The first had a straight, slender, conical shell, flattened on one side, and capped with a lid or operculum. The second was also slender and conical, but was round in section, and ornamented with transverse ridges. Both animals have been classed with or near the living pteropods or wing-snails, tiny molluscs which are pelagic in habit and thus swarm in the open sea. Neither really resembles any known pteropod, and their inclusion in this group is largely a matter of convenience.

Most characteristic and prolific of all Cambrian animals were the trilobites, a curious group of creatures long since extinct. They were true arthropods, and are classified with the Crustacea, of which the trilobites form an important sub-class. The sub-class is in turn divided into several orders, many families, and a great number of genera and species. They dominated the Cambrian seas, and myriads of them crawled on the bottom or swam about. They were not large: some were a centimetre or less in length, others grew to a few centimetres. They were more or less flat, oval or elongately oval in shape, the body being in three main parts: the head and the tail (both protected by a shield), and the thorax, which was divided into transverse segments varying from two in number to a great many. Two longitudinal furrows also divided the body lengthways into three regions, and one can see how, from such threefold division, the term trilobite was coined. Each segment of the body was provided with a pair of jointed legs, sometimes modified for swimming, but as these legs were right beneath the body they are rarely seen in the fossil specimens. Some species were quite blind, others had large, complex, many-faceted eyes. Some were quite smooth, others were armed with long spines, and at the height of their evolution many acquired the power of rolling themselves into a ball, and are fossilized in this position.

A phenomenal variety of trilobites have been found in Cambrian rocks. Though of somewhat primitive type compared with the later

Ordovician to Permian forms, they were sufficiently complex and diverse to suggest a previous long-continuing ancestral development in the Proterozoic. However, as mentioned in earlier chapters, because they probably were soft-bodied animals before the Cambrian, we have as yet no clue to their derivation.

In Australia they have been found in all mainland States and Tasmania. In other words, almost wherever Cambrian rocks occur there are also a host of trilobites. But it is in western Queensland and through the Northern Territory into the East Kimberley region that specimens have been obtained in the best condition and in the greatest numbers. One of the most prolific sources in western Queensland is on the Templeton River, about 30 kilometres west of Mount Isa. Shales full of trilobites were discovered by Campbell Miles in 1922, when he found the great silver-lead deposits of Mount Isa, and it is this rich assemblage of trilobites which became the subject of a now classic monograph by Whitehouse (*1936, 1939, 1941, 1945*).

A large and prosperous town now exists in this formerly isolated region, and the fossil beds are still an attraction for inhabitants and tourists alike. The collection of "beetles", as the trilobites are familiarly called, has for many become a pastime, and the upper reaches of the Templeton River are now known as Beetle Creek. In some of the beds trilobites are found in extraordinary profusion, and slabs of shale may be literally covered with their remains, notably those of a small species of *Xystridura*.

To the north of the Templeton River other collecting localities lie on the numerous tributaries of the Georgina River; and to the south and west, across the border into the Northern Territory, both from the surface and from wells sunk into the Barkly Tableland limestones, many fine specimens have been recorded. Still other localities lie in the hills of the Huckitta-Marqua region, along the Barkly Highway east of Tennant Creek, in the MacDonnell Ranges, and in the Daly River/Victoria River drainage areas in the northwest.

Over 500 species of Cambrian trilobites belonging to many different genera and families have been described from Australian localities and, according to Opik, at least another 200 are known and about to be described. Conspicuous among them are the agnostids—small, blind forms with relatively enormous head and tail shields, and only two or three thorax segments. They constitute one of the groups considered typical of Cambrian rocks, as they became extinct towards the close of that Period.

On Plate 4 a number of characteristic forms are shown, but for a more detailed description of the many species the monograph by Whitehouse, mentioned above, or the more recent papers by Opik (*1958a, 1961, 1963a, 1963b, 1967, 1975*) must be consulted.

Long ages were to elapse after the Cambrian Period before plants and animals left the sea and invaded the land. Though it is remarkable that so much should be known of this, the first fully diversified life on the earth, it may be again emphasized what a small fraction it must be of the whole, and what wonderful scope there still is for discovery, not only in the rocks of the Cambrian Period but also in those of other ages.

6 The ORDOVICIAN PERIOD

THE ORDOVICIAN, which followed the Cambrian, is the second of the six Periods of the Palaeozoic Eras. According to radiometric data the duration of the Ordovician was about 72 million years—that is, from 490 to 418 million years ago. In some parts of the world there is a clear unconformity between Cambrian and Ordovician formations and they can easily be separated from each other on a geological map. In Australia, however, the two series commonly have no obvious break or interruption of sedimentation between them. This means that the fundamental pattern of the distribution of land and sea in the Australian region remained the same from Late Cambrian into Early Ordovician times. Thus, looking at Map (11), and then at Map (12), we see that the general pattern is identical although there is more sea and less land on (12).

Under such circumstances one often finds that geologists differ on whether certain strata at the boundary between the two systems should be placed in the latter part of one period or the beginning of the other. As long as such strata are richly fossiliferous and also contain the recognized guide-fossils of the last Cambrian and the first Ordovician zone there can be no dispute, even if the boundary has to be drawn right through the middle of a particular limestone or sandstone formation. Nature, of course, seldom does us the favour of abruptly changing from one type of sediment to another in sympathy with our system of zones and stages based on the evolution of living organisms. The way nature erodes its lands and sheds its sediments into the sea has nothing, or little, to do with the way she evolves life on earth and the rate at which she does it.

There may, therefore, be one or more unfossiliferous, or poorly fossiliferous, formations between certain Cambrian and certain Ordovician beds which remain indeterminate. The correct way to treat this matter is to designate them as "Indeterminate: Cambrian to Ordovician" and enter them on the map and in the legend under an

Map 12:

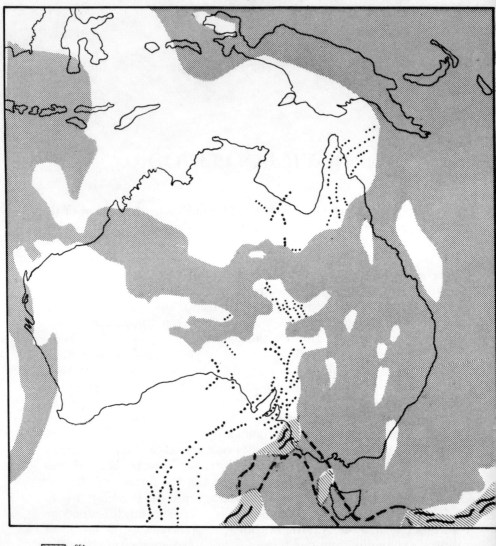

 SEA

 LAND

 ZONE OF CONTEMPORARY OR JUST COMPLETED MOUNTAIN-BUILDING

 STRUCTURAL TRENDS OF ANCIENT FOLD MOUNTAINS—NOW MORE OR LESS PENEPLAINED

The Australian region during the Early Ordovician, ca. 460 million years ago, showing the great east-west Centralian Straits and the beginnings of a marine transgression in coastal Western Australia which had been dry land since the Late Carpentarian.

appropriately different colour and symbol. More often, however, "convenience" steps in and one includes everything in one of the two systems, or simply splits it approximately down the middle. One simplifies the map; after all, map-making and printing cost time and money. Moreover, in many fields of Applied Geology such detail is of little use and therefore scarcely ever required and in most cases the situation is at least explained in the text.

This type of simplification is very common indeed, for few things in geology are clear-cut-and-dried for future generations of geologists. It also shows us that a geological map can be drawn up in various ways depending on its purpose. There are simple, and highly specialized, and complexly comprehensive maps, and the amount of information contained in each varies greatly. The geologist, confronted with nature's changeability and vicissitudes, must always make an intelligent selection from his innumerable, in part subconsciously accumulated observations before he can begin to interpret them in the light of his own experience and that of others. What he finally puts on the map and describes in an accompanying text is an intricate mixture of science and art, of facts, imagination, and reasonable probabilities, and therefore always to a greater or lesser extent disputable and unprovable.

This digression about the work of the geologist has been inserted deliberately because, protracted though the Ordovician Period was and in spite of the large development of rocks belonging to it, it is, especially in eastern Australia, one of the most difficult periods to study. For one thing, much of the system is hidden under great thicknesses of later rocks, with which it has been extensively crumpled and folded, and only in squeezed-up cores of such folds or in raised blocks and slices is it revealed on the surface. A large part is almost devoid of fossils, and the rocks themselves bear an often deceptive resemblance to the overlying Silurian. In many areas of southeastern Australia the geological structure is exceedingly complicated, which makes it very difficult to read records contained in the rocks in the right order. In spite of this, great progress has been made in our knowledge of the events which took place during the Ordovician.

Map (12) gives an idea of what things looked like about the turn from Cambrian to Ordovician times. In Tasmania the fairly strong Jukesian Orogeny which had ended in the early part of the Late Cambrian—and probably also affected western Victoria and New Zealand—left behind highlands in the west. These were, however, rapidly worn down again. As a result the basal formations of the next cycle of sedimentation, which is mainly an Ordovician story, are almost everywhere coarse conglomerates and pebbly sandstones, obviously derived from the western highlands, and resting unconformably on all older rocks. In some areas (*Webby, 1978*) the age of the earliest conglomeratic beds is still Late Cambrian, but in others their accumulation began only in the earliest Ordovician. Thus the great unconformity at the base of Tasmania's Owen Conglomerate does not invariably—and conveniently—mark the beginning of the local Ordovician System of formations, as was believed for many years.

Some of the water-worn boulders in the conglomerates of Tasmania are very large, a metre or more in diameter. The mass is very hard and compacted, and its resistance to erosion has given rise to much grand and rugged scenery. Mount Roland, St. Valentine's Peak, Mount Murchison, Mount Lyell and many other peaks are solid bastions composed of such conglomerates. It is difficult to imagine the exact conditions under which these deposits were laid down, but the most widely accepted theory is that they were formed as more or less water-worn slope scree, or fanglomerate, on the western slopes of a then existing Tyennan mountain range. The fanglomerate was caught in a graben-like valley. This valley was eventually invaded by the sea, whence came the deposition of marine sandstones and the Gordon Limestone with its rich Middle and Upper Ordovician shelly and corallian faunas.

The Ordovician System in Tasmania is known as the Junee Group. It is up to 2200 metres thick and includes considerable series of limestone, shales, and mudstones. Unlike southeastern Australia, where graptolites are the most common fossils, the Ordovician fauna here is shelly and graptolites are as rare as they are in central Australia. This is related to the depth of the seas. Mostly shelly animals are shallow-water dwellers and therefore rarely found as fossils in sediments that were laid down in deep waters, whereas the opposite is true for pelagic flotsam such as the graptolites.

The Ordovician geology of Tasmania is now incomparably better known than even just twenty years ago, when only sporadic occurrences of significant fossils had been described in a handful of papers. Until 1925 much of western and central Tasmania was almost uninhabited and geologically poorly explored. The pioneers were the prospectors, particularly the seekers after the rare osmiridium (an alloy of the two metals osmium and iridium). While trails were being cut through the mountains and the dense, cold, rain forests, rich deposits were found, and the little mining settlement of Adamsfield was established in the heart of the wilderness. The geological survey in this exceedingly difficult country was done by P. B. Nye (*1929*), who some 30 years later became the Director of the Australian Bureau of Mineral Resources, Geology, and Geophysics. Since then many geologists and palaeontologists have contributed much research; a fine summary of what has been achieved may be found in Banks (*1962*).

In Victoria, Ordovician rocks are of great extent, as practically the whole of the State was under the sea—though all parts were not covered throughout the Period, because the bed of this sea was gradually moving from the west to the east. The whole system of formations in Victoria has been closely studied, and divided into nine stages distinguished by different types of graptolites, an important group of extinct organisms of which more will be said later in this chapter.

If we draw an imaginary line north from Melbourne, most of the earlier Ordovician strata (stages 1–5) lie to the west of it, and all the later ones (stages 5–9) to the east. The latter extend into eastern and central New South Wales together with, at least as far north as Canberra, older Ordovician rocks. In other words, the western portion of the sea-bed rose above sea-level about half-way through the Period, while the eastern portion remained submerged and allowed the deposition of further series of strata.

The older Ordovician stages visible west of Melbourne are, from the lowermost up: (i) Lancefieldian; (ii) Bendigonian; (iii) Chewtonian; (iv) Castlemainian; (v) Yapeenian; (vi) Darriwilian. Apart from the differences in their fossil content, all these series are very similar in character. They consist mainly of very fine-grained black shales and slates, with a few beds of fine sandstone, and their deposition, commonly disturbed by submarine slope-creep and slumping, was probably fairly slow and took place in deep water. Although the Cambrian rocks to the west and south had been folded and elevated the coarser sediments from their erosion, because of north-south-trending swells and thresholds, were not carried into the central Victorian deep-sea.

Nearly all the once so rich goldfields of Victoria lie in Ordovician rocks or in alluvial gravels and sands derived from them. In Bendigo, Ballarat, Daylesford, Maldon, Dunolly, Poseidon, and Steiglitz the original location of the gold is in quartz lenses, or saddle-reefs, which are interspersed in the top of tight folds, often one above the other. These were formed at a later age of mountain-building. As the arches of the folds tended to open upwards under great lateral pressure, heated gold-bearing silica solutions filled the saddle-like cavities.

The four younger Ordovician stages are found throughout the eastern part of Victoria except where covered by comparatively thin remnants of Silurian, Devonian and later rocks. The four younger stages are in ascending order: (vi) Darriwilian; (vii) Gisbornian; (viii) Eastonian; (ix) Bolindian. It will be noted that the sixth stage, the Darriwilian—in fact the Middle Ordovician—has the widest extent of all as far as one currently knows. This shows that at this time the sea extended right across southeastern Australia as is shown in Map

(12). In the Snowy Mountains of southern New South Wales the Kiandra beds, which are of Darriwilian age and include substantial thicknesses of volcanic tuffs and lava, continue northwards, rapidly thinning, through Canberra (Pittman Formation—*Opik, 1958*) into the Goulburn District.

The maximum thickness of the three latest Ordovician stages in Victoria is about 750 metres. The rocks are very similar to those of the Lower and Middle Ordovician, fine black shales and slates are common, though a number of beds of coarse sand and grit suggest that land, probably islands, existed in the vicinity during part of the time. Northward through New South Wales Ordovician rocks are known as far as about Lat. 32° South in the Cobar and Dubbo districts, after which they are hidden under younger formations, or are in part probably absent, until they appear again in northern Queensland's Einasleigh District (*White, 1959, 1965*). Most of them are of Middle and Upper Ordovician age as far as sporadic fossil finds have indicated, but the presence of Lower Ordovician, and even Cambrian marine series is strongly suspected in a number of areas from Moruya on the South Coast (Mallacoota/Wagonga phyllites and quartzites) and the Snowy Mountains through the Australian Capital Territory to north of Orange. Thus, not only the Middle but also the Lower Ordovician sea may have extended over a large part of the eastern Australian region. Around Cobar probably the entire Ordovician is represented.

On the Shoalhaven River at Tallong, near Goulburn, and on Capitol Hill in the centre of Canberra the situation of the Ordovician formations is very interesting, for they are overlain by rocks of the next period, the Silurian, and there is a strong unconformity between the two, showing that the older rocks were folded, elevated, and to some extent eroded before they were submerged again to have the Silurian deposited on them. Evidence of such movements is found also in other places, and from one of these localities, the Benambra Highlands in Victoria, these movements have been called the "Benambran Orogeny" (*Browne, 1947*). However, as Opik (*1958*) has shown, this orogeny was merely one of several preliminary phases of a much greater mountain-building event which culminated towards the end of the Silurian.

Apart from the Einasleigh area, where Upper Ordovician has been proved by fossils (*White, 1965*), there is a large area along the east coast in Queensland, from Cape York Peninsula to the New South Wales border, in which Ordovician rocks are strongly suspected to occur, for there is a great thickness of more or less metamorphosed formations which are definitely older than Silurian. The Brisbane Schists, which are up to 6000 metres thick, have already been mentioned in connection with the Cambrian sea in eastern Australia. It is quite reasonable to assume that these easternmost regions of Australia were to a large extent under the sea during the whole of the Ordovician Period too. Other rocks, similar in character and position to the Brisbane Schists, occur farther north near the mouth of the Fitzroy River, and along the coast from Townsville to Temple Bay on Cape York Peninsula. The question of the exact age of these rocks is a very difficult one, and a great deal of work needs to be done even to solve one or two of the main problems.

But let us now return to Map (12) for an idea of what happened in the large and shallow seas in central and northwestern Australia and, strangely enough, also—for the first time again since the Mid-Proterozoic—in a newly opening seaway along the western coast from Perth to the Northwest Cape. The main formation on the west coast is known as the Tumblagooda Sandstone of Clarke and Teichert (*1948*). According to Condon (*1965*) this shallow-water formation is 6000 metres thick in some areas, but its exposures at the surface are few. Apart from animal trails similar to those found in Cambrian and Ordovician rocks of central Australia there is no clear indication of the formation's age.

Until 1949 there was no positive knowledge of Ordovician rocks anywhere in Western Australia. Today one knows that they occur at depth beneath younger beds throughout the Canning Desert Basin south of Broome, from where they emerge around the Proterozoic Kimberley Block in the Fitzroy River Valley in the west (Prices Creek) and towards the Bonaparte Gulf in the east (Pander Greensand). Southeastward from the Kimberley District various Ordovician formations continue, thinning for a time across a threshold along the Northern Territory-Western Australian border, into central Australia, and across into Queensland, where they link up with the eastern deep-sea series beneath the younger beds in the Great Artesian Basin. Two Ordovician embayments, which may at times even have been connected, are evident along the southern margin of the Centralian seaway, one extending westward from the Simpson Desert region along the southern side of the Mann-Musgrave Ranges, the other pointing south along Long. 124°East. The latter is an extension of the Canning Basin; the other is known as the Officer Basin. In all—except the last named—basins and platforms through the centre and the north of the continent a great deal of mapping, research, and drilling (mainly by oil search companies) has been carried out in recent years, and all the events of Ordovician times are now clearly outlined, at least in their main points.

In the Canning Basin inland from Broome Lower and Middle Ordovician beds measuring from 1500 to 2400 metres, mostly shales, have been cut by drills looking for oil. Northeastward towards Broome and the Fitzroy Valley this series thins on an old threshold to about 600 metres of mostly calcareous or dolomitic rocks. These rocks probably continue beneath the very deep Fitzroy Trough to surface against the Precambrian of the Kimberley Block where, as might be expected, the limestones and dolomites are frequently interbedded with sandstones and shales, the material of which was derived from the land to the north. A rich, shelly fauna of peculiar and long extinct ancestors of today's pearly nautilus from these beds was described by Teichert and Glenister (*1954*), but there are also trilobites, conodonts, brachiopods, gastropods, and even a few graptolites.

Towards central Australia too the Ordovician formations emerge from beneath the younger blanket rocks which hide them in the Canning Basin. Also, as we move into the Northern Territory, the Ordovician lies no longer directly on Proterozoic or Archaean rocks; over large areas it is underlain by the Cambrian series we have discussed in the previous chapter. This is seen particularly well along the MacDonnell Ranges near Alice Springs, and from there southward through the folded Amadeus Basin as well as to the east through the Huckitta region into northwestern Queensland.

Although there are some fairly extensive layers of limestone, most of the sediments through this region are detrital, that is, they consist of sandstones, sandy shales, and here and there conglomeratic formations. The thickness of the whole system varies greatly because there were constant irregular up-and-down movements of the sea floor which culminated in what is known as the Rodingan Deformation toward the end of the Ordovician Period (Map 13). After this the sea was pushed out of central Australia forever; all subsequent sedimentation in the area is non-marine. It is noteworthy that these important events coincide with the Benambran Orogeny in southeastern Australia as if they were a tele-effect. We will see a similar situation again some 50–60 million years later in Late Devonian times, i.e. during the Alice Springs Orogeny, the last major mountain-building event to affect the continent west of Long. 142°East.

The MacDonnell Ranges themselves are formed by the southern flank of the biggest of all folds thrown up in that Alice Springs Orogeny; of the northern flank nothing can be seen now, as erosion has removed it. Another, considerably smaller one of these folds, which trend northwest across the Western Australian border, east and west through the centre, and eventually northeast towards the Queensland border, forms the elongated breached dome known as the Waterhouse Range, 24 kilometres south of the main range. The rim of this structure consists of hard Ordovician sandstones, which rise several hundred metres above the plain, and in the core of the fold we find Middle and Upper Cambrian limestone, sandstone, and shale.

A most peculiar breached structure, only about six kilometres in diameter, with a near-circular rim standing on end or in the south even overturned, is Gosses Bluff (Plate 32), about 95 kilometres west of the Waterhouse Range. In a way it resembles an extinct volcano and there are, in fact, some minor indications of volcanicity in a rocky hill a few kilometres to the south. Like others of its kind in the world, Gosses Bluff is a so-called cryptoexplosive structure—crypto meaning secret or unknown—because one does really not know for certain how it originated. First it was thought to be a salt dome, then a rather peculiar extinct volcano, and later I suggested (*1969*) in this book it might even be a large extinct earth-gas and groundwater mud-volcano. However, these days the most widely accepted explanation of Gosses Bluff says it is the mark left by an extraterrestrial body which crashed onto the earth's surface maybe as recently as Tertiary times. It is most probably what is known as an impact crater very much like those which occur much more abundantly on Moon and Mars (*Crook and Cook, 1966*), that is, an astrobleme.

The MacDonnell Ranges are of great scenic beauty, a torn, rugged, and picturesque remnant of a once magnificent mountain chain. There is little vegetation to soften its outline in the desert air, and the vivid colouring has been made familiar in the paintings of such artists as Sir Hans Heysen and especially the late Albert Namatjira, who was a member of the local Arunta tribe.

The greatest thickness of the Ordovician System here is found immediately south of and along the western MacDonnell Ranges in the central part of the Amadeus Basin. Several thousand metres of it are exposed. Together with the already discussed older and also some younger systems it has been the object of many interesting studies, particularly by the geologists of the Commonwealth Bureau of Mineral Resources in Canberra, but also by geologists from Australian petroleum exploration companies who have found large natural gas reserves there. The Bureau's results have been released in a number of fine publications with 1 : 250 000 scale geological maps (*Prichard and Quinlan, 1962; Ranford, Cook and Wells, 1965; Forman, 1965, 1966; Wells, Forman, Ranford, and Cook, 1970*). Information on the much thinner Ordovician sequences to the northwest (Ngalia Basin) and northeast of Alice Springs (Georgina Basin) is also found in publications by the same institution (*e.g. Smith, 1972*).

Still another facet of the Ordovician story has been added by the petroleum explorers in the Great Artesian Basin. They have found that there was an embayment of the eastern sea into the Lake Eyre region via the Cooper Basin area in southwest Queensland. Slightly metamorphosed Ordovician formations also occur at depth in north-central Queensland, and it seems the Adelaide Geosyncline in its dying stages had an extensive but shallow northern branch which was folded only much later (Devonian movements), that is, long after the events of the Delamerian Orogeny in the south. However, studies made on bore cores from the Simpson Desert area indicate that the seaway which connected the ancient Pacific with the Amadeus and the Officer Basin directly via a depression in the Simpson Desert region (the Warburton Basin of Wopfner *1972*) was during both Cambrian and Ordovician times open only intermittently. For most of the remainder of these two long Periods the marine connections into central Australia seem to have been hindered at least by a prominent and barrier-like peninsular feature as depicted on Map (12).

LIFE OF THE ORDOVICIAN PERIOD

The Age of the Graptolites

Reference has been made to the graptolites, the interesting group of extinct organisms which are of particular importance and abundance in the Ordovician. Graptolites are not spectacular fossils. They occur as thin films or impressions on the bedding planes of shales and slates. Structurally they consist of a thin line or axis, on one or both sides of which is arranged a long line of cells somewhat like a bucket-dredge. They were thus compound animals of simple structure, related to the jellyfish, and in some ways not unlike the blue-bottle or Portuguese man-of-war, which in summer months is a menace to bathers on our beaches. Like the blue-bottle they were pelagic—that is, they floated in the ocean—and they must have swarmed in great numbers, being borne hither and thither by winds and ocean currents. Many of them as they died sank to the bottom, to be buried in soft muds which are prevalent in the open oceans. In shallow waters, however, wave action as well as untold legions of bottom-dwelling molluscs, trilobites, brachiopods and other animals, for which graptolites were food, left scarcely a trace of them even though they must have been as abundant in the shallows as in the deep open oceans.

There is evidence that the parts found as fossils were only portions of the animal, and that these parts were originally attached to a central disc or float that supported them on or near the surface of the water. They were represented by a great variety of genera and species, many of which have a worldwide range, and Australian scientists have contributed a great deal to the knowledge of these organisms. The late Director of the Geological Survey of Victoria, Dr. D. E.

Thomas, was a world-renowned authority on graptolites, and his work has become one of the classic publications on the subjects (*Thomas and Keble, 1933; Harris and Thomas, 1938*).

Unspectacular though the graptolites were, they are of tremendous value to the field geologist. They appeared at the end of the Cambrian, but attained their maximum in both numbers and variety during the Ordovician, and died out towards Middle Devonian times. Because the various species existed for a comparatively short time only, successively developing and becoming extinct, they are excellent for subdividing or "zoning" the whole sequence into stages, as well as for correlating these stages with subdivisions, based on the same fossils, in other parts of the world.

Long slender forms with the cells on only one side called *Monograptus*, for example, are typical of the Silurian Period. A branching form with small root-like attachments at the base and also with cells only on one side, *Dicellograptus*, characterizes the Upper Ordovician; *Diplograptus*, narrow and with cells on both sides, appears at the end of the Lower Ordovician. *Climacograptus*, narrow and with two root-like appendages at the base, comes in at the end of the Lower Ordovician too, but a little before *Diplograptus*. The four-branched *Tetragraptus* is characteristic of the Lower Ordovician, the broad and leaf-shaped *Phyllograptus* also, but appearing a little later. These are some of the commoner forms, but there are many others (Plate 5). Apart from graptolites fossils are very rare in the Victorian Ordovician, and only sporadically are trilobites or brachiopods found.

This is in striking contrast not only with other regions in Australia (except southern and central New South Wales) but also with many other parts of the world where a large part of the Ordovician faunas is shelly, although graptolites occur in many places too. It appears that the Australian graptolite sequence is the most complete and most varied in the world; it sets the standard for the rest of the world, and it is for this reason that Dr. Thomas and his collaborators' work was so important.

Information about shelly Ordovician fossils is available from the Kimberley District, the Amadeus Basin in central Australia and southeastern parts of the Northern Territory, the Toko Range in western Queensland, western New South Wales, and Tasmania. Late Ordovician bryozoa and corals are known from northern Queensland, central western New South Wales, central Australia, and Tasmania. From all these areas a considerable number of trilobites have also been reported, as well as echinoderms, conodonts, and others; in short, faunas as rich as any in the world.

Cephalopods of the present day are familiar to most people in the form of the octopus, the squid and the cuttlefish, and also the pearly nautilus. Some, like the cuttlefish, have an internal shell, and these have been found as fossils, but the geologist is more interested in those which, like the nautilus, secrete an external shell. These go back to the beginning of the Ordovician, perhaps even the late Cambrian, and the class has persisted right through the ages, being particularly important during the Mesozoic.

Examination of the shell of a pearly nautilus reveals that, unlike the shell of a snail, it is divided into numerous chambers. Only the outer chamber, the largest, is occupied by the body of the animal; the inner chambers are sealed off, but the partitions are pierced by a continuous, narrow tube or siphuncle. Some of the Ordovician cephalopods were involute like the nautilus—the shell was coiled in upon itself— and, since in other ways their shells were almost identical with that of the present-day nautilus, it may be that their habits and life history were much the same. This is a group that has persisted right through the ages with very little change, but offshoots from this main stock, as it were, evolved into many bizarre and even monstrous forms which flashed momentarily across the geological record and then disappeared.

The tendency for the cephalopods to evolve new and extreme forms with comparative rapidity will be noted frequently hereafter. In addition to the coiled type there were others with long, straight, slender, slightly tapering, more or less cylindrical shells. Which of the two types came first it is impossible to say, but whereas the nautilus and its relations survive to the present day, the straight shells were destined after several periods to become extinct. Both types are found among the shelly fossil faunas in the Australian Ordovician. A fine example of coiled ones is *Hardmanoceras lobatum* from Prices Creek in the Kimberleys, and of straight ones *Thylacoceras kimberleyense*, *Madiganella magna*, *Tasmanoceras zeehanense*, and others. Almost all of the descriptions of nautilid cephalopods—not only Ordovician ones— are by Teichert and Glenister (*e.g. 1952, 1954*).

Some of the Australian cephalopods were quite small while others attained a length of several metres. A specimen found in Ordovician rocks in North America is no less than 4.5 metres long. Such giant species generally flourished for a very short time only, and it would appear that the long straight shell was an encumbrance rather than an advantage. All cephalopods have a number of tentacles surrounding the mouth, and either lie in wait or dart through the water in order to seize their prey. In relation to their length, the body cavity of the giants was comparatively small. They were apparently extremely clumsy animals, dragging themselves laboriously over the sea-bottom, and unable to pursue any fast-swimming victims. Their very size would indicate that they were highly specialized for existing only under very particular conditions. As long as these conditions prevailed such specialized cephalopods flourished, but a small change would lead to their extinction, while other smaller and less specialized types adapted themselves and survived.

The brachiopods or lamp-shells were increasing rapidly during the Ordovician but had not become so varied and abundant as in subsequent Palaeozoic ages. A noteworthy feature is that by Ordovician times almost all lamp-shells had acquired solid calcareous shells with strengthening folds and ridges, a longer hinge-line, and more intricate internal processes. The simple horny forms common in the Cambrian persisted, however, and in fact still live almost unchanged in the present seas, whereas almost all of their more intricate and more handsome cousins of Palaeozoic times have disappeared. One of the handsome genera, very common in the Australian Ordovician, is *Orthis*. Several of its species occur in untold numbers at various levels in limestone and calcareous shale beds in the MacDonnell and other ranges of central Australia, northwestern Queensland, and Tasmania. Another genus, known especially from the Kimberley district, is *Spanodonta*.

An important event is the sudden forceful spreading of the bivalve molluscs through the early Ordovician although most of them still belong to the primitive types. The upper part of the Pacoota Sandstone in central Australia is full of them. The shell matter of the bivalves was formed of the more soluble, unstable, carbonate of lime known as aragonite. In sandstones and shales it is seldom preserved and the specimens are therefore preserved only as internal or external casts or moulds. A common Ordovician bivalve is *Isoarca* which, like the common mud-living ark-shells of the present day, had a hinge with many zip-type teeth. Gastropods, though not very varied, are also quite common (e.g. *Raphistoma*), and have now developed much larger forms than in the Cambrian. Echinoderms are still rather scarce but corals and other colonial organisms appear in force towards the end of the Period, heralding their prominence later during the Silurian.

Trilobites, of course, are still swarming and crawling in great hordes over the sea-bottom and, in fact, reach the peak of their development during the Ordovician. They have become much more complex, larger, almost all now have eyes with many facets, and many could roll themselves into a ball, perhaps as a protection against some larger predators. The common trilobite families in Australia are the asaphids, a rather simple type with rounded head and tail shields, and the typical genus *Asaphus* is common in central Australia. *Ogygites* and *Xenostegium* are found in the Kimberleys. Another family are the illaenids, with genera such as *Bumastus*, occurring throughout the non-graptolitic Australian Ordovician, and many others.

A peculiar type of very tiny fossils, appearing late in the Cambrian, but very widespread in the Ordovician and later systems (until the middle Mesozoic) are the conodonts. These look like minute single and compound teeth set up on a base, and one does not know to what animal they belonged. The tiny teeth of worms or gastropods of the present day look rather different. There is a phenomenal variety of these conodonts and, like the graptolites, they are excellent zoning fossils wherever they are abundant. Australian examples are shown on Plate 5. All these microscopic or generally very tiny fossils are of great importance to the geologist searching for oil thousands of metres below the surface because they can be identified, and therefore the age of the rock determined, from the small fragments which the drill-bit chews off at depth and which the drilling-mud lifts to the surface. There are many kinds of such small fossils and, as we realize now, a lot of them are not at all simply single-celled protozoans but remains from otherwise destroyed, much larger, organisms, even such things as spores and later pollens of large plants.

Still another common fossil, persisting from the Cambrian to the present day, are the tiny bivalved crustaceans known as ostracods, and again many are known here from the Ordovician in Australia although few have been described yet.

The First Vertebrates

Among the most exciting palaeontological news of the past decade was the discovery, in the mid-Ordovician Stairway Sandstone of the Amadeus Basin, of very early vertebrates. They are jawless ostracoderms, now known as *Arandaspis* and *Porophoraspis*, the first forms of that kind found in the Southern Hemisphere (*Ritchie and Gilbert-Tomlinson, 1977*).

7 The SILURIAN PERIOD

THE CLOSE of the Ordovician Period witnessed great changes in the topography of the Australian region. There was considerable elevation in central and northwestern areas as well as in western areas of New South Wales, parts of Victoria, and central Queensland. The great central seaway disappeared, and its place was taken by mountain ranges and lake-filled valleys, especially in central Australia. The thousands of metres of sediments which had been deposited during the Cambrian and Ordovician were squeezed as if in a great vice; they were bent and folded, twisted and shattered to various degrees, and eventually pushed up into mountain chains. Then began another period of demolition; the mountains crumbled and were worn down to lower levels or, because of continuing crustal unrest, simply subsided again in parts below sea-level. In some regions, of course, the sea remained as it was from Ordovician into Silurian times, just as it did in many areas from Cambrian into Ordovician times. All these names originated in Wales, by the way, because it was there that the English geologists of the early 19th century first described and studied the respective rock formations. In Roman times Wales was known as Cambria, the Ordovices were a tribe in northern, the Silures in southeastern Wales. Because the differing rock sequences to the south in Devonshire overlie those of the country of the Silures, the next younger Period received the name Devonian.

Map (13) gives an idea of how our region looked some time after the Benambran Rodingan orogenetic movements had quietened down and the sea repossessed many areas in eastern Australia which it had lost for a time. Along the Western Australian coast the extent of the Silurian sea is similar to that of the Ordovician. However, Silurian rocks nowhere appear on the surface and are known only from drill holes. At depth beneath the Canning Desert the Ordovician series is followed by rocksalt beds of very considerable thickness, but these apparently contain no fossils which could tell us whether the beds

Map 13:

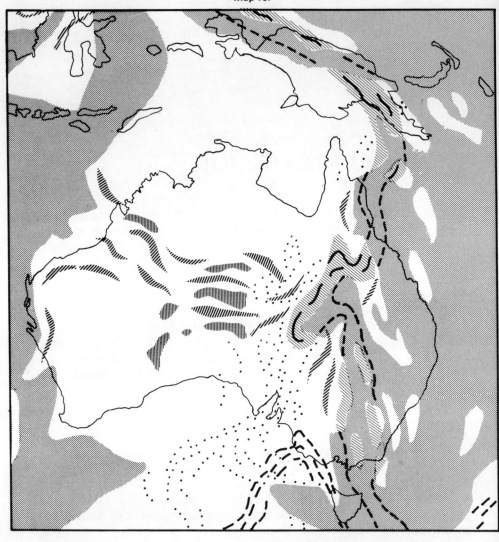

 SEA

 LAND

 FRESHWATER LAKE

 ZONE OF CONTEMPORARY OR JUST COMPLETED MOUNTAIN-BUILDING

 AXIS OF UPWARPING —IN PLACES AS AN EFFECT OF DISTANT MOUNTAIN-BUILDING FORCES

 STRUCTURAL TRENDS OF LAST COMPLETED FOLD MOUNTAINS—NOW UNDER INTENSE EROSION

 STRUCTURAL TRENDS OF ANCIENT FOLD MOUNTAINS—NOW MORE OR LESS PENEPLAINED

After the sea had been pushed out the welding-together of southern and northern Australia was complete by the time of the Rodingan crustal movements in central Australia. Orogenic events there— including the Rodingan which are related to the Benambran Orogeny in eastern Australia—are now usually tele-effects of mountain-building in the Tasman Geosyncline of eastern Australia. Although the major structural patterns in central Australia are now well established further, at times very strong, folding and thrusting in later Palaeozoic times will reinforce the older structures. The sketch shows the situation in early Silurian times, about 410 million years before present.

were laid down in the Silurian or later in the early part of the Devonian. The same uncertainty applies to some non-marine formations, mostly sandstones and conglomeratic rocks, known from various localities in central Australia and interior Western Australia. The most prominent and widespread of these formations, commonly about 1000 metres thick, is the Mereenie Sandstone in the Amadeus Basin of central Australia. It is also this sandstone which forms the upturned rim of the Gosses Bluff cryptoexplosive structure.

In eastern Australia the sea had retreated further towards the open Pacific, and no Silurian rocks are known to occur west of the 143rd meridian of longitude in the south, or the 138th degree farther north. By this latter northern area we mean the Palaeozoic Coopers Greek Basin which is lying hidden beneath the Mesozoic infilling of the Great Artesian Basin.

In the early Silurian the sea had probably not yet penetrated as far inland as is shown on Map (13), but apart from island arches it certainly covered the eastern half of Victoria, most of Tasmania, much of southern and probably central western New South Wales, and northeastern Queensland. By Middle Silurian times, however, the extent as shown on the sketch was pretty well established. This quick subsidence shows that the "Benambran" movements had not, as real orogenies do, consolidated anything much in what is known as the "Tasman Geosyncline", and this is probably due to the fact that there was little or no substantial injection of solidifying granite magmas into the folding region. As Opik (*1957*) has shown, the Ordovician and earliest Silurian is over large areas singularly free from igneous activity, especially from granitic intrusions. This was changed, however, from the Middle Silurian onward, particularly in New South Wales and Victoria, though not in Tasmania or Queensland.

At this point attention must be drawn to the fact that the Silurian Period had a duration of no more than 20–25 million years, that is from 418 to about 394 million years ago; it is the shortest of all the periods recognized in the geological time-scale. When comparing the sequence of events of the Silurian with those in the Ordovician or other periods, we must realize that the entire Silurian was probably no longer than say the Middle Ordovician or half of the Upper Cambrian. We will then not be surprised when we find that there were perhaps two or three significant orogenetic events in the Middle and Upper Cambrian alone, and only one in the whole of the Silurian.

Another important notion arising here is that the rate of deposition of sediments into the Silurian marine troughs was apparently three or more times faster than that prevailing during the Ordovician. The thicknesses of the two Systems are of the same order—but the duration of the Ordovician is three to four times longer than that of the Silurian. In spite of this the character of the two series is quite similar throughout the southeast Australian part of the Tasman Geosyncline's deeper troughs. Here again, however, "deeper" means in fact "faster rate of subsidence" because, as in the earlier Adelaide Geosyncline, evidence of deep water is at best intermittently observable. Most of the series was laid down in reasonably shallow water, although the sea was alternately shallower and deeper, and there was a corresponding difference in the coarseness and type of the sediments and in the living things which inhabited the waters.

Fossiliferous Silurian in Queensland is found from the Palmer River (Chillagoe) northwest of Cairns southward intermittently as far as the Clarke River west of Townsville. From there along the coast to the New South Wales border—that is, from near Rockhampton and Gladstone to Brisbane and Neranleigh-Fernvale—rocks believed to be of Silurian age contain no clearly identifiable fossils; they are more or less metamorphosed and their age is inferred from the field relationships with younger sequences of Devonian age.

The Chillagoe formation with its once rich mineral deposits is many thousands of metres thick and consists mainly of corallian limestones, cherts, sand- and silt-stones, and some minor intercalations of volcanics. It is of Upper Silurian age. The similar, but more complete sequence on the Burdekin and Clarke rivers is many tens of thousands of metres thick, but includes also Lower and Middle Silurian rocks, all laid down in a rapidly subsiding trench or graben which started to "cave in" towards the end of the Ordovician period.

In Victoria the thickness of the Silurian series may be as much as 6000 metres in places. It is only found east of a line from Melbourne to Heathcote and is divided into three stages from bottom to top: the Keilorian, the Eildonian, and the Melbournian. These correspond to the Llandoverian, the Wenlockian and the Ludlovian of the Silurian System in Great Britain, all distinguished by their fossil content, e.g. graptolites, cephalopods, and trilobites.

In the Walhalla district, about 160 kilometres east of Melbourne, Silurian series lie in a broad, north-northwest trending belt of folds with Ordovician rocks appearing in the cores of the folds and then, to the east, as a broad belt themselves. The disturbances at the end of the Ordovician Period had, in fact, been such here as to raise the latter belt above sea-level and prevent the deposition of Silurian beds on top of the Ordovician. Only much farther to the east, about the watershed between the Murray (up here known as Indi) and the south-flowing Buchan River, and on the Bonang Highway near Delegate (N.S.W.) do Silurian sediments appear again in the form of outliers of folded troughs which extend southward from New South Wales.

Silurian rocks are well developed near Melbourne, mainly to the northeast in the Upper Yarra district and near Lilydale. Most of the country consists of them, but to the north of Melbourne they are masked by lavas which in much more recent times spilled over the land. Melbourne itself lies well out over what were once deeper parts of the Silurian sea, and the rocks exposed in this area show a complete sequence of the three stages. They are well folded and consist mostly of sandstones and shales, the latter being used for making bricks and tiles. Many Silurian fossils were found in city excavations.

Another large development of Silurian rock occurs in the Heathcote district, about 110 kilometres north of Melbourne. These beds were formed nearer the shore-line than those around Melbourne, and though the sequence with its characteristic fossil zones is much the same the thickness is reduced by about one-quarter. At the base are the Costerfield mudstones, which contain scarcely any fossils and are at least 600 metres thick; then follow the mudstones of the Wapentake beds with the so-called "*Illaenus* band" at their base. *Illaenus* and other trilobites such as *Dalmanites*, *Calymene*, *Encrinurus*, and *Proetus*, together with brachiopods of the orthid, strophomenid, and athyrid families, and the wing-snail relative *Conularia*, are typical in this sandstone band. The Wapentake beds are about 1500 metres thick. Above them lie the 1500-metre-thick Dargile beds which in various horizons contain many species of trilobites, brachiopods, and graptolites. At the top is the McIvor sequence of 1500 metres, of sandstones chiefly, which are richly fossiliferous (trilobites, brachiopods, cephalopods, pelecypods).

In Tasmania there was a change from limestone to sandstone deposition at the beginning of the Silurian because the source area of the sediments in the northwest was further uplifted at the time of the Benambran movements. Boulders of Ordovician Gordon Limestone are found in the lower Silurian sandstones of central western parts of the island. Moreover, this uplift resulted in the inundation of northeastern Tasmania which had received little or no sedimentation in the Ordovician. Several thousand metres of silt-, clay-, and sandstones (Mathinna beds) were laid down here, whereas the slates and sandstones of the Eldon Group in central western Tasmania, near Queenstown, reach as much as 3600 metres; but some of the latter are already of Devonian age. While the Mathinna beds contain few fossils, the lower (Silurian) part of the Eldon Group has been subdivided by means of graptolites (*Monograptus* and *Cyrtograptus* species) and ostracods (*Gillatia*). Especially in northeastern Tasmania it is rather difficult to separate the Silurian from the Devonian rocks.

The Tasmanian region is still a difficult one to analyse. One may ask, for example, how far to the west of Tasmania did the sea extend during Ordovician and Silurian times? There is little doubt that during much of the Proterozoic and probably also Cambrian times a continental region of great extent displaced the Southern Ocean, close to, and at times connected with, Australia. From its mountains must have come at least some of the great glaciers or ice-sheets of the Late Proterozoic ice ages. Perhaps the Silurian sea which inundated northeastern and western Tasmania was in the form of a great gulf separating this southern land from southeastern Tasmania. More likely, though, there were groups of larger and smaller islands in the southeastern Australian region as shown on Map (13). It was only when the sea-bed rose about Middle Devonian times that Tasmania again became part of this southland and remained so until the beginning of the Permian Period, as is shown in Maps (15) and (16).

In New South Wales conditions in Silurian times differed from those in Victoria in that for the early part of the period most of the northern and eastern regions remained above sea-level. Only in the south and the central-west of the state are Lower Silurian rocks known to occur. Middle and Upper Silurian, however, occur over most of the state. The sea entered from the south along the south coast and the Monaro Tableland and may even have spread inland through the Central West into the Coopers Creek Basin of northeastern South Australia and southwestern Queensland. Rivers from the highlands in the west as well as from islands in eastern Australia brought down sand and mud, and this accumulated in beds of sandstone and shale. In many areas the water must nevertheless have been clear enough to allow a rich growth of coral, and many reefs are to be found, comparable in extent with those of the present day.

It is interesting from the point of view of climate to consider what a wide geographical range these ancient corals had. Reef-building corals are confined today to shallow water in tropical and sub-tropical regions; that is, with 25° latitude to both sides of the equator. Yet corals are found in Silurian rocks in Australia from latitude 15° south to beyond latitude 42°. Studies of palaeomagnetism in Silurian rocks seem to confirm that these Australian corals indeed grew in tropical latitudes, but the whole question of climates in past geological ages is one about which very little is known for certain. Some conclusions can be drawn from the present distribution of life in the sea, but it is not certain whether similar organisms, corals for example, were subject to the same climatic restrictions in the past. The problem is one for physicists and astronomers as well as geologists. Many questions remain unanswered. Was the climate of the earth always divided into polar, temperate, and tropical belts? Or has the degree of tilting of the earth's axis, which is the primary cause of the alternation of summer and winter, varied from age to age, with a corresponding increase or decrease in the extremes of seasonable temperature? Have the positions of the poles themselves altered? Or did the position of the continents relative to the poles and to each other change?—did they wander? It is all very perplexing, but data are slowly accumulating which should eventually answer most of these questions.

One of the best areas to study the corallian type of Silurian is at Yass, 65 kilometres northwest of Canberra. Here the country is open and undulating, and the gently folded strata are well exposed on the sides of the low hills and in the beds of the Yass River and tributaries such as Derrengullen Creek. Interstratified with the sedimentary rocks are two massive beds of volcanic rock, ancient lavas which flowed from volcanoes in the vicinity. The other rocks consist largely of shales and limestones. It is a wonderful thing to be able to walk for hours over eroded surfaces of some of the earliest coral reefs in the world. Some of the individual colonies of coral are very large. In the bed of Derrengullen Creek is one continuous mass of a large branching reef-builder, over five metres across, which lies undisturbed, just as it grew over 400 million years ago. The shales between the coral limestones are also crowded with fossils, lamp-shells for the most part, and the uppermost part is a bed rich in trilobites. Similar conditions may be found at many other New South Wales localities, notably at Fernbrook near Bathurst, and at Borenore and Wellington. At the Jenolan Caves the rocks are so greatly folded and, in some places, altered that many contained fossils have been destroyed.

LIFE OF THE SILURIAN PERIOD

The First Big Coral Reefs

Among the lower marine forms the graptolites had decreased considerably in number and variety at the end of the Ordovician, but some types, for example *Monograptus*, are very typical and quite common in the Silurian System. Other members of the phylum to which the graptolites belong existed in Silurian times, but their soft bodies are rarely preserved. One fine specimen of a jellyfish, found in a brick pit at Brunswick in Victoria, is in the National Museum in Melbourne.

Mention has already been made of corals and coral reefs, and their development was a conspicuous feature of the period. Corals, it is true, had appeared in Ordovician times, but even in the early part of the Silurian they were rather small and insignificant. It was only in the latter part of the Silurian that they increased enormously in size, number and variety.

The individual coral animal or polyp has a rather simple anatomy, with a body wall, a mouth surrounded by tentacles, a gastric cavity with internal, radiating, fleshy partitions, and little else. The food, consisting of very small organisms, is taken in and waste matter ejected through the mouth. Sea-anemones, which are related to the corals and are common on all seashores, have a similar anatomy. In the corals the soft parts secrete and are supported by a hard calcareous skeleton which is typically cup-shaped, but may be very varied in shape and internal structure. Corals may be simple, consisting of one individual polyp, or compound. Compound corals consist of many individual polyps united together by living tissue, so that in a way the whole colony may be considered as a single organism. It is the compound corals which form the bulk of a coral reef, though calcareous seaweeds, worm tubes, shells, and many other organisms contribute.

All Silurian types of corals have long since become extinct, but their descendants may be seen in the coral reefs of the present time, and there are reasons to assume that the conditions under which they lived and grew were not greatly different from those of today—clear warm seas, with no muddy sediments from rivers. The water must also be fairly shallow, for corals grow most profusely from below low-tide level to a depth of a few fathoms. It was probably under these conditions that coral reefs flourished in Silurian times in northern Queensland, New South Wales, Victoria, and even Tasmania.

One of the characteristic Silurian reef-building corals was *Halysites* or chain coral, so called from its resemblance, when viewed from above, to a series of intersecting chains. It is not found at Yass, but at Borenore it constitutes a large part of the reefs, and it is abundant in many other localities. *Favosites* (organ-pipe coral) is another common genus, and consists of small, closely packed, hexagonal columns connected by pores. *Heliolites* is like *Favosites* but there are larger columns at intervals between the smaller ones. *Syringopora* consists of a mass of small, very irregular, ramifying tubes, connected at intervals by cross rods. A large massive coral, *Spongophyllum*, is, like *Favosites*, composed of hexagonal columns, but they are very large, their internal structure is quite different, and the genus belongs to a different family. *Zenophila* is another compound coral in which there is no wall between the corallites, and in cross section the septa, or radiating internal partitions, run from one corallite to another. *Tryplasma* and *Cyathophyllum* are two of many large branching forms which formed considerable masses in the reefs.

Of the many simple corals *Mucophyllum* is shaped like a broad, flat, flower calyx several centimetres across, while some genera are long and cylindrical, and others shaped like horns. An extraordinary little coral called *Rhizophyllum* is found in abundance in the shaly limestone immediately above one of the main reefs at Hatton's Corner near Yass. This is shaped like a slipper and had an operculum or lid.

These are but a few of the many types found, the description of which has filled many lengthy monographs and scientific papers, but they will give some idea of the profusion and variety corals attained before the close of the Period.

Sea-lilies or crinoids had increased greatly by the Silurian Period. These animals, akin to the starfish and sea-urchins, had the body enclosed in a calyx composed of thick stony plates. This was attached to the bottom by a long jointed stem, and was surmounted by a crown of branching jointed tentacles, giving the whole a flower-like appearance. After death the calyx, stem and tentacles almost invariably disintegrated and fell to the bottom, so that complete specimens are very rarely found. These animals must have lived in countless numbers, for accumulations of the stem joints or ossicles are sometimes so great that by themselves they form considerable thicknesses of the rock known as crinoidal limestone. Such deposits are fairly common in Australia, a notable example being the crinoidal limestone at Rockley, New South Wales. Polished slabs of this may be seen in some Australian city buildings.

Space does not permit description of the many genera and families of fossil crinoids. They are generally similar in appearance, and their identification, which depends mainly on the shape and arrangement of the plates composing the calyx, is a matter for the specialist. Mention may be made, however, of a magnificent specimen found at Brunswick and preserved in the National Museum in Melbourne. It is a *Heliocrinus* with a long slender stem, the end coiled for attachment to seaweed, and the small calyx surmounted by graceful plume-like tentacles which give the whole a very palm-like appearance (Plate 7).

Specimens of starfish are not common in Australian Silurian rocks, but that does not mean that the animals were rare, for one bed in the Upper Silurian of the Heathcote district contains them in abundance. Their classification, like that of the cystoids and blastoids, which are very scarce in Australia, is a matter for the specialist.

Lamp-shells and true molluscs were increasing, though the latter still played a minor part in the life of the Australian seas. The lamp-shells had become very abundant, and many new families and genera had appeared with a great diversity of forms. One large genus, *Pentamerus*, had a strong, overhanging beak, and both valves were divided by an internal partition which itself had a central cavity, so that the whole shell was actually divided into five chambers—hence the name. This genus is sometimes so abundant that it gives a distinctive character to masses of limestone, such as the black and white "marble" from Spring Hill, New South Wales. In another genus, *Chonetes*, one valve was very flat and was covered with long spines. Others, such as *Atrypa* and *Spirifer*, had intricate internal spiral processes which supported the brachiae or breathing organs. Related to the lamp-shells are the bryozoa or lace-corals, beautifully preserved specimens of which are found at Yass.

Amongst the true molluscs the nautiloids, so prolific in the Ordovician, had declined somewhat in number and size, but *Actinoceras*, a straight, long, cylindrical form, is abundant in a number of localities, including Yass, where *Ormoceras* is also common. Bivalves were fairly abundant, and although most of them, like the Ordovician forms, occur mainly as moulds and casts, species with stronger shells began to appear. Univalves or sea-snails were on the whole not very common, but species of *Murchisonia* and *Omphalotrochus*, simple spiral forms, are often found. One curious genus, *Bellerophon*, had an involute shell with an expanded mouth, and in appearance was not unlike a miniature nautilus, except that the shell was not divided into chambers. Good specimens of such univalves have been found at Wellington, Orange, and Yass in New South Wales, and in the top part of the Dargile beds in the Heathcote district of Victoria.

Near Yass a specimen of a chiton or coat-of-mail shell has been found. Chitons are true molluscs in which the back is covered with an almost flat shell consisting of eight transverse shield-like valves. When detached from the rocks to which they cling the animals are capable of coiling themselves into a ball. Chitons are familiar creatures at the present day, living for the most part under stones on the foreshores.

The trilobites, which had attained a great development in Ordovician times, were still very common in the Silurian. Specimens have been found in a great many places. At Rainbow Hill, near Yass, and at Bowning, in the same area, they are quite abundant, as also in a number of beds in the Heathcote and Melbourne sequences, and in the Eldon Group of Tasmania. At Borenore, near Orange, one species of 15–20 cm or more in length has been discovered, one of the largest so far found in Australia apart from the very large asaphid forms in the Middle Ordovician of central Australia. Forms with big compound eyes, such as *Phacops*, are typical of the period; *Dalmanites* had a long pointed tail and, like *Calymene*, could roll itself into a ball. It must be remembered, though, that many of the trilobite fossils we find are not the remains of a whole dead animal, but only of its chitinous carapace or shell which, as it grew, it needed to shed like all arthropods of that type—or, for that matter, like snakes.

One other crustacean group may be mentioned—the eurypterids. To this group belonged a remarkable creature called *Pterygotus*, found near Melbourne. Magnificent specimens of eurypterids have been found in Europe and America. They averaged less than 30 centimetres in length, but a few gigantic forms were from 1.8 to 2.8 metres. The head was covered with a bony shield, on the upper surface of which were the eyes, while beneath were eight short, jointed legs (or appendages). Behind these were two other, larger and paddle-shaped appendages, probably used for swimming, and then a long tapering body covered with large, transverse plates and terminating in a long, spine-like tail. The present-day king crabs, some of which attain an equally large size, and which are found in the depths of Australian and Indonesian seas, are distantly related to the extinct eurypterids, and may in some way be descended from them. The eurypterids had appeared in the Cambrian Period and specimens have also been found in Ordovician rocks, but they attained their maximum development and widest distribution in Silurian times.

Before concluding this chapter we must draw attention to a most important evolutionary event which took place towards the very end of the Silurian—the appearance of vascular land plants, but we will hear more about this in the chapter on life in the next younger Period, the Devonian.

8 The DEVONIAN PERIOD

THE DEVONIAN Period, like many which came before and after it, is vast in terms of time; it lasted about 40 million years, from 394 to 354 million years ago. Such time spans and the colossal events which took place during them, compared with human affairs, are on such a scale that the human vocabulary and imagination are entirely inadequate to convey their immensity.

In Australia, especially eastern Australia, it was a rather unruly period, in which great changes took place. Even in the relatively stabilized western parts of the continent the sea managed to make considerable inroads by the beginning of the Upper Devonian, and towards the end of the period there were some crustal movements also between Perth and the Northwest Cape.

Of the three subdivisions of the Devonian—Lower (early), Middle, and Upper (late)—the first, consisting of the three stages Gedinnian (bottom), Siegenian, and Emsian is only represented in eastern Australia, and even there the record is in some areas fragmentary. Most of the Middle Devonian (Eifelian and Givetian) is also absent in the west, but in eastern Australia there is a fairly complete record. Only the Upper Devonian, that is the Frasnian and Famenian stages, is found in both the west and the east. In the centre the landscape of the Silurian—ranges, rivers and lakes—continues. The same occurs in the whole of the Northern Territory except for a small area on the Joseph Bonaparte Gulf which was perhaps reached by the sea in late Devonian (Famenian) times.

WESTERN AUSTRALIA. The Devonian story in Western Australia is at first sight similar to, or almost a repetition of, that we have seen in the Ordovician. In fact, there is a fundamental difference. There is no longer a marine connection from the west and northwest through central Australia to the geosynclines in the east. The Devonian inundations of parts of Western Australia represent significant marginal or epicontinental transgressions of an oceanic feature, which from now on is rapidly increasing in its importance to the geography of the Australian region—the Indian Ocean, a feature that belongs to the relatively youthful traits on the face of the earth.

In the Joseph Bonaparte Gulf basin Devonian formations are known only in the Western Australian part, although the sea may have reached into the Northern Territory at times. The oldest deposits are plant-bearing sandstones, a few thousand metres thick, which also contain marine bivalves. The sandstones are overlain by richly fossiliferous limestones, some 1200 metres thick. The main elements of the fauna are brachiopods and gastropods, but at some horizons cephalopods, including nautiloids, goniatites, and clymenioids, and at others trilobites, ostracods, corals, bryozoans are very strongly represented. The main outcrops of these Upper Devonian series are from 50 to 90 kilometres east of Wyndham in the Burt Range, but smaller outliers are also known in patches west of the Ord River.

In the West Kimberleys Middle and Upper Devonian formations are widespread at depth in the Canning Desert where they were discovered by the drill. At the surface they can be studied only along the northern margin of the Fitzroy Basin from King Sound in the north to Balgo Mission in the south. The sequence is very complex, but particularly interesting because much of it represents an exhumed fringing reef which runs for about 300 kilometres along the western slopes of the King Leopold Ranges. The individual growth stages (fore-reef slope sediments, main-growth coral reefs, still-water backreef beds) can be clearly distinguished and studied. It is one of the finest and largest outcrops of reef-type Devonian rock in the world. As Veevers and Wells (*1961*) said and Playford (*1980*) demonstrated, this area is ideal for detailed studies on that kind of sedimentation and on distribution of marine faunas in such environments. It is quite an experience to look up at some of the sheer reef-limestone cliffs, well knowing that the scene is almost exactly the same as it was some 350 million years ago, save for the fact that the reef's colourful life is dead and that we stand on dry land instead of on the sea bottom.

Away from the reefs, that is to the southeast and southwest, shales and sandstones were deposited—some in brackish or freshwater lagoons or lakes, as for instance in the Balgo Mission area at latitude 20 degrees south near the Northern Territory border. Oil drillers have also shown that Devonian rocks are absent in a southeasterly trending zone from Broome inland. This, as shown on Map (15), was probably an island consisting of Ordovician and older rocks which had been folded during the Benambran-Rodingan orogeny, Map (13), at the end of Ordovician times.

Another important region in the West where Devonian sediments were laid down is that between Northwest Cape and Murchison River. It is known as the Carnarvon Basin. The outcrops are some 300 kilometres inland from the coast and a number of drill holes have cut Devonian rocks in many areas nearer to the coast. Like all other Devonian formations in the West those in the Carnarvon Basin were first recognized and described by Teichert, the father of modern stratigraphy in Australia, whose varied and often fundamental contributions to our knowledge, especially of Palaeozoic rocks, remain milestones in Australian geology. We cannot list them for there are too many, but in many of the reference works mentioned at the end of this book one will find references to Teichert's papers. The summary prepared by McWhae and collaborators (*1958*) for the Geological Society of Australia may be cited in particular.

The Devonian in the Carnarvon Basin consists largely of sandy sediments, but with a substantial number of limestone and dolomite beds interstratified. Of the total 1200–1500 metres only about 500 metres are fossiliferous (Gneudna Formation). Corals do occur but there is no reef-building of the kind we have just seen in the West Kimberleys. Most fossils are shelly, and conodonts are also common. Their age is early Upper Devonian and there may be some Middle Devonian too.

CENTRAL AUSTRALIA. An important non-marine sedimentary area during the late Devonian is in central Australia. In the Amadeus Basin along the MacDonnells the sequence measures up to 3500 metres and is known as the Pertnjarra Group. East and southeastward it thins out considerably and is called the Finke Group. The most conspicuous formation, up to 2500 metres thick, of the Pertnjarra sequence, which otherwise consists chiefly of sandstones and shales, is its youngest part, the Brewer Conglomerate, a large southward spread of boulder and cobble beds derived from the then again rising MacDonnell Ranges (*Jones 1972, 1973; Playford et al., 1976*). These movements are known as the Alice Springs Orogeny. The raising of the Range can be read in reversed order, so to speak, from the change in the type of the conglomerate's boulder, cobble, and pebble components. One can see how ever older beds became eroded from the steadily growing southward overfold of the MacDonnells. In the lowest, that is oldest, 1000–1500 metres of the Conglomerate the components are all derived from the Ordovician, Cambrian, and topmost Pertatataka formations which formed the cover of the rising fold, whereas the younger conglomerate horizons contain up to 50 per cent of material from the Heavitree Quartzite and the metamorphics of the Arunta System, indicating that erosion

Map 14:

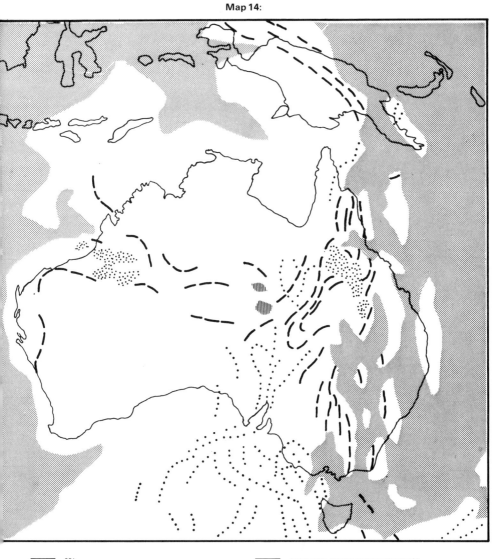

The Australian region in the Early Devonian, about 380 million years before present. The salt lakes are shown considerably larger than they probably were, but it is interesting to note that their position foreshadows the large marine embayments which will be developed in the Late Devonian (see Map 15).

Map 15:

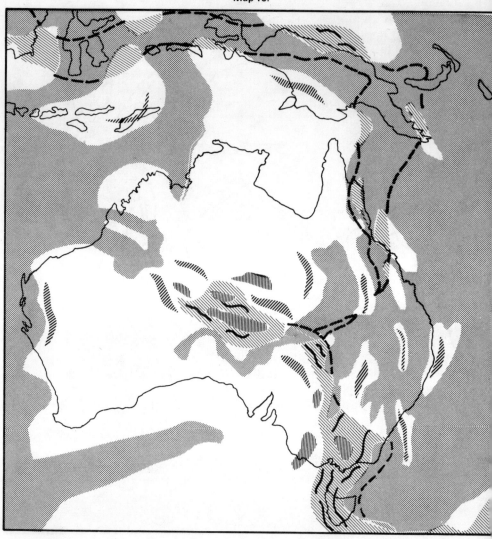

	SEA		AXIS OF UPWARPING —IN PLACES AS AN EFFECT OF DISTANT MOUNTAIN-BUILDING FORCES
	LAND		
	FRESHWATER LAKE		
	ZONE OF CONTEMPORARY OR JUST COMPLETED MOUNTAIN-BUILDING		

The Australian region about 360 million years ago during the Late Devonian and after the strong Alice Springs Orogeny in central Australia which was connected with the Tabberabberan Orogeny in Victoria and strong movements in central eastern Queensland. The minor tectonic events in Western Australia took place a little later towards the end of the Devonian and led to a neat unconformity between Devonian and Carboniferous sequences on the eastern margin of the Carnarvon Basin.

by then had cut right down into the crystalline basement formations.

Before discussing the Devonian in eastern Australia let us briefly look at Maps (14) and (15), for when describing the scene in Western Australia we did jump the gun a little; we started with the late Middle Devonian there because earlier Devonian beds are absent. These are known only in the eastern Australian region, and it is this early to middle Devonian geography which is shown on Map (14).

EASTERN AUSTRALIA. The beginning of the Devonian saw Tasmania still mostly below sea-level; from the central west coast northeastward past Launceston to Flinders Island and beyond there was an embayment into which sediments were shed from the west and south. Shallow water deposition is evident in the southwest where at least 900 metres of sandstones and shales with a rich fauna of brachiopods, pelecypods, crinoids, some trilobites and corals are preserved. In the deepwater environment in the northeast fossils are rather scarce and it is difficult to draw the boundary between Silurian and Devonian rocks. Towards the end of the Middle Devonian a strong mountain-building event, which also brought widespread intrusion of granitic magmas, lifted the whole of Tasmania out of the sea, together with other parts of southeastern Australia and much of New Zealand, and it was to remain land for about 100 million years, until Permian times.

In much of southeastern Australia the Devonian was ushered in by important crustal movements known as the Bowning Orogeny. These movements resulted in the consolidation of a number of regions which had already been folded by the later phases (Early Silurian or "Quidongan") of the otherwise Ordovician Benambran Orogeny. Granite magmas were intruded, and their associated volcanic effusions are spread widely into the shallow-sea Silurian deposits in the Australian Capital Territory, the Southern Tablelands of New South Wales, and the Monaro and Snowy Mountains. Although somewhat modified by later erosion cycles, such areas represent very ancient terrestrial environments, even if the idea (*Opik 1958a*) of Canberra being built on a re-exhumed Devonian surface can no longer be upheld (*Jennings 1972*). Other such ancient land surfaces are what were in Silurian times a chain of mountainous volcanic islands, for instance the "Capertee Islands" of Webby (*1972*) shown approximately on Map (14).

In Victoria, the early Devonian sea covered the central areas (Melbourne-Heathcote) in the form of a gulf extending from Bass Strait northward, and this gulf was separated by the Benambran peninsula from another gulf or meridional strait across East Gippsland. The latter gulf finds its continuation on the north side of the Great Dividing Range in the Tumut and Murrumbidgee River valleys and from there northward in the Young, Bathurst, Mudgee and Wellington districts of New South Wales. Through these areas—and some others in New South Wales and Queensland which will be mentioned later—sedimentation went on through most or all of Lower and early Middle Devonian times, and there are rich faunas which testify to this. On the other hand, there is no doubt that during the same period there was much crustal unrest, and that it led to the extrusion of large masses of volcanics within, and in the vicinity of, all these marine gulfs and straits.

The biggest volcanic masses of early Devonian age are found in the Snowy Mountains region where they are several thousand metres thick and cover more than 3000 square kilometres. Chemically they are mostly of the acid or granite-to-porphyrite type. Many fine mineralogical and petrographical studies of them have been made during the tens of kilometres of tunnelling work carried out for the Snowy Mountains Scheme. Their Lower Devonian age can be neatly demonstrated by the fact that the marine Middle Devonian limestones of Buchan in eastern Victoria have been deposited over an eroded surface on these volcanic masses.

Throughout central New South Wales, New England, and northward into southern Queensland as far as the Rockhampton district there is almost rhythmical volcanic activity during the Lower and early Middle Devonian, so that the marine sediments are everywhere interstratified with lavas and volcanic tuffs, and the thickness of the whole series reaches in a number of places 3000 metres and more. North of Rockhampton, however, there is no evidence of volcanicity in Lower Devonian time; there are no tuffs or lavas for example in the respective formations on the Burdekin River west of Townsville. On the other hand, tuffs—but no lava—occur in the far west of New South Wales in the Cobar district, where the sea also persisted from Silurian into Devonian times. These tuffs evidently originated from huge ash clouds that were carried westward from centres of volcanicity in eastern Australia.

Much work has been done since 1960 in Victoria to elucidate the complex happenings in Late Silurian and Early Devonian times. Talent (*1965*) and Douglas and Ferguson (*1976*) have described a fascinating mosaic of marine and non-marine sedimentation depending in the various areas on the vicissitudes of many crustal movements. There are many unconformities, breaks in sedimentation, and shifting of shallow-water and deep-water environments from one place to another almost continuously. The main point to remember is that the sum total of all this is the gradual withdrawal of the sea from the central Victorian gulf during the Lower Devonian and its final complete withdrawal from Victoria before the end of the Middle Devonian, as the result of a great mountain-building event at that time, the Tabberabberan Orogeny, which, as we have already seen, was also strong in Tasmania.

The complexity of the record in Victoria has led to much confusion about the position in the sequence locally, and the age of various formations in general. Until a few years ago the Silurian in Victoria included five main stages—above the Melbournian was the Tanjilian and the Yeringian. Both of these are now regarded as Lower Devonian, or simply as different aspects or facies of sediments that were laid down about the same time. The Yering group of formations in the Lilydale district includes series which are of the same age as the Tanjil formation around Mount Matlock farther east and the Norton Gully Sandstone of the Eildon Dam-Jamieson district. All of these are apparently of Siegenian (late Lower Devonian) age.

The Yering Group consists of over 2400 metres of siltstones and sandstones followed by the 200 metres of Lilydale limestone, and finally again by 60 metres of sandstone. The marine fauna is shelly and, in the limestones, also corallian. The Tanjil Group starts off with varicoloured silt- and sandstones (900 metres) without fossils, on which lies a 6–30 metres sequence of mudstones, black shales, and some sandstones containing bivalves (*Panenka*), pteropods (*Styliolina*), and straight-shelled nautiloids. This is overlain by 150–300 metres of fine-grained, graded, current-bedded sediments with some brachiopods in the lower part and then by black slates, about 30–60 metres thick. Unlike the Tanjil Group, where there are several graptolite horizons of Lower Devonian age, the Yering beds are not graptolitic. Such differences in faunal content and character of sedimentation within short distances were until recently interpreted as due to different age of the beds rather than to local effects of land and sea topography working at the same time in different areas.

There is thus considerable variation in the character and thickness of the Middle and early Upper Devonian sequences found in a number of areas, from Lilydale in the west to Buchan in the east. The long-standing controversy among geologists as to the age of the uppermost beds in the Walhalla district is now about to be settled. There is little doubt that they are Lower and Middle Devonian and that fossiliferous Silurian is absent in that area. The famous Cave Hill beds on top of the Lilydale Limestone are also Middle Devonian and were laid down about the same time as the lowest part of the plant-bearing red-bed shales and sandstones in the Cathedral Range near Alexandra and the similar Koala Creek series in the Eildon Dam region. At about the same time limestones were laid down in the Waratah Bay and the Buchan area. The Buchan limestone has long been quarried and sold as marble because of the beautiful patterns created by the millions of fossil corals and brachiopods.

Within its rather narrow confines of straits and embayments the eastern sea must have been comparatively calm, warm, and—in spite of numerous rivers bearing mud and silt—clear enough in many parts to allow the formation of coral reefs. These are even more extensive than those of the preceding period. The reefs in the vicinity of Tamworth in New England are particularly noteworthy. A kilometre or so to the north of the town they can be seen on the Common. At one point on the hillside is a flat mass, about 60 centimetres thick, of a branching coral called *Disphyllum*, which can be traced continuously for over 30 metres. This was apparently one single colony. Northward from here the pine-clad hills at Moore Creek are practically composed of fossil corals.

Still farther to the north the limestone beds continue in a great arc in the foothills of the Moonbi Range, past Attunga, Manilla and Barraba. They are found again at Bingera and Warialda, the general trend being at first north and then northeast, with the main massive of the New England tableland always to the east. This belt continues across the Queensland border almost as far as Toowoomba. After a gap here the beds may be picked up again and followed to the coast at Rockhampton and Gladstone where they are known as the Mount Etna series and show a thickness of at least 4500 metres. Coral reefs occur throughout this series but most prominently about the centre. There are many collecting localities for fossils in the district, notably at Mount Etna itself, and at Morinish, Cawarral, Ula, Raglan, Mount Larcom, Kroombit, and Yarrol. To petroleum explorationists these areas are known as part of the Yarrol Basin of Upper Palaeozoic marine deposition.

An interesting discovery made by oil exploration companies is the occurrence of rocksalt formations of Devonian age at depth below the younger series of the Great Artesian Basin in Queensland. Perhaps these salt beds, Map (14), were formed at the same time as those in the Canning Basin of Western Australia, for palaeomagnetic studies indicate that then, as today, these areas were about the same latitude—between zero and 10 degrees south in the Devonian, and between 18 and 23 degrees south today.

About the end of the Middle Devonian considerable geographical change occurred in Australia under conditions which varied in different regions. We have already seen how much of Western Australia was invaded by the sea at that time, Map (15). In Victoria

THE DEVONIAN PERIOD 133

and Tasmania the general unrest of the Period culminated in an important mountain-building event, the Tabberabberan Orogeny. All rocks up to and including the Middle Devonian (e.g. the Wentworth Group at Tabberabbera, the Buchan Group, the Bell Point Limestone at Waratah Bay), which had just been deposited, were strongly folded, compressed, elevated above sea-level, and again partly eroded over vast areas. Subsidence soon lowered the level of the land once more, but instead of the sea a series of large freshwater lakes and brackish lagoons covered most of the eastern portion of the State. In these were deposited conglomerates, sandstones and shales, but nowhere are they very thick. On the Mitchell River in Gippsland and other places these Upper Devonian beds are found resting unconformably upon the upturned edges of the Middle Devonian formations, showing thus clearly that major crustal movements must have taken place in the interval of time between the deposition of the youngest Buchan beds and the oldest beds of the Upper Devonian non-marine series.

At Bindi distinct impressions of raindrops may be seen on the surfaces of the shale, made no doubt by a passing shower when the shale was merely an expanse of soft mud. Elsewhere networks of fossilized suncracks caused by the shrinking of the drying mud are also preserved. When so much that was tangible has disappeared it is awe-inspiring to contemplate this simple evidence of sunshine and rain in a remote age. Ripple marks made by waves and currents in shallow-water sand, or by the wind on the surface of dry sand have been preserved in the same way.

Volcanic activity was renewed and especially towards the end of the period eruptions took place on a large scale. In the valley of the Yarra between Warburton and Healesville the lava flows are over 1500 metres thick. In one place, in a lake which had been dammed by a lava flow, remains of fish have been found. In western Victoria a separate area of subsidence seems at times to have admitted the sea, for within the freshwater series one finds in places beds containing marine fossils.

In central and southern New South Wales conditions were quite different. Although shown on Map (15) as passing through New South Wales, the folding movements of the Tabberabberan Orogeny are scarcely evident in many areas there. The western seaway remained open throughout the Upper Devonian, now with embayments into central Queensland to the north, towards Lake Eyre in the west (Pedirka Basin) and beyond to the Officer Basin, and southward to the Murray River. There was, in fact, a great increase in the area which was invaded by the sea, particularly west and northwestward;

also the far southeast subsided. Much of southern New South Wales, however, remained above sea-level.

The western and southern seas were shallow and received much sandy and shaly sediment from the high lands to the south as well as from larger and smaller islands which had risen as a consequence of the Tabberabberan Orogeny. The shallow embayments frequently silted up. Saltwater was replaced by brackish lagoons and freshwater lakes. There is evidence of dense vegetation around such places, and many remains of freshwater fish have been found. As might be expected under such circumstances the coral reefs so characteristic of the earlier Devonian stages are gone too.

There are many areas in New South Wales where these rocks can be studied. The type locality is Mount Lambie near Rydal on the main western line between Lithgow and Bathurst. Mount Lambie is composed of Upper Devonian sandstones and quartzites with beds of red shales and some conglomerates at intervals. Marine fossils as well as plant remains are abundant. Another large area is north of Goulburn, where the escarpment of the Cookbundoon River is composed of Upper Devonian quartzites resting on nearly vertically up-ended Ordovician slates (*Naylor, 1936*). Farther to the east and south they can again be picked up in the gorges of the Shoalhaven River, Yalwal Creek, Clyde River and other places. On the far South Coast the Upper Devonian series consist largely of light coloured, acid lavas and tuffs, known as the Yalwal Group, Merimbula Formation, etc.

In the Tamworth area of central New South Wales the sea remained in much the same trough as it was during Middle Devonian times, but sedimentation was not entirely continuous. The effects of the Tabberabberan Orogeny are noticeable in that the Upper Devonian mudstones, radiolarian cherts, and tuffs of the Barraba Group, which all together are about 4000 metres thick, lie more or less unconformably upon the late Middle Devonian coralline limestones of the Moore Creek Formation. There are also pebbles of earlier Devonian rocks embedded in the Barraba Mudstones, thus indicating emergence and erosion in part of the region within Devonian times.

It is rather difficult to picture the exact conditions under which the Barraba sequence was laid down, for with the exception of the remains of microscopic radiolaria at some levels, and the occasional occurrence of stems of plants which had evidently drifted to their present position, there is an extreme scarcity of fossils. Coral reefs are absent too in spite of the fact that, according to palaeomagnetic studies, the area may have been well within the tropics. The most likely explanation is that these peculiar sequences, which are best studied in the vicinity of Barraba and run in a wide belt from Manilla in the south to beyond Bingara in the north, are sediments laid down in a deep-sea trough which kept on subsiding throughout the Upper Devonian and, in fact, well into the next Period, the Carboniferous.

Conditions in Queensland were similar to those in New South Wales, and waters also spread far inland. However, at times part of the country was not low enough to allow the sea to penetrate. Thus lakes existed in the Drummond area near Clermont, and in them were deposited red sandstones containing fossil plants. Similar plant-bearing red-beds are known from the southern part of Cape York Peninsula at Gilberton. At the same time, over 6000 metres of marine and freshwater sediments were laid down west of Cairns (Hodgkinson beds). In general, this alternation of marine with non-marine formations is characteristic of the Upper Devonian in Queensland and, in fact, these conditions persisted into the Carboniferous, so that many of the beds which do not contain fossils clearly assignable to one or the other Period must be called transitional in age, that is Devonian-to-Carboniferous.

LIFE OF THE DEVONIAN PERIOD

The First Fish and the Conquest of the Land by Plants

LAND PLANTS. An event of great significance to the future development of life on earth occurred in earliest Devonian times—the first appearance, to our knowledge, of undoubted vascular land plants (somewhat doubtful ones occur in late Silurian rocks in Europe). In Australia they were first found some decades ago in slatey beds, then believed to be of Silurian age, in a cutting on the slopes of Mount Pleasant on the old road between Alexandra and Thornton, about 150 kilometres northeast of Melbourne. Following this, similar discoveries of vascular plants have been made in other areas of Victoria, in Tasmania, and near Mudgee in New South Wales. In some areas the fossils were found in great abundance, and though not always well preserved they show sufficient details to prove that they were undoubtedly land plants of lowly organization, akin to the club-mosses and also related to algae or seaweeds. They are classed partly with the psilophytes and partly with the lycopods. Several quite distinct forms are known under names such as *Baragwanathia*, *Yarravia*, and *Hostimella*, the latter occurring also in the Mathinna beds of north-eastern Tasmania. All were first described and identified by Cookson (*1935*).

The discovery of these plants is a matter of absorbing interest. It is hard to imagine a land without any vegetation at all; yet, through eons, vast areas of mountain, valley, and plain must have been stark and lifeless, with nothing to relieve the bare landscape of rock and rushing river. If these early Devonian plants were the first to find a footing on land, they are the type which might reasonably be expected, lowly forms derived from the seaweeds. Creeping at first into the Zone between the tide-marks, they would be exposed to the air for a varying period each day, would develop roots and would become gradually adapted for a new existence out of water. Spreading to river-banks, swamps, and low-lying areas, they at last became established on the land itself. In their new environment competition would at first have been negligible, the soil virgin, unconsolidated, but rich in mineral plant food, and they would multiply rapidly. Under varying conditions of climate, altitude, and humidity, evolution must have proceeded apace; new and more advanced species would gradually appear. Thus, later in the Devonian, the land flora had reached a comparatively high level.

The coming of the plants to the land is a major milestone in the history of the earth. With them, or closely behind, came animal life, which without them could not exist. Worms and snails were among the first to start a terrestrial existence, also the ancestors of the insects, spiders and scorpions. Later came higher forms, until the whole land was populated with a multitude of creatures, developing in complexity as time passed, and culminating in man himself.

After establishing their first hold on land, plants were well established towards the end of the Devonian. Real forests were probably still absent, but there were certainly dense thickets which covered what had been bare and desolate regions in earlier ages. Whether high land was covered with vegetation is not known, for most of the fossil plants seem to have grown in swamps or in low-lying country adjacent to rivers, lakes, and estuaries. By modern standards the vegetation was drab and monotonous and had a striking similarity throughout the world.

In Australia plant remains are common in the early and again in the late Devonian formations, but scarce or very poorly preserved in Middle Devonian rocks. The Upper Devonian plant remains are commonly found in sandstone which, unfortunately, is a rather coarse medium in which to preserve such fragile structures. For that reason, we can obtain scarcely more than a glimpse of the flora of that time. Most of the specimens consist of branches and stems which seem to have been carried far from their source; they are just fragments and seldom show structural details.

Among the lowest groups, the psilophytes, which may be most

easily described as something between seaweed and moss, were still prevalent, especially in the early part of the Devonian. They grew in masses in damp situations to a height of perhaps 50 centimetres. Primitive fungi were also present. Although their chances of preservation were slim they are occasionally found inside the stems and on the roots of the larger plants of the period, but they are microscopic forms. The larger woody types found on logs and forest trees had not yet evolved.

Other plants, though still primitive, showed a considerable advance. Among these, two groups became predominant in the Upper Devonian: the earliest true ferns, and the lycopods, the latter being popularly known as scale-trees or club-mosses. The earliest fern in Australia, *Archaeopteris*, has been found in several localities—at Genoa near the border of New South Wales and Victoria, at Nauguda in Victoria, and elsewhere. In appearance it was not unlike many modern ferns, and apparently grew thickly on the low-lying banks of streams.

The club-mosses were the largest and most striking plants. Though not yet the immense size they were to attain in the next period, they were still approaching the status of trees, fifteen to twenty centimetres in diameter and probably up to six metres in height. The trunks were straight and surmounted with a small crown of branches, at the termination of which were the seed-bearing cones. Both trunks and branches were covered with closely packed and spirally arranged rows of leaves not unlike pine needles, the shedding of which left a characteristic pattern of scars. It is these scars which have led to the use of the term "scale-trees", while the terminal position of the cones has suggested the term "club-mosses".

The common club-moss, *Lepidodendron* and its close relative *Leptophloeum* have been found at many places—from Gilberton and the Burdekin River in northern Queensland to the Avon River in Victoria, and it occurs also in the Devonian of Western Australia and central Australia. The plant can be easily recognized even in small fragments by the regular, rhomboidal pattern produced by the leaf scars, in appearance not unlike the scales of a large snake; indeed it is often sent as such by correspondents to the various museums.

Though the lycopods became almost extinct at the end of the Palaeozoic, a few still survive to the present day. One of these, "mountain moss", a small erect plant less than 60 centimetres high, is fairly common in the heath country around Katoomba in the Blue Mountains of New South Wales.

LAND ANIMALS. Little is known of the animal life which followed plants from the water to the land in Devonian times. Organisms such as spiders and insects disintegrate rapidly after death and need rare and peculiar circumstances for their preservation as fossils. There are thus many gaps in our knowledge, and our evidence for Devonian land animals comes from a few widely scattered sites. Fossils of scorpions, spiders, and insects have not yet been found in the Devonian rocks of Australia but are known from other continents, e.g. wingless insects from the Rhynie Chert of Scotland. More importantly, the Late Devonian saw the first appearance of four-legged land animals—labyrinthodont amphibians. Most of the known skeletal remains (*Ichthyostega* for example) come from Late Devonian rocks of east Greenland, but recent discoveries reveal that similar animals were then, or even earlier, present in Australia. A possibly amphibian jaw, *Metaxygnathus* (*Campbell and Bell, 1977*), comes from the Jemalong Range, west of Forbes; and well-preserved and undoubtedly tetrapod footprints occur in Late Devonian rocks along the Genoa River in eastern Victoria (*Warren and Wakefield, 1972*). The potential for more Devonian tetrapod discoveries appears good.

INVERTEBRATE MARINE LIFE. The most noteworthy features of Devonian times throughout the world were the increase in lamp-shells, the decline of the trilobites, and the extinction of the graptolites. Otherwise life did not differ greatly from that of the Silurian Period, though there were considerable changes in detail. There were great developments of coral reefs, but it is interesting to note that in Eastern Australia these took place in Lower and Middle, but in Western Australia in Upper Devonian times.

A great variety of corals has been found in Tasmania, Victoria, New South Wales, and Queensland, particularly in the two last-mentioned States. *Favosites*, *Heliolites*, and *Syringopora* had survived from the Silurian, but the chain-coral, *Halysites*, and many others had become extinct or evolved into distinctly different forms. Simple corals, that is those consisting of a single polyp, had become very abundant, including the curious little slipper-shaped *Calceola*. A large form, *Actinocystis*, was shaped like a cow's horn; another smaller horn-shaped genus, *Zaphrentis*, appeared and survived until the Permian Period, and there were many others. Amongst the compound corals, *Sanidophyllum*, was a handsome genus in which the individual corallites were connected by broad, continuous platforms. It is found in the Tamworth district and on the Burdekin River. A branching coral, *Disphyllum*, formed colonies of great extent, and it has already been mentioned that one of these, at Tamworth, is traceable for over 25 metres. An order of hydrozoan coelenterates closely related to the corals, but with minute tubular corallites, is particularly characteristic of the Upper Devonian in Western Australia, although it is also quite common in eastern regions. It is known as Stromatoporites. Whole reefs are composed of forms such as its typical genus *Stromatopora*, which has consolidated into extensive beds of limestone, especially in the West Kimberleys. A curious sponge, *Receptaculites*, is also common in the Australian Devonian and has been found in many places.

In some areas, notably the Tamworth district, fossils other than corals are rather rare. A characteristic rock in such areas is radiolarian chert, a hard and dark-coloured siliceous rock, which when sectioned very thinly for the microscope is seen to be composed largely of the minute shells of radiolaria. These humble unicellular animals secreted shells of silica, many of beautiful geometrical design, and ornamented commonly with long slender spines.

Crinoids remained abundant. These creatures seemed to be gregarious, forming large colonies on the sea floor because, when found at all, they are in great numbers, but it is very difficult to obtain complete specimens. The calyx plates and the joints of the stems and tentacles fell apart after death and accumulated generation after generation, until they built up considerable thicknesses of glittering crystalline limestone. One such bed at Nemingha, near Tamworth, is red in colour, and when polished makes a beautiful ornamental marble. A similar red crinoidal marble is found on the banks of the Murrumbidgee River near Michelago south of Canberra, but this is of Silurian age.

Lamp-shells or brachiopods are abundant in some places in Queensland, at Lake Bathurst and on the Murrumbidgee near Yass in New South Wales. In Upper Devonian rocks they constitute the majority of marine fossils found, and several species, among them *Spirifer disjunctus* and *Camarotoechia* (*"Rhynchonella"*) *pleurodon*, though unspectacular fossils, are of particular interest because they have been found in many other parts of the world. They are what are called worldwide type, or guide, fossils, and characteristic of the late Devonian beds everywhere. A great variety of lamp-shells is also known from northern Western Australia, but most species are different from those in eastern Australia (*Veevers, 1959*) and even from those in the Devonian of the Carnarvon Basin in Western Australia. In the West Kimberley district, around Mount Pierre, lamp-shells are so abundant that they give the name "brachiopod limestone" to considerable rock masses. Beds are composed sometimes of the remains of very few, even of only one species, showing that the animals lived gregariously, like the crinoids, in large colonies on the sea-bed.

With the brachiopods lived numerous true molluscs, including many bivalves. Among these many of the *Pteria* type had long hinges and a wing-like expansion of the shell, not unlike the pearl oyster of the present day. Univalves, which were not so common, included long, slender, spiral types such as *Loxonema*, as well as the involute *Bellerophon*. Devonian cephalopods are particularly interesting. These molluscs have always been subject to rapid change, and already the evolution had begun which was to culminate long ages after in the development of many extraordinary families. For purposes of classification the chief feature to be noted is the form of the numerous partitions by which the shell is divided. In *Nautilus* and in the straight-shelled forms such as *Orthoceras*, these partitions were smooth and made an almost straight or but slightly wavy line at their junction with the outer shell.

In the Devonian there appeared numerous small nautilus-shaped shells called goniatites. This is a family name and the group includes many genera. In these cephalopods the edges of the partitions or septa are puckered, and their junctions with the outer shell, called sutures, appear as a series of smooth folds and saddles. The suture cannot be seen from the outside, but on internal casts or in specimens whose shell is partly broken away it is often visible. Various species of goniatites are abundant in Western Australia, and are very important time markers in the series of Devonian rocks. Most of the species survived for only a limited, often rather short time, and they are therefore characteristic of particular zones or levels within the sequence (*Teichert, 1941*).

The trilobites were further declining, although in the early Devonian they were still fairly common, but their remains are rare in Australia. Near Lake Bathurst railway station in southern New South Wales, the curious compound eyes of a species of *Phacops* are found quite separated from other parts, and to anyone unfamiliar with the whole organism would be very puzzling. *Harpes*, a curious little genus with a large rounded head shield, and *Bronteus*, which had a large tail, shaped like an inverted rising sun, have been found near Yass. Fine collections also come from a particular bed in the lower part of the Upper Devonian in the southeastern portion of the Fitzroy Basin in Western Australia, where they are very well preserved.

The largest known trilobite is a Devonian species, but was not found in Australia. This form, a *Dalmanites*, attained a length of no less than 73 centimetres. After the end of the Devonian the order of the trilobites declined even more, but it lingered on almost to the end of the Permian Period.

THE FIRST FISH. One of the most fascinating parts of the study of historical geology or stratigraphy is to find out how each period sets its milestones in the advancement of life on earth. We have seen that through Cambrian times a large and diverse marine fauna had already developed, and that the Ordovician was notable for the development of graptolites and cephalopods, and the Silurian for that of the corals and other interesting marine forms of life, as well as the coming of the first land plants. Now, in the Devonian, there is evidence of another great step upwards in the ladder of life with the first appearance, in abundance, of vertebrates with jaws, animals with both a backbone and a cranium. The first chordates were probably soft-bodied animals, unlikely to be fossilized. They may have resembled the living lancelet, *Amphioxus*, a small and primitive, superficially fish-like animal with a cartilaginous notochord (but no backbone) which lacked jaws, eyes, bones, and paired fins. A possible fossil lancelet, *Pikaia*, is known from Middle Cambrian rocks in Canada. The earliest fish-like animals, jawless forms known as ostracoderms, have been found as bone fragments in Late Cambrian and Early Ordovician rocks of the U.S.A., but better-preserved material, for example *Arandaspis*, have been described (see p. 112) from the Middle Ordovician in Central Australia. No ostracoderms are known from the Silurian of Australia, and in the Devonian thelodonts are represented only by isolated scales.

It was not until bony plates and scales as well as internal bones were developed that the vertebrates could leave definite evidence of their existence. Fortunate discoveries of such plates and scales have been made in Devonian rocks in many parts of the world which allowed, after much patient research, something like a complete reconstruction to be made. Even single plates may be identified and referred to one of several groups.

The last two decades have seen rich and diverse discoveries of Devonian fish fossils in many parts of Australia. Acid treatment of Lower Devonian marine limestones from Buchan, Vic., and Taemas/ Wee Jasper, N.S.W., has yielded many well-preserved specimens. An Early Devonian freshwater fish fauna occurs in the base of the Mulga Downs Group in the west of New South Wales. Middle Devonian faunas have been recovered from near Burrinjuck Dam, N.S.W., and from the Bunga beds on the New South Wales south coast. The richest and most widespread fish-bearing rocks, however, are those of Late Devonian age, particularly the freshwater fauna from Mount Howitt in central Victoria and a marine fauna of about the same age from the Gogo Formation in Devonian reef complexes of the West Kimberley region, W.A. Many other rich sites are now known and under investigation in Victoria, New South Wales, Queensland, and the Northern Territory.

Most abundant are representatives of an extinct class of armoured fish known as placoderms. Most had a head and trunk shield encased in bony plates and articulated with a ball-and-socket joint. The flexible trunk and tail was only rarely covered in scales. The two most important placoderm groups were the arthrodires and the antiarchs. Some arthrodires reached nine metres in length, for example the Late Devonian *Dunkleosteus* (Ohio, U.S.A), but most were much smaller. East Australian genera include *Buchanosteus* and *Wittagoonaspis* (Early Devonian) and *Groenlandaspis*, *Phyllolepis* (Late Devonian). The Late Devonian in the west contains *Eastmanosteus* and *Coccosteus*. The related antiarchs, armoured fish with strange, paddle-like pectoral fins, (*Bothriolepis*, *Remigolepis*) occur widely in Late Devonian rocks. Many of these genera are closely related to forms found in Devonian sediments in North America, Greenland, Europe, Asia and Antarctica. In Australia, as elsewhere, they became extinct by the end of the Devonian.

Somewhat more advanced in organization than the placoderms were the bony fish or Osteichthyes which include the air-breathing, lobe-finned sarcopterygians (sometimes called Choanichthyes) and the ray-finned actinopterygians. The sarcopterygians are of particular interest, for palaeontologists have long agreed that from them evolved the first land vertebrates, the amphibians. The two orders of the Sarcopterygii are the Crossopterygii and the Dipnoi.

Of the two main families in the Crossopterygii the earlier and more primitive were the rhipidistians (*Osteolepis*) which were long-bodied and covered with thick rhomboidal scales. Their curious fleshy-lobed pectoral fins had an internal bony structure similar in many ways to the fore-limbs of higher vertebrates. Representatives of the group, which survived from Devonian to Permian times, have recently been discovered in the Upper Devonian of Canowindra, N.S.W. and of the Kimberleys, W.A. (*Miles 1978*) It is from this group, possessing the ability to breathe air and with the potential limb structure of tetrapods in their fins, that higher animals adapted to a terrestrial existence developed. Out of them evolved, in the Late Devonian, the first amphibians and later the reptiles, mammals and birds.

The other crossopterygian family are the coelacanths, rather deep-bodied fish with a three-lobed diphycercal tail and a calcified swim bladder. The coelacanths were thought to have become extinct in late Mesozoic times, until 1939, when the scientific world was thrilled by the discovery of a coelacanth still living in the sea. The first specimen was trawled by fishermen in deep waters off East London on the east coast of South Africa. Unfortunately only the skin was kept but it was sufficient to allow classification of the strange animal as a new genus of coelacanth, which was named *Latimeria* after its discoverer. Another seventy or so specimens have been caught and carefully preserved since 1952; most have come from deep waters off Madagascar. The detailed analysis of this "living fossil" by Millot and Anthony (*1958*) has contributed a great deal to our understanding of the group from which the tetrapods sprang. An actual specimen may be seen in the Australian Museum.

In the second order of Choanichthyes are the Dipnoi, or true lung-fish. This order is particularly interesting to Australians because dipnoans have been found both living and as fossils in this country. As fossil teeth and scales have been found in the Devonian rocks near Taggerty associated with the remains of placoderms, and they occur also in the far west of New South Wales (*Fletcher, 1964*). Single plates and a fine head of a genus called *Dipnorhynchus* were also found on the Murrumbidgee River near Yass (Plate 10). More is known of the Dipnoi than of the other groups of Choanichthyes, for the world-famous lung-fish of Queensland is a member of this order. It is characterized by, among other things, the possession of organs closely approximating to lungs, which enable it when necessary to breathe out of water. The lung-fish is, in fact, the closest relative of the higher vertebrates among all existing fish.

The second major division of the bony fish or Osteichthyes are the ray-fins or actinopterygians. Early members, often called ganoids from the enamel coating on their interlocking scales, are rare in the Devonian but predominate in the Carboniferous, Permian and Triassic. Surviving representatives include the sturgeon of eastern Europe, the bony pike of North American rivers and lakes and *Polypterus* of the rivers Nile and Senegal. Fossil forms have recently been found in the Late Devonian of Western Australia, New South Wales and Victoria.

The earliest of the cartilaginous, shark-like fish or Chondrichthyes are known from Middle Devonian beds in the U.S.A., but only isolated teeth and scales have so far been reported from Devonian rocks in Australia.

9 The CARBONIFEROUS PERIOD

IN MOST of the Northern Hemisphere this Period is known as the Age of Coal—hence "Carboniferous"—but in Australia and other southern land masses the great coal deposits belong to later ages, particularly to the Permian and early Mesozoic.

Because of the continued growth and consolidation of the Australian continent through the earlier Palaeozoic we find now that important, and especially orogenetic events have become more and more confined to the periphery of our land mass; in fact, to the eastern perimeter alone, as far as true mountain-building is concerned. Not that elsewhere things were absolutely still; but the broad warping, occasional gentle folding, and slight slipping along stress lineaments, while often quite considerably altering the shoreline and producing various smaller and larger gulfs and embayments, could no longer effect large-scale and enduring changes to the main block of continental crust that had by now developed.

In Western Australia, because of such slight crustal movements about the close of the Devonian, we notice considerable changes in the extent, though not the general geographical position, of the shallow marine incursions onto the continent (*compare Maps 15 and 16*). While there is no obvious change in the Joseph Bonaparte Gulf Basin in the north, we see a substantial reduction in the size of the Canning Desert marine gulf; by early Carboniferous times it has, in fact, shrunk to what is known as the Fitzroy Trough of Late Palaeozoic sedimentation. In the Joseph Bonaparte Gulf Basin marine sedimentation which had begun in the Late Devonian (Famenian) continued uninterrupted into and until the end of Lower Carboniferous (Dinantian) time. There is a formation of sandy limestone here which contains a great number of brachiopods, conodonts, and corals of early Carboniferous age (Spirit Hill Limestone).

In the Fitzroy Trough there seems to have been some interruption in sedimentation between the latest Devonian and the Laurel Beds of Lower Carboniferous age. The latter are also calcareous and, in addition to a marine fauna of similar character to that of the Spirit

Hill Limestone, contain many shark teeth. The age of the Laurel beds is late Lower Carboniferous, that is younger than the Spirit Hill Limestone, and the earliest Carboniferous has not been found here in outcrop although it is likely to be present in the hidden central portions of the Trough.

In the Northwest Cape region (Carnarvon Basin) there is a similar reduction in the extent of the marine invasion; instead of the general submergence of the area during the Late Devonian we notice now a northwesterly opening embayment. Like the Fitzroy Trough this feature, after further shrinking, will become the main area of subsidence and sediment accumulation in this region during the following period, the Permian. The movements which brought about the changed shoreline produced slight folding and some erosion of the Devonian formations, so that the oldest of the Lower Carboniferous beds, the locally richly fossiliferous Moogooree Limestone, lies with a slight discordance or unconformity upon the Devonian formations.

About the end of the Lower Carboniferous further slight, or epeirogenetic, movements dried up all but one of the Western Australian embayments. In the Carnarvon Basin we find plant-bearing freshwater sandstones overlying the early Carboniferous marine beds, while in the Joseph Bonaparte Gulf Basin there is no evidence at all of later Carboniferous sedimentation. Only in the Fitzroy Trough we find the sea persisting probably throughout Middle and Upper Carboniferous times, but it must often have dried up too because the sandstones and shales of that age are interbedded with evaporites such as anhydrite. None of these beds appear on the surface. They were cut by the drill at several thousand metres depth in the Grant Range and Fraser River oil bores.

The first of the significant Carboniferous events in eastern Australia was the gradual elevation of the whole of Victoria and southern New South Wales, parts of which had been under sea in later Devonian times. Early Carboniferous sedimentation in these areas, and particularly in Victoria, is similar in character and extent to that of the Late Devonian; it is represented by largely unfossiliferous freshwater sandstones and siltstones, very sparsely interstratified with brackish water, or even marine sediments. The sandstone series of the Grampian Mountains in the west of the State, for example, include one or two thin beds of shales which contain small species of the simple lamp-shell *Lingula*, a creature that cannot live in freshwater environment. The Grampian formations are now tilted at an angle and exposed in gorges and precipitous escarpments of considerable scenic beauty.

The Mansfield area, northeast of Melbourne, was also the scene of a large, low-lying lake, which stretched far to the east and southward, and in which sandstones, conglomerates, and shales were accumulated. Here, too, occasional marine incursions seem to have taken place, as is shown by the occurrence of "sea-spider" echinoderms near Mansfield. North of this town, at Bridge Creek, is also one of the best known localities for collecting Carboniferous fish and plants; over 300 specimens of freshwater fish were found here many years ago (*Woodward, 1906*).

In southern New South Wales there is no evidence at all of Carboniferous sediments. The same applies to central western and northwestern parts of the State, unless one wants to connect uppermost parts of the Mootwingee and Mulga Downs beds west of Cobar with the Grampian series of western Victoria.

In northern New South Wales the seaway which had existed in Middle to Upper Devonian times still persisted in slightly different form, as can be seen by comparing Maps (15) and (16). The main outlines of this strait, though not its actual coastline, can be clearly traced from the open ocean to the east through the country between Newcastle and Kempsey northwestward past Gloucester, across what is now the main Dividing Range at Murrurundi, past Gunnedah and Bingera towards Warwick in southeastern Queensland, where the Carboniferous series pass beneath younger rocks. North of this gap they may be picked up again in the vicinity of Mundubbera; thence they extend in a continuous belt past Cania and Diglum to the coast at Rockhampton, where they pass again beneath later rocks, only to appear again farther north on the coast near Repulse Bay and Whitsunday Island. It is noteworthy that the only marine sediments of Upper Carboniferous age known in Australia are found in this part of Queensland, from Mundubbera through the Stanwell-Mount Morgan district to Rockhampton—in what is now known as the Yarrol Basin, an area of persisting strong subsidence during the later Palaeozoic.

In recent years another area of occasional marine incursions during the Lower Carboniferous has been found in the region between Townsville and Cooktown (*Wyatt and White, 1960*), and it seems that it is through this region, where outcropping Carboniferous rocks appear, that the Carboniferous part-marine part-freshwater sequences cut by deep bore holes beneath younger sediments in central and southwestern Queensland were connected with the Pacific, see Map (16). At times, though, a connection may also have existed from northern New England northwestward to these inland regions. In the far north of Queensland Carboniferous rocks are known only in plant-bearing freshwater facies, for instance the Pascoe River beds on Cape York Peninsula.

Although these various seaways remained more or less open for perhaps one-third of the Period—which is, with 74 million years,

Map 16:

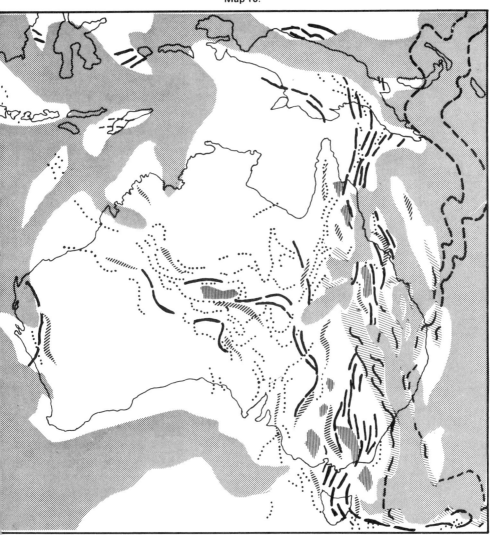

	SEA		STRUCTURAL TRENDS OF LAST COMPLETED FOLD MOUNTAINS—NOW UNDER INTENSE EROSION
	LAND		STRUCTURAL TRENDS OF ANCIENT FOLD MOUNTAINS—NOW MORE OR LESS PENEPLAINED
	FRESHWATER LAKE		
	ZONE OF CONTEMPORARY OR JUST COMPLETED MOUNTAIN-BUILDING		
	AXIS OF UPWARPING —IN PLACES AS AN EFFECT OF DISTANT MOUNTAIN-BUILDING FORCES		

The Australian region during the Kanimblan Orogeny in southeastern Australia, about 310 million years ago in mid-Carboniferous times. The sketch also shows the structural trends which were formed earlier during the Palaeozoic Era (for Precambrian trends see Map 9).

almost twice as long as the Devonian—the sequence of events differed considerably in various regions. Thus, a better portrayal of the sequence of events is obtained of the areas within New South Wales and those of the Yarrol Basin if they are discussed separately, in spite of the fact that both of them seem similarly constituted of two main subdivisions of Carboniferous rocks, each of them very distinct, especially in New South Wales.

The lower and older of the two subdivisions in New South Wales is known as the Burindi series, the upper as the Kuttung. The two together are over 6000 metres or six kilometres thick. Many papers dealing in detail with this sequence have been published.

The Burindi series, which is about 3000 metres thick, is a marine sequence laid down in a steadily subsiding trough which had received much sediment already in Devonian times. Professor A. Voisey (*1959*) has called it the New England Eugeosyncline, that is a type of marine trough in which the waters are comparatively deep and the sediments on the whole rather fine-grained. There was, however, repeated shallowing in places, and then coral reefs and other abundant marine life are found. The deeper water sediments are commonly fine mud with extensive beds consisting to a large extent of microscopic siliceous shells of radiolaria. There was also much volcanic activity, some of it submarine; in other words, not only were lavas poured onto the sea-bed but showers of volcanic ash also became mixed with the other sediments.

The shallow-water beds in the Burindi, both in New South Wales and Queensland, are often characterized by developments of a curious rock called oolitic limestone. This is composed of small granules of carbonate of lime which resemble the roe of fish—hence the term oolitic, which means "egg-stone". The densely packed granules are formed by the deposition of the lime around minute nuclei of sand or shell particles.

In Queensland the equivalent of the Burindi series was long simply known as the Rockhampton Series, but in recent years a great deal of work has been done in many areas, and the sequence has been subdivided into a number of formations which are characterized both by typical rock associations and certain fossil faunas. In the Yarrol-Monto district, for example, the earliest Carboniferous beds, consisting of cherts and greywacke (a kind of "dirty" sandstone), are known as the Tellebang Formation, and its characteristic fossils, brachiopods such as *Dictyoclostus* and *Balanoconcha*, appear in the upper part. In the Mount Morgan-Stanwell area beds equivalent to the Tellebang Formation are siltstones and sandy shales, with brachiopods such as *Tenticospirifer* and *Dimegelasma* below, *Chonetes* and *Brachythyris* at higher levels, and they are known as the Pond Argillite. We cannot go into much detail here, but those interested may consult the relevant chapters in the "Geology of Queensland" published as Volume Seven of the Journal of the Geological Society of Australia (*Hill and Denmead, 1960*).

The upper part of the Rockhampton Group corresponds in the Mount Morgan district to the Neil's Creek sequence (greywackes, cherts, many limestone layers), which there follows upon the Pond Argillite; and in the Yarrol-Monto district to the Baywulla Formation (shaly sandstones, limestones, and conglomerates) with its corals (e.g. *Palaeacis*).

The Rockhampton Group as a whole is then followed by the Neerkol Formation in Queensland, in the same way as the Burindi series is followed by the Kuttung Group in New South Wales. The town of Rockhampton itself is in a large area of exposures of these Carboniferous rocks, which consist largely of marine shales with layers of oolitic limestones similar to those in New South Wales. There seems to have been an absence of lava flows in this region, but many of the rocks consist partly of volcanic ash fragments, which shows that the intense volcanicity which marked the whole Period was never far away. Many of the rocks in the Rockhampton district contain abundant marine fossils, particularly the oolitic limestones, but also some of the shales. One of the best localities is at Lion Creek, about 30 kilometres to the southwest of Rockhampton, where the limestone contains many corals, lamp-shells and other fossils (*Hill, 1934*). Other fine fossil localities, farther to the south, are Station Creek, Mount Morgan, Mount Grim, Diglum, Cania, Many Peaks, Cannindah, Riverleigh, and Mundubbera.

The thickness of the Rockhampton Group is over 1800, that of the Neerkol Formation about 600 metres. It should be noted, however, that especially in the area from Rockhampton southward to Cania there is no continuity in sedimentation from one series to the other. During the Namurian (the uppermost stage of the Lower Carboniferous) the area had emerged. When it subsided again at the beginning of the Middle Carboniferous the first of the new deposits in the advancing sea were large sheets of conglomerates of coastal shallow water type. The first marine fossils appear only several hundred metres above the conglomerates. This fauna is characterized by the lamp-shell *Levipustula*, a form often mistaken for *Productus*, with which we will deal later. Another bed, higher up in the Formation, is distinguished by the abundance of a bryozoan or lace-coral called *Protoretepora*, and these two beds or "marker" horizons wherever found, give an infallible clue to the understanding of the sequence and structure of the region. Towards the top of the Neerkol Formation there is a thick flow of andesite lave which is probably of Upper Carboniferous age.

Farther to the west in Queensland, conditions remained much the same as during the later Devonian, but the extent of these shallow inland seas and lakes decreased considerably during Lower Carboniferous times. One of the largest features of that type was "Lake Drummond" which lay some 300 kilometres west of Rockhampton, covering the area now occupied by the Drummond Ranges. Whether it was always a freshwater lake is questionable because oolitic and algal limestone beds occur among the sediments laid down in it. Probably the sea did repeatedly find access to it, perhaps in the way shown in Map (16). On the whole Lake Drummond must be regarded as having been part of the shallow-water, estuarine and lagoonal environments which, at times reduced to strings of large lakes, characterize the Carboniferous landscape of central and southwestern Queensland as well as that of eastern Cape York Peninsula, and of marginal parts of the Rockhampton-New England-Newcastle seaway—in the latter, especially from the late Lower Carboniferous onward.

While the deposition of marine rocks in the Yarrol Basin of Queensland was proceeding quietly and with but few and short interruptions, there began in New South Wales another of those great convulsions in the earth's crust which completely changed the face of the land. This is known as the Kanimblan Orogeny. The folding movements began at the close of the Viséan—that is, a little more than half way through the Early Carboniferous epochs or, in local terms, at the end of the deposition of the Burindi Group—and they continued intermittently and with varying intensity in different districts for a long time. The first effect was the elevation of the floor of the seaway northwest of Newcastle above sea-level, and its conversion into a series of lakes. The rocks which had been deposited within the seaway were also compressed and folded, and in places worn down and eroded, so that the new lake sediments were laid unconformably upon the upturned edges of the older marine beds.

More marked, however, was the folding and elevation of the lands which had bordered the Early Carboniferous southeastern seaways. The island areas in the New England region—the "Ancestral New England" of Andrews (*1938*)—must have been particularly affected and elevated into mountains well above the snow-line. From these mountains glaciers pushed far down the valleys and may have even reached the lakes and estuaries which lay at the feet of the ranges or, if they failed to do so, swift-flowing streams resulting from the melting of the ice carried masses of glacial debris into the lakes as well as into

PALAEONTOLOGICAL PLATES

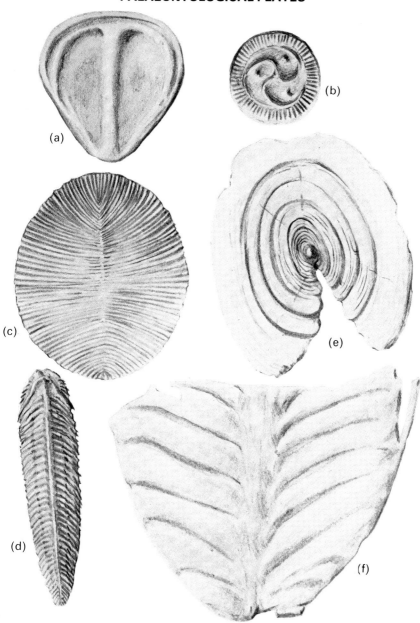

PLATE 1—**PROTEROZOIC FOSSILS**
CLASS UNKNOWN: (a) *Parvancorina* ×2.9, (b) *Tribrachidium* ×1.4
WORMS: (c) *Dickinsonia* ×0.7, (d) *Spriggina* ×1.4
JELLYFISH: (e) *Spriggia* ×0.7
SEA-PENS: (f) *Rangea* ×0.4
(Pencil drawings by R. O. Brunnschweiler)

PLATE 2—**PROTEROZOIC FOSSILS**
ALGAE: (a) and (b) *Algal limestone* (Carter, Brooks & Walker 1961);
(c) *Osagia* ×2.7 (Edgell 1964)
UNICELLULAR MICROFOSSILS: (d) *Huroniospora* ×1500 (Muir 1976)

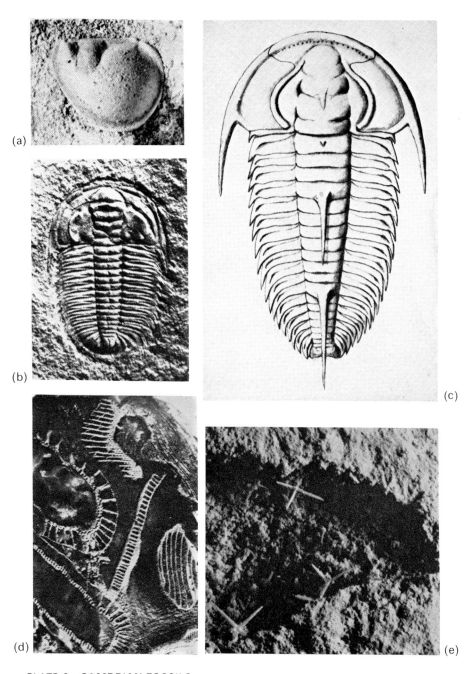

PLATE 3—**CAMBRIAN FOSSILS**
BIVALVED CRUSTACEANS: (a) *Aristaluta* ×7.3 (Opik 1961)
TRILOBITA: (b) *Xystridura* ×1.4 (Laseron 1955); (c) *Redlichia* ×2.3 (Opik 1958)
PLEOSPONGIA: (d) *Archaeocyathina* ×0.6 (Laseron 1955)
SPONGIA: (e) *Spicules of Pleodioria* ×8.4 (Opik 1961)

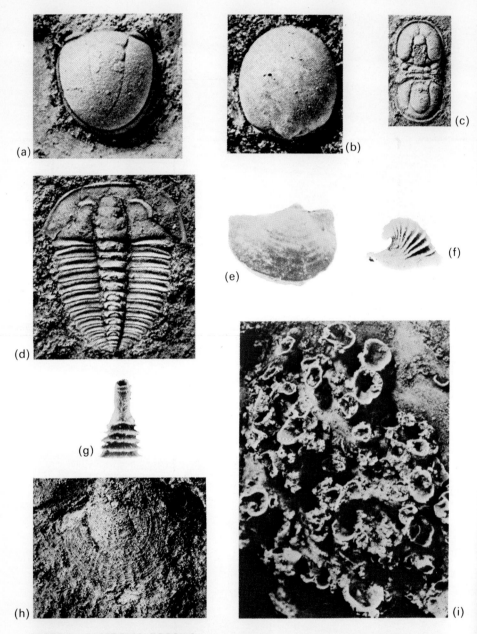

PLATE 4—**CAMBRIAN FOSSILS**
TRILOBITA: (a) *Leiopyge* ×14, tail, and (b) *Leiopyge* ×12, head (Opik 1961);
(c) *Agnostus* ×4.2, (d) *Olenus* ×11 (Opik 1963)
PROBABLE ANCESTORS OF PELECYPODA: (e) *Myona* ×24 (Runnegar & Jell 1976)
PROBABLE ANCESTORS OF GASTROPODA: (f) *Helcionella* ×5 (Runnegar & Jell 1976)
ANCESTORS OF CEPHALOPODA: (g) *Yochelcionella* ×19 (Runnegar & Jell 1976)
BRACHIOPODA (LAMP-SHELLS): (h) *Micromitra* ×3 (Opik 1961)
ANTHOZOA (CORALS): (i) *Cothonion* ×1.5 (Jell & Jell 1976)

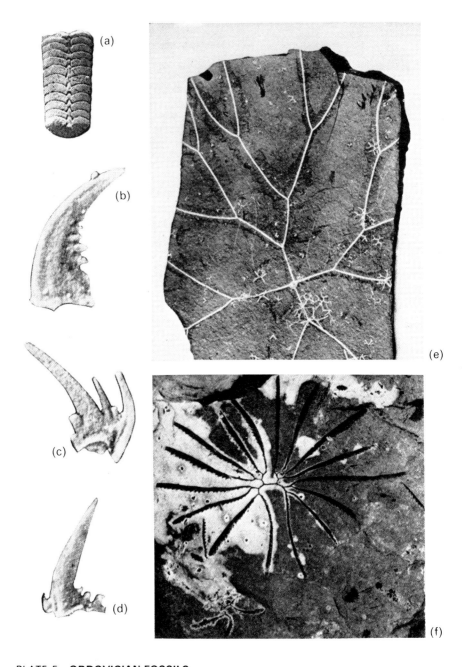

PLATE 5—ORDOVICIAN FOSSILS
CEPHALOPODA: (a) *Ventroloboceras* ×1 (Teichert & Glenister 1954)
CONODONTA: (b) *Belodina* ×50, (c) *Phragmodus* ×50,
(d) *Dichognathus* ×50 (Philip 1966)
GRAPTOLITES: (e) *Tetragraptus* and *Loganograptus* on same slab ×1.9 (Laseron 1955),
(f) *Clonograptus* ×1.9 (Laseron 1955)

PLATE 6—**ORDOVICIAN FOSSILS**
CEPHALOPODA: (a) *Hardmanoceras* ×1 (Teichert & Glenister 1954)
GASTROPODA: (b) and (c) *Raphistomina* ×0.66 (David 1950)
BRACHIOPODA: (d) *Ectenoglossa* ×2 (Fletcher 1964);
(e) *Gunningblandella* ×3 and (f) *Durranella* ×5 (Percival 1979)
BRYOZOA: (g) *Stictopora* ×10 (Ross 1961)

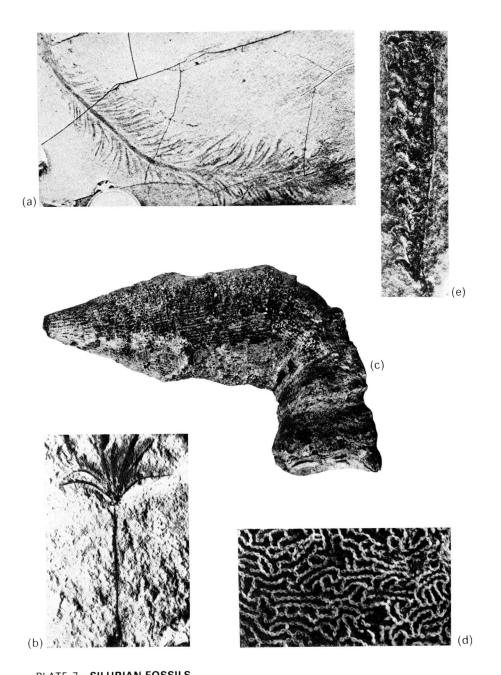

PLATE 7—**SILURIAN FOSSILS**
PLANTS: (a) *Baragwanathia* ×0.4 (Williams 1964) appeared just
before the end of the Silurian. It is more common in early Devonian rocks.
CRINOIDEA: (b) *Heliocrinus* ×0.3 (Laseron 1955)
CORALLIA: (c) *Phaulactis* ×1.7 and (d) *Halysites* ×1 (Laseron 1955)
GRAPTOLITES: (e) *Monograptus* ×8.4, also occurs in earliest Devonian rocks (Berry 1965)

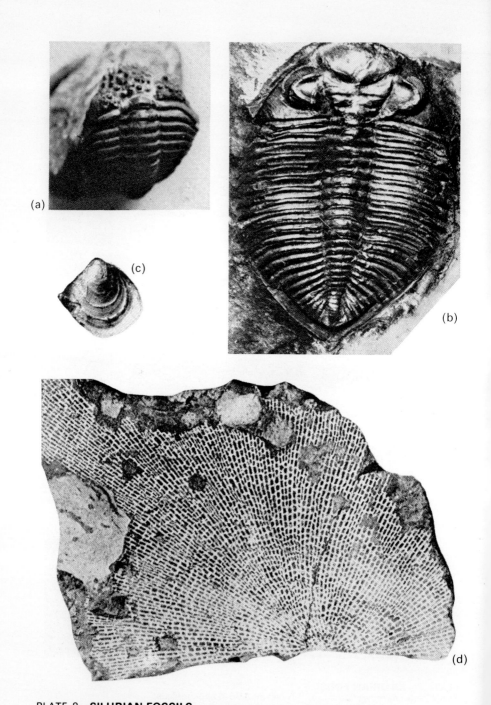

PLATE 8—**SILURIAN FOSSILS**
TRILOBITA: (a) *Encrinurus* ×1 and (b) *Dalmanites* ×1 (Laseron 1955)
PELECYPODA: (c) *Pterinea* ×1.4 (Talent & Philip 1956)
BRYOZOA: (d) *Fenestella* ×2, also occurs in Devonian and Carboniferous rocks (Ross 1961)

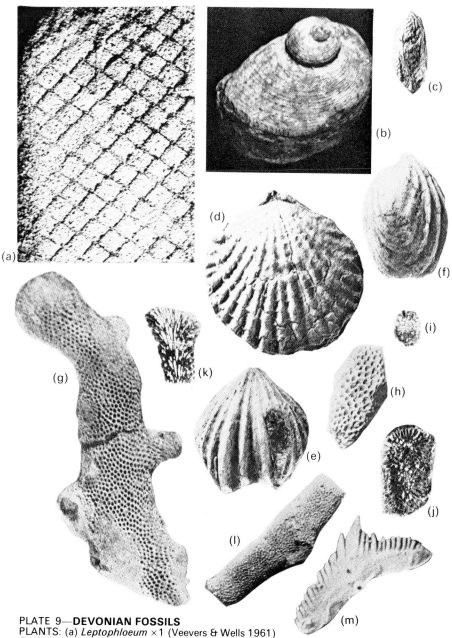

PLATE 9—**DEVONIAN FOSSILS**
PLANTS: (a) *Leptophloeum* ×1 (Veevers & Wells 1961)
GASTROPODA: (b) *Phanerotrema* ×1 (Laseron 1955)
BRACHIOPODA: (c) *Spinatrypa* ×2 and (d) *Spinatrypa* ×4 (Veevers 1959);
(e) and (f) *Fitzroyella* ×6 (Veevers 1959)
BRYOZOA: (g) *Percypora* ×3.4 (Ross 1961)
CORALLIA: (h) *Alveolites* ×1 with (i, j, k) various cross sections of same species ×1;
(l) *Thamnopora* ×1 (Hill 1954)
CONODONTA: (m) *Ozarkodina* ×35 (Philip 1965a)

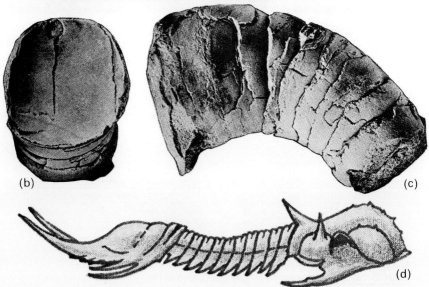

PLATE 10—DEVONIAN FOSSILS
VERTEBRATA: (a) *Dipnorhynchus* ×1 (Laseron 1955)
CEPHALOPODA: (b) and (c) *Lytogyroceras* of Nautilidoidea ×0.9
(Teichert & Glenister 1952)
TRILOBITA: (d) *Acanthopyge* ×8 (Edgell 1955)

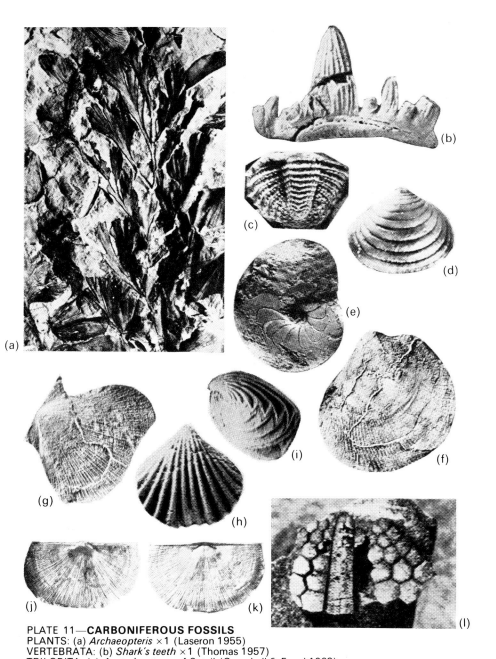

PLATE 11—**CARBONIFEROUS FOSSILS**
PLANTS: (a) *Archaeopteris* ×1 (Laseron 1955)
VERTEBRATA: (b) *Shark's teeth* ×1 (Thomas 1957)
TRILOBITA: (c) *Australosutura* ×4.3, tail (Campbell & Engel 1963)
BRANCHIOPODA: (d) *Cryptophyllus* ×9 (Jones 1962)
CEPHALOPODA: (e) *Muensteroceras of the Goniatitina* ×1.5 (Campbell & Engel 1963)
PELECYPODA: (f) *Panenka* ×1 and (g) *Euchondria* ×3 (Campbell & Engel 1963)
BRACHIOPODA: (h, i) *"Camarotoechia"* ×2.7 (Veevers 1959) and
(j, k) *Schuchertella* ×1 (Roberts 1964)
ECHINOIDEA: (l) *Oligoporus* ×1.5 (Thomas 1965)

PLATE 12—**PERMIAN FOSSILS**
PLANTS: (a) *Glossopteris* ×0.8, (b) *Noeggerathiopsis* ×0.8
(c) *Phyllotheca* ×0.8 (Laseron 1955); (d) *Gangamopteris* ×0.8 (Veevers & Wells 1961);
and PLANT SPORES: (e) *Potonieisporites* ×820,
(f) *Parasaccites* ×820 (By courtesy of Eliz. Kemp)
ECHINOIDEA: (g) *Calceolispongia* ×0.8 (Etheridge 1914)
FORAMINIFERA: (h) *Reophax* ×57, (i) *Frondicularia* ×46,
(j) *Ammodiscus* ×43 (Crespin 1958a)

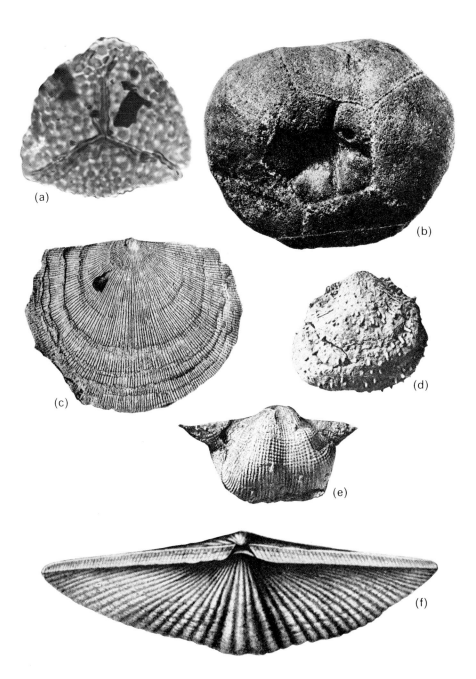

PLATE 13—**PERMIAN FOSSILS**
PLANTS (SPORES): (a) *Verrucosisporites* ×700 (Eliz. Kemp)
CRINOIDEA: (b) *Phialocrinus* ×1.6 (Laseron 1955)
BRACHIOPODA: (c) *Permothetes* ×0.8 (Thomas 1958);
(d) *Strophalosia* ×0.8, (e) *Dictyoclostus* ×0.4 (Coleman 1957);
(f) *Spirifer* ×0.8 (Laseron 1955)

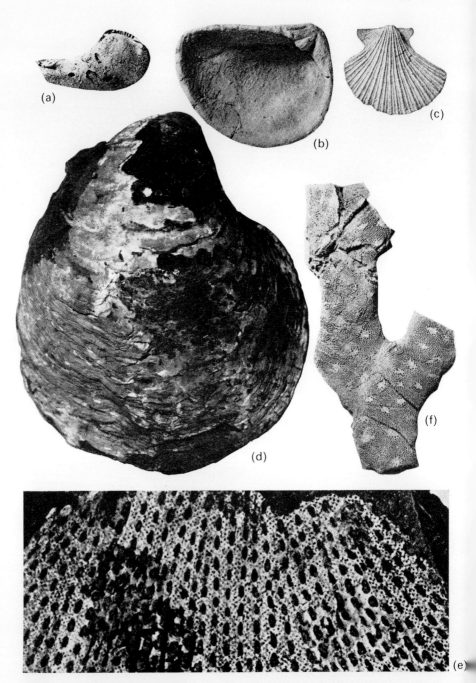

PLATE 14—**PERMIAN FOSSILS**
PELECYPODA: (a) *Nuculana* ×1, (b) *Schizodus* ×1 (Dickins 1956);
(c) *Aviculopecten* ×1 (Dickins 1963); (d) *Eurydesma* ×0.9 (Laseron 1955)
BRYOZOA: (e) *Polypora* ×2.6 (Laseron 1955); (f) *Hexagonella* ×0.9 (Crockford 1957)

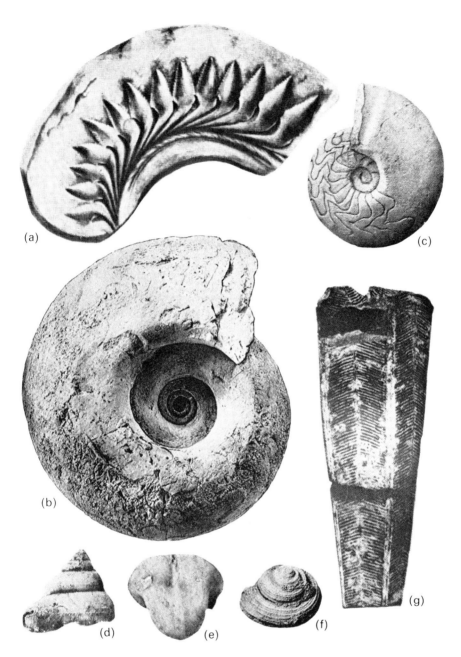

PLATE 15—**PERMIAN FOSSILS**
VERTEBRATA: (a) *Teeth of shark Helicoprion* ×0.4 (David 1950)
CEPHALOPODA: (b) *Metalegoceras* ×0.8 (Teichert & Glenister 1952);
(c) *Pseudogastrioceras* ×1 (Teichert 1953)
GASTROPODA: (d) *Mourlonia* ×4, (e) *Bellerophon* ×1 (Dickins 1963); (f) *Ptychomphalina* ×2 (Dickins 1957)
CONULARIDA: (g) *Conularia* ×1 (Laseron 1955)

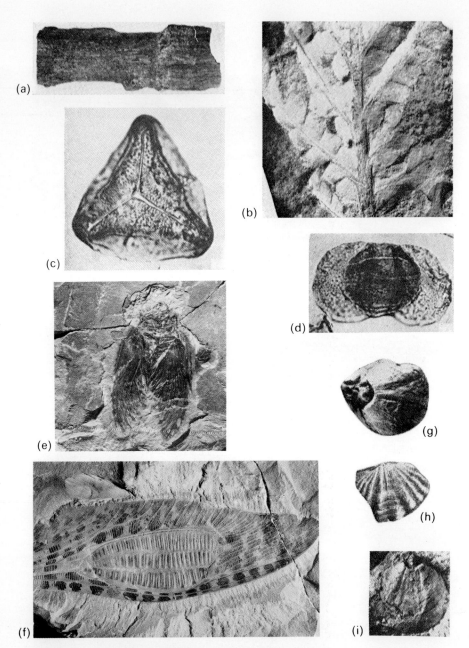

PLATE 16—**TRIASSIC FOSSILS**
PLANTS: (a) *Equisetites* ×0.8 (Laseron 1955), (b) *Thinnfeldia* ×0.8
(Veevers & Wells 1961) and SPORES: (c) Concavisporites
×500, (d) *Protohaploxypinus* ×380 (Playford 1965)
INSECTA: (e) *Cicada*, (f) *Clathrotitan* ×0.8 (Laseron 1955)
CEPHALOPODA: (g) *Subinyoites* ×2 (Dickins & McTavish 1963)
PELECYPODA: (h) *Claraia* ×2, (i) *Bakevellia* ×2 (Dickins & McTavish 1963)

PLATE 17 — **TRIASSIC FOSSILS**
PISCES: (a) *Brookvalia* ×0.9, (b) *Cleithrolepis* ×0.8 (Laseron 1955)

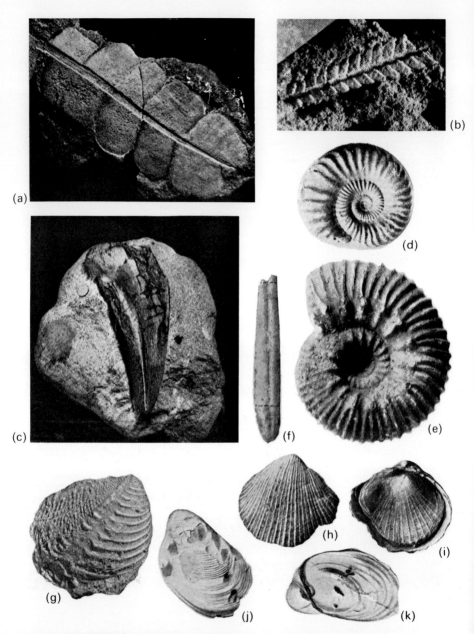

PLATE 18—**JURASSIC FOSSILS**
PLANTS: (a) *Nilssonia* ×0.8, (b) *Otozamites* ×1 (Veevers & Wells 1961)
DINOSAURS: (c) Claw bone × 1 (Laseron 1955)
CEPHALOPODA: (d) *Fontannesia* ×0.8, (e) *Otoites* ×0.8 (Arkell & Playford 1954);
(f) *Belemnopsis* ×0.8 (Brunnschweiler 1960)
PELECYPODA: (g) *Trigonia* ×0.8 (Skwarko 1963); (h) left valve and
(i) right valve of *Meleagrinella* ×1.2 (Teichert 1940b);
(j, k) left and right valve of *Buchia* ×0.8 (Fleming 1959)

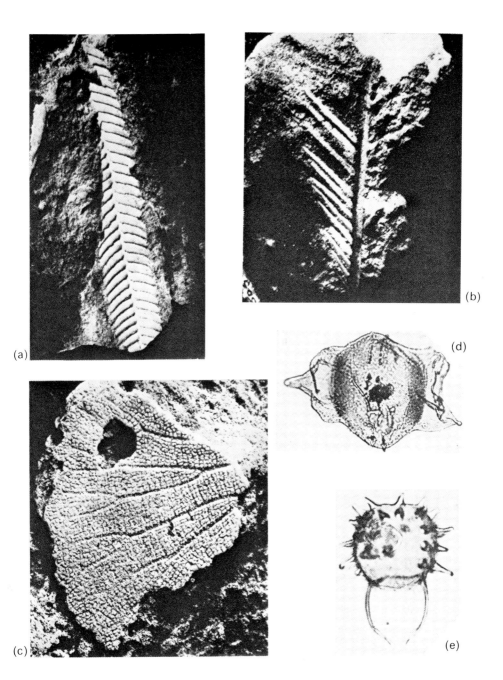

PLATE 19—**CRETACEOUS FOSSILS**
PLANTS: (a) *Ptilophyllum* ×0.9, (b) *Zamites* ×1,
(c) *Hausmannia* ×1.9 (Veevers & Wells 1961)
DINOFLAGELLATA: (d) *Deflandrea* ×500 (Cookson & Manum 1964),
(e) *Diphyes* ×670 (Cookson 1965)

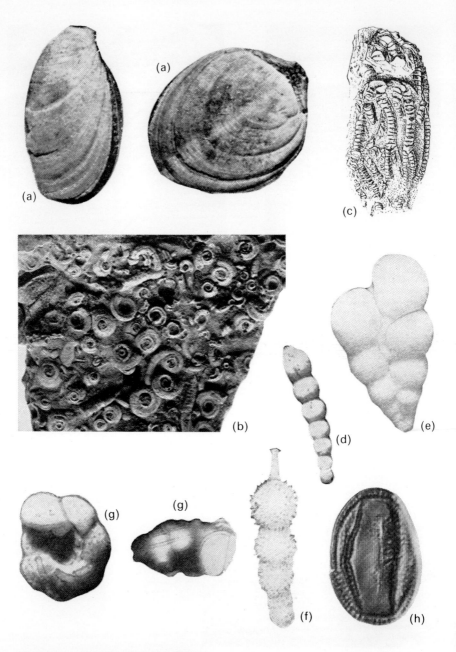

PLATE 20—**CRETACEOUS FOSSILS**
BRACHIOPODA: (a) *Australiarcula* ×3 (Elliott 1960)
ANNELIDA (WORMS): (b) Curled tubes of *Spirulaea* ×0.5 (Laseron 1955)
CRINOIDEA: (c) *Isocrinus* ×1 (Etheridge 1901)
FORAMINIFERA: (d) *Dentalina* ×13, (e) *Guembelina* ×97,
(f) *Stilostomella* ×47, (g) *Globotruncana* ×48 (Belford 1960)
PLANT POLLEN: (h) *Phimopollenites* ×1000 (by courtesy of D. Burger)

PLATE 21—**CRETACEOUS FOSSILS**
PELECYPODA: (a) *Pseudavicula* ×3 (Cox 1961); (b) *Syncyclonema* ×1.5 (Feldtman 1951);
(c) *Pycnodonta* ×1.25 (Feldtman 1963); (d) *Inoceramus* ×1 (Brunnschweiler 1960);
(e) *Pterotrigonia* ×1, (f) *Nototrigonia* ×1 (Skwarko 1963)
GASTROPODA: (g) *Globularia* ×1 (Cox 1961)

PLATE 22—**CRETACEOUS FOSSILS**
CEPHALOPODA: (a) *Crioceras* ×0.8, (b) *Labeceras* ×0.8 (Laseron 1955);
(c) *Turrilites* ×0.8, (d) *Scaphites* ×0.8 (Wright 1963)
PLANT POLLEN: (e) *Nyssapollenites* ×800 (D. Burger)

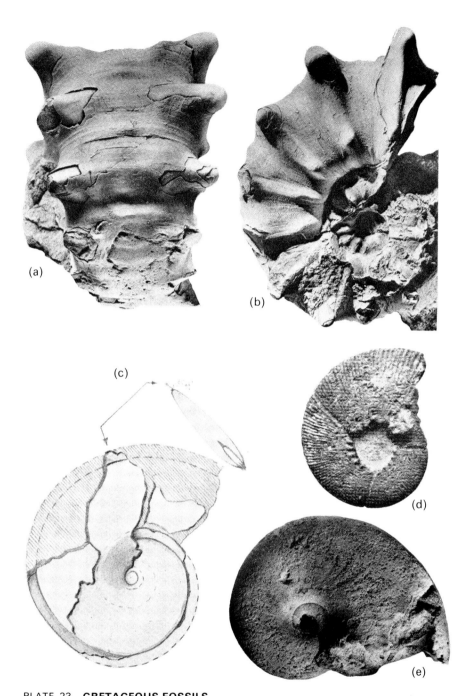

PLATE 23—**CRETACEOUS FOSSILS**
CEPHALOPODA: (a, b) *Euomphaloceras* ×0.8 (Wright 1963);
(c) *Aconeceras* ×0.8 (Brunnschweiler 1959); (d) *Kossmaticeras* ×0.8 (Spath 1940);
(e) *Pseudophyllites* ×0.8 (from the author's own collection in Cmwlth Bur. Min. Resour.)

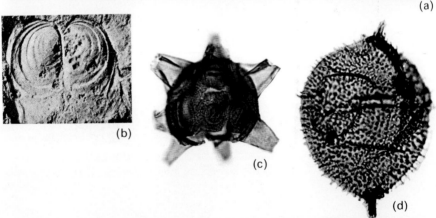

PLATE 24—**CRETACEOUS FOSSILS**
VERTEBRATA: (a) *Ichthyodectes* ×0.4, paddle foot of fish lizard (Laseron 1955)
CONCHOSTRACA (BIVALVED CRUSTACEANS): (b) *Cyzicus* ×7 (Talent 1965)
DINOFLAGELLATES: (c) *Litosphaeridium* ×580 and
(d) *Diconodinium* ×580 (Eliz. Kemp)

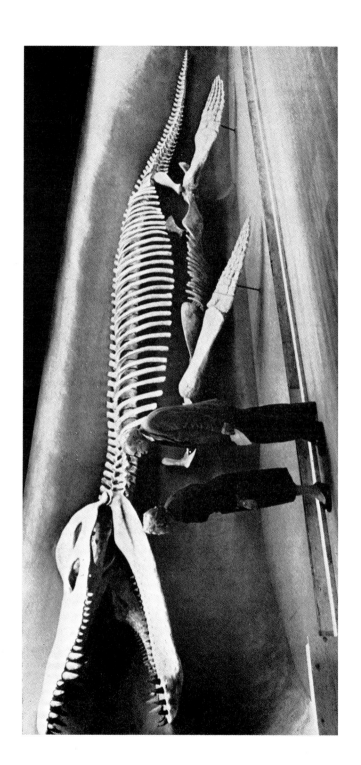

PLATE 25—**CRETACEOUS FOSSILS**
VERTEBRATA: *Kronosaurus queenslandicus* Longman. The complete skeleton, 12.6 m long, of an extinct marine reptile found in Queensland, housed in the Mus. Zool. Univ. Harvard, U.S.A.

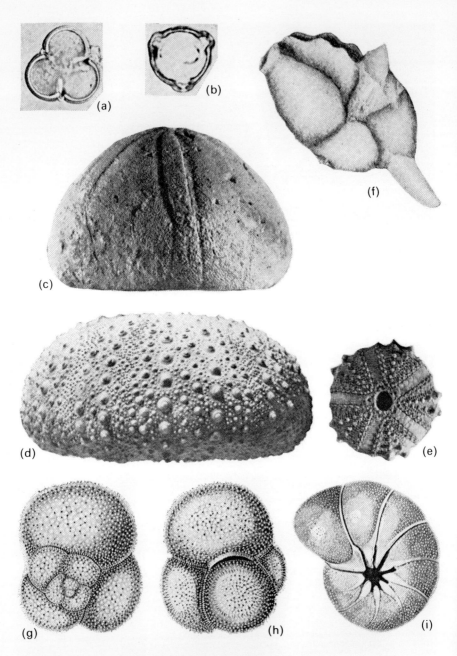

PLATE 26—**TERTIARY (CAINOZOIC) FOSSILS**
PLANTS (POLLEN): (a) *Gunnera* ×600, (b) *Casuarinidites* ×600 (Cookson & Pike 1954)
CRINOIDEA: (c) *Hypsoclypeus* ×0.9 (Crespin 1943);
(d) *Heliocidaris* ×0.9, (e) *Murravechinus* ×1.8 (Philip 1965)
FORAMINIFERA: (f) *Hantkenina* ×100 (Crespin 1958);
(g, h) *Globigerina* ×114, (i) *Astrononion* ×65 (Carter 1958)

PLATE 27—**CAINOZOIC FOSSILS**
PELECYPODA: (a, b) *Eucrassatella* ×0.63 (Darragh 1965);
(c) *Neotrigonia* ×1 (Laseron 1955); (d, e) *Zenatiopsis* ×1.7 (Gill 1963)
GASTROPODA: (f) *Diala* ×9, (g) *Notomella* ×4 (Valentine 1965)

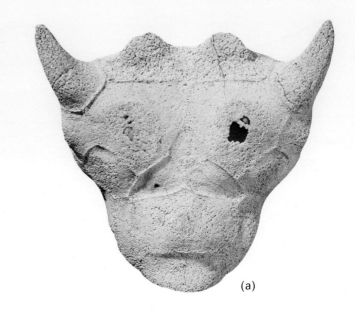

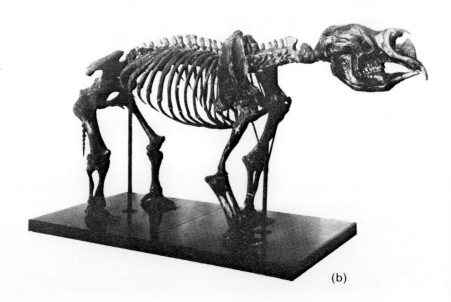

PLATE 28—**CAINOZOIC FOSSILS**
VERTEBRATA: (a) *Miolania* ×0.5, a horned turtle, and
(b) the marsupial *Diprotodon* (about ×0.1), both from the original edition of this book
(Laseron 1955)

GEOLOGICAL PLATES
The following plates are presented in chronological sequence designed to show outstanding features of the major Geological Periods.

Plate 1
Mica-schist showing original bedding of lower Precambrian Willyama Series in the Barrier Ranges, Western N.S.W.

Plate 2
Chewings Range thirty kilometres west of Alice Springs, a few kilometres north of the western Macdonnell Ranges. A Pertaknurran (Heavitree Quartzite) wedge rising from the Nullaginian Arunta Complex. **(J. A. Dulhunty)**

Plate 3
Tilted Mt Isa shales (early Carpentarian) which contain large silver, lead, zinc and copper deposits.
(D. F. Branagan)

Plate 4
Extensive exposure of early and mid-Proterozoic (Nullaginian and early Carpentarian) rocks in the Mt Isa district. The mineralised beds lie just west of the town in the area of the dark ridges where the high smoke-stacks are.

Plate 5
The Flinders Ranges consist of a broad series of folds formed during earth movements which were active between 450 and 520 million years ago. Slightly tilted layers of hard sedimentary rocks are interbedded with softer, less resistant layers.
(Bureau of Mineral Resources)

Plate 6
Unusual shapes produced by weathering of granite on Kangaroo Island, S.A. Running water and wind have produced the surface features.
(Douglass Baglin)

Plate 7
Gosses Bluff, a peculiar, circular mountain structure, about six kilometres in diameter, seventy kilometres west of Hermannsburg Mission in Central Australia. The origin of this structure, which consists of Ordovician and Silurian rocks, is no longer a mystery; it was caused by the impact of an extra-terrestrial missile.
(Bureau of Mineral Resources)

Plate 8
The Grand Arch which cuts through the 275-metre-thick limestone at Jenolan, N.S.W.; the limestone was originally formed in the Silurian period. The Jenolan Caves occur in this limestone and were formed in late Tertiary times. **(A. Healey)**

Plate 9
Cathedral Rocks, a Silurian reef of limestone near the Kowmung River, N.S.W. Running water has dissolved the limestone, producing the interesting surface pattern and a series of underground caves.
(B. Byles)

7

9

10

11

Plate 10
Westerly dipping sandstones of the Grampian Range, Western Victoria. These were probably deposited in a large lake during the upper Devonian-lower Carboniferous age, 380 million years ago. They now form a prominent range because of later faulting, folding and erosion of the surrounding rocks.

Plate 11
Edge of the Bathurst granite intrusion, N.S.W., of Carboniferous age. Heat from the granite has altered the surrounding rocks and has made them more resistant to erosion. The dark hills consist of these altered rocks, the low-lying foreground of weathered granite.
(Government Tourist Bureau of N.S.W.)

Plate 12
Fossil shells of the early Permian cold water genus **Eurydesma** in coarse conglomerate. Railway cutting near Allandale, N.S.W. **(C. F. Laseron)**

Plate 13
Triassic sandstones at North Avalon, N.S.W., showing contortions in the sandstone beds. **(D. F. Branagan)**

Plate 14
Cradle Mountain in Central Tasmania is carved out of Jurassic dolerite which here lies on conglomerates of early Permian age. Note the prominent columnar jointing of the lava rock. **(Australian News and Information Bureau)**

Plate 15
The Warrumbungles, Northern N.S.W., an excellent example of Tertiary volcanic rocks, Trachytic lavas and intermediate intrusions associated with basalts. The Breadknife in the foreground is a resistant dyke rock. **(A. Healey)**

Plate 16
A sink-hole in the Nullarbor, which is composed largely of Miocene limestone (middle Tertiary). The wide extent of limestone is one reason why the area lacks much vegetation and has no surface streams. **(B. J. Watts)**

Plate 17
The flat-lying Tertiary rocks of Port Campbell off the west coast of Victoria. Cliffs of similar Tertiary formations form the coast of the Great Australian Bight. **(A. Healey)**

16

17

18

19

Plate 18
Foliated granite tors near Station Hill, Tennant Creek. Fine particles have been eroded during millions of years to expose the granite boulder surface. The landscape now consists of low ridges of resistant rock separated by wide expanses of dirt.
(Bureau of Mineral Resources)

Plate 19
Gibber Plains, near Woomera, S.A., of Pleistocene to Holocene age. These have formed by erosion of hard surface cappings, which are common in the arid regions of Australia.
(Australian News and Information Bureau)

Plate 20
Angular unconformity between two sandstone sequences (near top of cliff) in the early Adelaidean system of the eastern Bangemall Basin, W.A. **(Geol. Soc. Aust.)**

Plate 21
The famous Devonian reef complex fringing the northern margin of the Fitzroy Trough, W.A. Buried under Carboniferous and Permian formations of the plains to the left, oil drillers have found still more such reefs. **(Geol. Surv. W. Aust.)**

Plate 22
Windjana Gorge cuts a most instructive natural cross-section through the Devonian reefs of the Fitzroy Trough. The fore-reef dips away to the left, the nearly horizontal back-reef beds are on the right and the reef bioherm in the centre. **(P. E. Playford)**

marine environments. In fact, while in some areas the marine sediments of the upper part of the Burindi Group were deposited, there was already glacial debris accumulated in nearby regions; in other words, the latest beds of Burindi type are of the same age as the earliest beds of the Kuttung or, in geological language, the two are laterally interfingering.

This shows that the story of the Kuttung-time glaciers had nothing to do with the great glaciation of the following period, the Permian. It happened several tens of millions of years earlier, chiefly during the Viséan and the Westphalian Epoch, that is, during the late Middle and the early Upper Carboniferous. More importantly, it was a story of local mountain glaciers, not of glaciation of continental extent like the Permian one, as we shall see later. Some of these glacial Kuttung beds nevertheless are of astounding volume, and individual series such as the main glacial sequence at Seaham (*Sussmilch, 1935*) can be as thick as 550 metres.

A factor which contributed to the formation of very thick beds of tillite and tillite-like materials was the combination of glacial and volcanic action. In a temperate climate glaciers from the high mountains may push well below the snow-line but do not reach the sea. The rocky detritus borne by them normally accumulates as moraines on the flanks, underneath, and in front of the creeping ice-sheet, though much or most of it is eventually carried by the streams away to the sea. This normal process of destruction and deposition may, however, be enormously accelerated if there are active volcanoes in the vicinity. Such an event actually took place in southeast Iceland during the last century, when a volcano erupted immediately below a large glacier. Fortunately it was in a sparsely inhabited district, but the destruction within a few hours was appalling. The glacier was literally lifted from its bed, and an enormous flood of broken ice, water, and rock, with a front of 50 kilometres, swept across the coastal plain and into the sea. From this one catastrophe alone thick beds of tillite-like sediments (not, however, true tillites) were deposited on the land, and also undoubtedly on the sea bottom for some distance from the shore. While there is no evidence that such a combination of glacial and volcanic action took place in Kuttung times, it is a distinct possibility, and it might well account for some of the astoundingly thick and rather localized deposits of tillitic material, deposits that are in fact up to ten times thicker than the thickest glacial deposits remaining from the last ice age which ended some 10 000 to 20 000 years ago. The reason for this distinct possibility is that throughout Lower and Middle Carboniferous times, and especially during the Kuttung, volcanic activity was exceptionally strong in southeastern Australia.

It was during this time that deep within these great mountain ranges were intruded the great granite masses which now form the larger part of the New England plateau. The mountains have long since disappeared, and it is only their slow removal by erosion that has exposed on the surface the granite masses in their core. The great volcanicity of those times was, of course, due to the intrusion pressure of these magmas from below. Thus, intermingled with the glacial deposits are the ash and scoriae ejected during eruptions, while large lava flows form a considerable portion of the total thickness of the Kuttung Group.

No greater contrast can be imagined than that between the present landscape and the terrible grandeur of earlier and middle Carboniferous times. Today the journey northwest from Newcastle, through the Hunter River valley, across the low main divide at Murrurundi to Barraba and beyond, is for the most part through a peaceful undulating country, rich in farm lands, with low rounded hills and long stretches of alluvial plain. In those far away times, however, the country was dominated by magnificent mountains towering to the skies, covered with large glaciers which sent their debris-laden masses into deep valleys with lakes large and small. The sky was lit at night by lines of active volcanoes erupting through their caps of snow and ice; the whole land was unstable, rising and sinking, and shaken continuously by earthquakes.

Conditions were not always so unceasingly turbulent during the period. Between the glacial and volcanic beds deposition of sand and mud continued in many regions quietly and built up sandstones and shales which now contribute considerably to the 3000 metres thickness of the Kuttung Group. In these formations are to be found numerous plant remains, most of which were probably washed into their present positions during times of flood. Amongst the numerous localities where fossil plants may be found should be mentioned Paterson on the Paterson River, north of Maitland, many places in the Gloucester district, Currabubula, Werris Creek, the Barraba area, and the Kempsey district.

From some point during the Middle Carboniferous epoch onward, things began to quieten down again. The mountain-building movements slowed down and almost ceased entirely, the volcanic fires died, and the mountains were soon eroded down to levels too low to permit the formation of glaciers. Lakes, estuaries and lagoons silted up, became smaller and smaller, and—except in the Yarrol Basin of Queensland—eventually disappeared.

In one way the period between the end of the Kuttung time and the beginning of the Permian, that is, the Upper Carboniferous, is of special interest—during this interval of time, which may have lasted as long as 25 million years, Australia was a very much larger continent than it had ever been before, and would ever be thereafter. Let us look at Map (16). Push the sea out of all the embayments and seaways around western and northwestern Australia, out of the central and southwest Queensland gulf and most of New Guinea; lock out the sea from New England, most of the Kempsey and Newcastle districts; narrow the big gulf south of the Great Australian Bight; broaden the connection from Tasmania to New Zealand; and one can imagine what was there. It must have been a very large continent indeed—land from Timor to New Zealand, from the Moluccas through New Guinea to Tasmania and beyond! In all this huge area we know of only one or two small marine provinces of Upper Carboniferous age, the Yarrol Basin in southeastern Queensland between Rockhampton and Brisbane and perhaps a small embayment in the Kempsey and Newcastle districts. However, as we shall see, this astonishing dominance of land in the Australian region was to disappear again very soon.

It had, however, a very remarkable effect on the evolution of life on our continent. During this Upper Carboniferous epoch the evolution of an entirely new and peculiar type of flora took place, a flora which characterizes the continents of the Southern Hemisphere together with India and Madagascar. The interval also produced a distinctive marine fauna of southern "provincial" character which we will get to know when discussing the Permian Period.

LIFE OF THE CARBONIFEROUS PERIOD

The First Forests

PLANT LIFE. Knowledge of Carboniferous plant life in Australia is comparatively meagre. In Europe and America the main coal measures are of Carboniferous age and have yielded a host of plant fossils. There has been much intensive research over many decades, and consequently more is known of the Carboniferous flora than of that of many later ages. In the Northern Hemisphere it is possible to form a fairly comprehensive picture of vast coal swamps; of forests of gigantic club-mosses 20 to 25 metres high; of horse-tails which also attained the size of large trees; of ferns 15 metres high with fronds 1.5 metres and more in length; of the curious plants called cordaitales, which may have been the ancestors of the pines; and of many smaller plants which were mainly ferns and seed-ferns.

In Australia we have no Carboniferous coal measures, and the plant fossils consist mainly of fragments which were washed into the lakes, lagoons and inlets. They afford but a glimpse of the vegetation of the time. Among these specimens are numerous fragments of the club-moss *Lepidodendron*, although not the same species as lived in the Devonian Period. With *Lepidodendron* is the horse-tail *Calamites*, but it is impossible to say whether these plants formed large forests as they did in other parts of the world. In the Middle Carboniferous (Kuttung), and very characteristic of our rocks, are two small primitive ferns, *Rhacopteris* and *Cardiopteris*, which though not spectacular are exceedingly useful for determining the age of the formations in which they occur. *Rhacopteris* (now called *Pseudorhacopteris*) had small rhomboidal pinnules alternating on either side of the stem; in *Cardiopteris* the pinnules were larger and rounded. Specimens of this type have been brought from the far north in the Cape York Peninsula, and from many other localities in Queensland and New South Wales. A well-known collecting locality is a road cutting within the boundaries of the small town of Paterson in the Hunter River district. These two forms, especially *Rhacopteris*, are so common in the late Lower and Middle Carboniferous of Australia that the name *Rhacopteris* flora has been given to the vegetation of the time (*Walkom, 1944*).

INVERTEBRATE MARINE LIFE. Although many of the genera and most species of Carboniferous marine animals were quite different from those of the Devonian, the relative proportions of the main classes remained much the same, and there were few spectacular changes among the lower invertebrates. Of these changes the most notable was the further decline of the trilobites which, rare in the Devonian, are now just barely lingering on. A few genera have been found, and there are not many species of these. Why the trilobites should have declined so much and, in fact, died out entirely before the close of the Permian, is not known. They seem to have been well adapted to their environment; possibly the development of predacious fish led to their final destruction.

In eastern Australia most of the Carboniferous rocks have been strongly folded and intruded in many areas by great masses of granite and porphyries, and often so altered by heat and pressure that their fossil contents have commonly been destroyed or made unrecognizable in detail. However, in some localities well preserved fossils are still abundant, notably the Clarencetown and Somerton districts in New South Wales and the Rockhampton district in Queensland. A number of strata in the Western Australian Carboniferous, both in the Carnarvon Basin and the West and East Kimberleys, are richly fossiliferous, but apart from a number of brachiopods and some ostracods the faunas have not yet been described in detail, although their composition is by now well known.

Coral reefs still occurred in Australia but they do not form such spectacular features as in the Devonian. A very distinctive coral is *Lithostrotion*, a massive, compound type, consisting of fairly large hexagonal columns packed closely together. In New South Wales, a reef composed of another coral, *Aphrophyllum*, occurs at Taree, and another of the same type at Hall's Creek, 30 kilometres south of Bingera. Practically all the spectacular corals found in the Silurian and Devonian sequences had become extinct, but the tubular, ramifying *Syringopora*, which we first met in the Silurian, and the small horn-shaped *Zaphrentis*, first appearing in the Devonian, still survived, the latter even into the next Period.

Lamp-shells are the most abundant Australian Carboniferous fossils, and a great variety has been found. Conspicuous among them are genera of the Productidae, such as *Productus* itself, *Marginirugus*, *Levipustula*, and the Chonetidae, including *Chonetes* itself, in all of which the smaller valve was flat or even concave, while the other was swollen and very convex. Most brachiopods of this type were ornamented with spines, often of great length. *Spirifer* and its relatives of the Spiriferacea superfamily, ranging from the Silurian into the earliest Mesozoic, had spiral internal processes, and the shell often had lateral wing-like expansions. The internal casts of lamp-shells of this type are in shape not unlike butterflies, for which they are sometimes mistaken by the uninitiated.

True molluscs, both bivalves and univalves, were increasing in numbers and variety, and were beginning to compete with brachiopods, which they were eventually to overshadow. Amongst the cephalopods the goniatites were prevalent, but the clymenioids and Nautiloids were also common. Again, the short-lived species among all these are most useful in dividing the Period into epochs and ages.

Crinoids were still abundant. It is unfortunate that we cannot give more space to the description of these complex and beautiful creatures. Australian Carboniferous specimens are, however, generally fragmentary and it is impossible to be sure of their details. So far no crinoids have been found that approach the magnificently preserved specimens from Iowa and Indiana in the United States, which are world-famous.

FISH AND OTHER VERTEBRATES. Carboniferous fish show a steady evolutionary advance over those of the Devonian Period. In Australia, knowledge has been mainly derived from some three hundred specimens found near Mansfield in Victoria. These were described by Woodward (*1906*). In addition to these there are the fish beds of the Drummond Range in central Queensland, one of which is said to be literally packed with remains of fish. This bed is a layer of sandstone about 30 centimetres thick, and has been traced over an extensive area. A series of specimens was collected by the Queensland Museum as long ago as 1892, mainly from the railway cutting at Hannam's Gap, near Bogantungan railway siding. Here the rock is largely made up of innumerable scales and teeth, complete specimens being very rare. Like those from Mansfield, the fish lived in fresh water and are mainly ganoids; the commonest species is a small fish, about the size of a haddock, belonging to the group known as the Palaeoniscidae (*Etheridge and Jack, 1892*).

The ganoids were the dominant group of fish in late Palaeozoic and early Mesozoic times. Sharks were also abundant and the dipnoids and crossopterygians (lobe-fins) still survived. A dipnoid is among the specimens found at Mansfield. Reference has already been made to the hard, enamelled, external plates of the ganoids, but they differed from the modern true bony fish in details of the skeleton also. In most of the orders comprising the large ganoid sub-class, the skeleton is cartilaginous, but in some of the more advanced groups it had become hardened into bones. The vertebrae of the backbone were, however, more or less articulated, unlike those of the bony fish which are separate and concave at each end. The tail is heterocercal like that of the sharks—that is, the backbone is prolonged into the upper lobe. In the homocercal tail of the bony fish, the backbone ends at the base of the tail, which is symmetrical.

With the fish ends this brief account of Carboniferous life in Australia. Details are meagre because the Australian fossils in no way compare in richness or variety with those from many other parts of the world. Such a lack will often be found when the general advancement of life throughout the ages is under study, but it is fortunate that one region will generally provide the information that is lacking in the others, and so a comprehensive world picture may also be built up.

It is possible that vertebrates higher than fish did not exist in the Carboniferous in Australia as they undoubtedly existed elsewhere; if they were present, they may not have been preserved or we simply have not found them yet. Nevertheless, through all the following periods there will be seen this peculiar lag in the appearance of the higher forms of life in Australia. The continent would seem to have been, if not isolated, at least remote, from those areas where evolution of the higher classes originated, and it generally took them several tens of millions of years to migrate by probably devious routes to our shores. Thus, elsewhere in the world, many lands were already being invaded by air-breathing vertebrates—by amphibians first and then by reptiles—yet evidence of their presence in Australia is sparse before the end of the Permian; none are known here in the Carboniferous.

10 The PERMIAN PERIOD

THE FOLDING movements during the Carboniferous Period had the effect of pushing the sea right off the continent to leave no more than a narrow inlet in central eastern Queensland's Yarrol Basin, and we have already shown that the largest extent this continent ever had was during Late Carboniferous times. However, the "giant" period of Australia was short lived, and by the very beginning of the ensuing period, the Permian, more than half of the giant continent was almost suddenly drowned again as shown in Map (17). From this map we can see that the pattern of subsidence is a new one in that it now combines several of the older subsidence trends into one large-scale submergence. In the west and in the east we still notice the previously predominating meridional, that is roughly north-south trending, seaways but the centre of the region shows strong northwest and northeast-directed seaways and embayments. Admittedly, as far as we know none of these Permian seas were deep at any stage, but the newly arisen geographical pattern is nonetheless a surprise.

The Permian Period is subsequently of great economic significance to Australia, for during it not only most of our important coal seams accumulated, but most currently known oil and gas fields in Queensland and South Australia are also intricately connected with Permian rock series. In Western Australia, too, rocks of that age have shown promising signs of the presence of oil and gas, especially in the northern part of what is known as the Perth Basin.

The period as a whole may be described, particularly as far as concerns eastern Australia, as one of alternating subsidences and elevations, during which the sea encroached on the continent for varying, and at times great, distances. In the epochs of stability or elevation large areas were converted into freshwater lakes and swamps, and in many of these the slow accumulation of vegetable matter produced beds of coal thick enough to be of great economic importance. It is difficult to say just how much time it took to accumulate the material for one thick coal seam, and there are, of course, many of them. We know that the duration of the Permian Period was about 50 million years, from, say, 280 to 230 million years ago; and that is considerably shorter than other Palaeozoic Periods, except the Silurian and Devonian—but it is still a very long time.

It was at the dawn of the Permian Period also that a great climatic change took place, and there was inaugurated throughout the world one of those general refrigerations which brought about another ice age (*Dickins, 1978*). At the height of this ice age huge glaciers and perhaps even continental ice-caps similar to those of Greenland and Antarctica at the present day covered many lands, particularly in the Southern Hemisphere. Enormous areas in South America were so covered, as was much of southern Africa, parts of central Africa now almost on the equator, much of India, part of the eastern United States, Alaska, and northern Asia. In Australia there was extensive glaciation over Tasmania, Victoria, New South Wales, South Australia, and central and Western Australia as far north as the Kimberley districts. On Map (17) the pattern of this glaciation is shown in the form of numerous highland-glacier areas merging southward past Tasmania with a northward advancing continental ice-sheet. Some geologists would probably extend the glaciated areas on the Australian mainland considerably to form one contiguous continental ice-cap. This writer does not favour such an interpretation of the evidence but, be that as it may, the glaciation as shown on our sketch-map is surely remarkable enough.

Lowered temperatures seem to have continued throughout the Permian Period and, in fact, well into the Mesozoic Era thereafter, but the cold must have varied considerably in intensity. As in the last, recently concluded, glacial period there were various interludes when the climate became milder and the glaciers shrank or disappeared completely. The cause, or causes, of such marked variations in the earth's climate are still unknown. Many theories have been advanced, but so far none is adequate to account for all the facts. All that can be said is that such glacial periods did occur, that they appeared and disappeared within relatively short times, and that their origin has to be sought largely in extra-terrestrial influences. From the geologists' point of view they serve as useful events, fixed in points of time, by which to correlate the geological records of different regions.

Until a few years ago it was thought that the Permian seas had covered only the fringes of the present outline of the continent. The drilling records of the search for oil have changed this, even though most of the subsurface Permian beds in the inland basins seem to be of non-marine origin (e.g. Cooper Basin). As shown on Map (17), areas of sea existed for varying lengths of time in several regions in Western

Map 17:

- SEA
- LAND
- FRESHWATER LAKE
- GLACIATED AREA

The glaciers of the very earliest Epoch (Sakmarian) of the Permian Period, and the great marine transgression which followed immediately afterwards in the Artinskian Epoch, about 275–260 million years ago.

Australia, in northwestern and southeastern South Australia, in central and eastern Queensland, as well as in eastern New South Wales, Tasmania, and southwest of Melbourne in the Werribee district.

WESTERN AUSTRALIA. The four regions to which the sea had access during the Permian in Western Australia are the Perth, the Carnarvon, the Canning-Fitzroy, and the Joseph Bonaparte Gulf basins. In addition there were areas in the south where lakes and swamps existed.

PERTH AND CARNARVON BASINS. The Perth Basin is a deep trough or graben, filled in with Permian and younger sediments, stretching along the southern 650 kilometres of Western Australia's west coast. Although attaining a width of no more than 100 kilometres the pile of sediments is known to be in places more than 1200 metres thick. The existence of Permian beds in this basin is known from drilling in its northern part, that is to the south of the small areas where Permian rocks come to the surface around the ridge of early Palaeozoic and Proterozoic rocks between Geraldton and the Murchison River. Only about 1200 square kilometres of Permian outcrops are known in this basin.

To the north of the Murchison River as far as the Lyndon River, and from 100 to 300 kilometres inland, Permian formations are exposed over an area of 40 000 square kilometres. To the east they lie in some places on Devonian and Carboniferous rocks, but for the most part directly on the Archaean and Proterozoic of the Yilgarn Shield. This area is known as the Carnarvon Basin (previously as North-West Basin). It is almost entirely sheep country rising gradually from the coast, with ranges of flat-topped hills rising sometimes in escarpments 150 to 300 metres above sea-level. It is dry country, sparsely covered with spinifex, and the main rivers, the Gascoyne, Minilya, and Lyndon only flow for short periods in the wet season.

A great deal of drilling in the search for oil has been done in this area, and it was found that, from about Carnarvon southward, Permian formations are absent beneath the younger, Mesozoic and Tertiary, rocks along the coast. In the main, therefore, Permian sedimentation was limited to a northwesterly opening gulf as is indicated on Map (17). The Perth Basin, on the other hand, was a separate embayment, opening in a westerly and southwesterly direction. Much of the Carnarvon Basin is artesian, and one of the principal intake beds, the sandstones of the Wooramel Group, some 100 to 200 metres thick, lie just about the middle of the Permian system. These sandstones form a conspicuous horizon where they come to the surface and are readily recognized when pierced by bores.

THE PERMIAN PERIOD

In the Carnarvon as well as the Perth Basin the Permian was ushered in by widespread glaciation, and the lowermost beds—the Lyons Group and the Nangetty Formation respectively—show every sign of that event. In the valley of the Wyndham River and some other localities farther north, some 500 to 1000 metres of Lyons Group sediments are exposed, and an oil boring at Rough Range found the formation at depth to be 900 metres thick. The Nangetty glacials in the Perth Basin are also at least 450 metres thick in many places.

Particularly in the Lyons Group there are a number of large boulder beds, some several hundred metres thick, marking times when the glaciers had retreated and left mighty rivers to roll out the blocks, boulders, and gravels into the sea, since the shaly material in which the boulders are embedded, contains marine fossils such as lamp-shells and molluscs. The size of an odd boulder can be quite astonishing. One granitic erratic weighing 50 tonnes may be seen where the Moogooree to Williamsbury road crosses the Lyndon River; another, four metres in length, lies in the bed of the Arthur River. Boulders up to two metres in length are not rare in many of the conglomerate beds. There is no need, however, to invoke the help of icebergs in order to explain the presence of odd large blocks among much smaller material. It is well known that powerful rivers, especially in times of flood, can transport huge boulders over great distances. The smaller material, pebbles and gravels, act like roller-bearings underneath the big blocks, and the slope gradient of the river bed need not even be steep. In other words, most or all of these famous boulder beds were formed as delta deposits where the rivers entered the sea. They are not true moraines, deposited by the glaciers themselves, but river deposits consisting largely of outwash from the higher country where the glaciers with their moraines had retreated.

This is not to say that true glacial deposits do not occur; but they are not bouldery as are the delta materials. They consist of very finely bedded mudstones or "varved shales", each varve consisting of a thin, light-coloured winter layer, and a thicker, dark summer layer. The boundary from light to dark is knife-sharp; from dark to light-coloured mud it is gradual. These peculiar, annual varves, which correspond to annual growth rings of trees, are the result of sedimentation in a lake in front of the tongue of a glacier during times when glaciation was at its height, the glaciers and ice-caps large, reaching far down into the lowest reaches of the valleys. Varved shales are very seldom preserved because with the retreat of the glaciers the wild rivers take over and usually wash out most or all of the finer features which the quiet glaciers had left behind. In Australia there are probably no more than half a dozen localities where remains of varved sediments of Permian age occur. One or two of them are in the Carnarvon Basin—for example, at Coordewandy Homestead, 250 kilometres east of Carnarvon, near the margin of the Permian basin. Glacier-scratched or striated rock pavements have also been observed in that area (*McWhae et al., 1958*).

Above the Lyons Group in the Carnarvon Basin lies a succession of shales and sandstones, with a richly fossiliferous limestone at the base. All together they reach several thousand metres in thickness in most places, but well over 3000 metres in a deep trough to the east of Wandagee Hill on the upper Minilya River. All were deposited in the Permian sea. This succession has been subdivided into a dozen or so formations characterized by rock type and fossil contents. Collecting localities are innumerable, and a large part of the rich marine faunas still remains to be described.

In the Perth Basin the glacial beds of the Nangetty Formation are also followed by a richly fossiliferous limestone, the Fossil Cliff Formation. Thereafter the regime is mostly continental, and coal measures are a strong feature of the sequence, for instance in the isolated Collie Basin south of Perth and on the Irwin River southeast of Geraldton. After another marine interlude, in which a monotonous series of siltstones were laid down on top of the coal measures, the area was elevated again, and apart from a series of freshwater sandstones there are no Upper Permian sediments known, at least not in the northern part of the basin. In the south the Permian lies at great depths which have not yet been probed by borings.

In both west coast basins, as indeed everywhere else except in southeastern Australia, glacial environments disappeared again in the early part of the Permian. The climate, however, remained cool and perhaps there were still some glaciers around, but no evidence of their existence can be found in the younger parts of the Permian sequences in the West. In many places and at various times the sea here teemed with life, and for that and other reasons, such as the fact that the strata are in many places gently folded and contain both impervious and very permeable beds, these basins will prove to be economically very important. Conditions are in many areas favourable for the accumulation of oil and gas. Oil, the life blood of our modern industries, seems mostly to have been formed from the chemical alteration of plant and animal matter in marine sediments, and it is commonly, though not exclusively, where these sediments are folded into dome-like structures that it is concentrated—because it is lighter than water, which stays below it—upwards beneath the summits of the domes. Several borings have already proved the existence of oil in the area, and off the northern opening of the Carnarvon Basin, on Barrow Island, a commercial oilfield has been proven, although the oil there is contained mainly in rocks of Mesozoic age. Oil and gas have, however, been encountered in Permian beds beneath impervious Triassic shales near Dandaragan in the northern portion of the Perth Basin, and here too a commercial field has been established.

It should be noted that marine sedimentation in the two western basins probably ceased before the end of Middle Permian times. Even in the two northern embayments, of which we will hear in the next paragraph, there were only minor incursions of the sea in Upper Permian times. Thus, as in eastern Australia, the latter part of the Period is characterized by lake and swamp environments. There is one big difference, however—nowhere in the west and north do we find the mighty series of lavas and tuffs which characterize the Permian sequences in New South Wales and Queensland (but not Victoria and Tasmania). Moreover, it should be noted that the size of Australia in later Permian time was considerably smaller than during the Upper Carboniferous, when it was in its giant stage.

CANNING-FITZROY BASIN. This large area, formerly called "Desert Basin", covering over 400 000 square kilometres, stretches from the Fitzroy River valley southwest for over 320 kilometres to the Oakover and De Grey Rivers in the Port Hedland area. The Canning stock-route traverses its southern extensions which reach southward almost to the Nullarbor Plain. This huge area of mighty sand dunes and salt flats, occasionally broken by low buttes and cuestas of flat-topped hills, has only in recent years been systematically explored with the help of helicopters, special multi-drive desert vehicles, and other paraphernalia used in large-scale assaults on such remote and difficult regions. Roads have now been laid into this uninviting desert country, and mighty oil drilling bits are churning thousands of metres down below the bone-dry surface in search of the "black gold". A fine account of most of what is known now about this country and its rocks is given in Veevers and Wells (*1961*).

The northernmost portion of the Canning Basin is known as the Fitzroy Basin or Fitzroy Trough, and the outline of this geological feature coincides roughly with the area of the Fitzroy River valley. During the Permian and part of the Triassic times this area was one of particularly strong subsidence, much stronger than elsewhere in the Canning Basin, and although the sea there was never deep because sedimentation kept pace with subsidence, the maximum thickness of Permian and Triassic sediments beneath the Fitzroy valley reaches between 4500 and 6000 metres, many thousands of metres more than in the large basin areas to the south and west. This colossal piling-up of sediments in restricted areas of narrow, subsiding, troughs or grabens is characteristic of the Western Australian Permian scene.

Between the Kimberleys and Darwin, along the eastern shores of the Joseph Bonaparte Gulf, and trending out towards the Tanimbar Islands, there is probably another one of these Permo-Triassic troughs, but most of it is beneath the sea now, and little is yet known about it.

Throughout the large shallow-water offshore areas of northern Australia the oil explorers are now very active. Even more so than the sedimentary basins exposed on land, the extensions of these basins on to the shallowly submerged continental shelf have great potential as petroleum-producing areas.

As in the western Permian basins the sequence in the Canning (and Fitzroy) Basin begins with formations of glacial character: the Braeside Tillite on the Oakover River, the Paterson Range conglomerates farther east—both of which are at the most a few hundred metres thick and may, in fact, be true glacier-deposited morainic beds—and the up to 2700-metre-thick Grant Formation in the Fitzroy Basin. Unlike the western regions there was scarcely any limestone deposition in the Canning Basin after the retreat of the glaciers, but the thin and localized development of calcareous beds and sandy limestone layers (such as the Nura Nura Member of the Poole Sandstone and the thin limestones in the Noonkanbah Formation) are as richly fossiliferous as the limestones along the west coast basins, and some of the marine faunas are very similar. However, interstratified with these fossiliferous beds are many formations of shallow-water or even freshwater origin, and in them there are no fossil remains apart from plant fragments, worm tracks and other animal trails. The post-glacial rocks attain a thickness of several thousand metres in some areas, and they end with the more or less ferruginous sand- and silt-stones of the Liveringa Formation. In this formation we find a bed, only about 30 metres thick, that contains the only marine fauna of Upper Permian age known in Australia, a fauna that also occurs in the Joseph Bonaparte Gulf Basin.

During the Permian Period what is now the Kimberleys, therefore, must have been a prominent and probably high peninsula jutting out into the growing Indian Ocean. There is little doubt that during the earliest part of the Period glaciers crowned the mountains of this region. Most of their load was dumped into the Fitzroy Trough, but some was carried out by rivers into the Joseph Bonaparte Gulf area.

JOSEPH BONAPARTE GULF BASIN. Almost all of the Permian in this basin lies either in the Northern Territory or offshore in the Timor Sea. The best area for a study of the sequence is around Port Keats, some 240 kilometres southwest of Darwin, where it has been found that the exposed beds are roughly equivalent to the

Liveringa Formation, and that older Permian rocks are largely hidden away beneath the younger beds. In this basin, too, some drilling for oil has been carried out in recent years and has resulted in a great deal of valuable new information and a big gas blow-out.

EASTERN AUSTRALIA. We shall discuss the Permian of eastern Australia in terms of two major groupings, the Queensland-South Australia seaways with their southward extension through the Sydney Basin to Tasmania, and the strongly glaciated land areas of southeastern Australia. Both features represent essentially the story of the earlier part of the period. Except for small areas in Western Australia the later Permian is characterized throughout Australia by non-marine and commonly coal-bearing series.

THE GLACIER LANDS OF SOUTHEASTERN AUSTRALIA. Although only small parts of southeastern Australia were under the sea at one time or another during the Permian, the events of the period left abundant evidence of their passage. The story begins again with a chapter of intense glacial activity. Glacial deposits occur in many parts of Victoria, in some places evidently filling valleys carved out of pre-Permian rocks. Near Bacchus Marsh they are upwards of 600 metres in thickness, and a cross section of one of the ancient valleys may be seen in Werribee Gorge. It is typically U-shaped, and thus characteristic of the action of glaciers. Apart from the glacial deposits of clay with large boulders, evidence of ice action can be seen in the polished and striated pavements of Ordovician rocks over which the glacier moved. The best of the glaciated pavements are at Darrinal, near Heathcote, and the direction of the scratches indicates that the ice was creeping north and northwest from mountainous lands whose place is now taken by the Southern Ocean. Glacial deposits can be traced at intervals right across Victoria into South Australia. In the western districts they pass under younger rocks, but they have been cut by bores from Nhill to the mouth of the Murray River.

In South Australia glacial beds are also extensive, though in many areas hidden under later deposits. Large parts of the area from Spencer Gulf in the west across the southern Yorke Peninsula to the lower Murray were covered by glaciers. A bore near Kingscote, on Kangaroo Island, penetrated 335 metres of what are believed to be Permian glacials before reaching the valley bottom of much older rocks. In the Inman Valley south of Adelaide the old landscape may still be traced beneath the glacial beds, and hills, ridges, gullies, and valleys in the older rocks are smoothed down, even polished, and striated by the passage of ice. However, Permian boulder beds have been found far in the north in the Peake-Denison Ranges, on the

Finke River across the border into the Northern Territory, as well as westward beneath the younger formations of the Nullarbor Plain whence, no doubt, they link up with the boulder beds in the Warburton Range region along the southern margin of the Canning Basin.

As in Victoria, the main glaciers seem to have come from the south, from mountains which probably extended far out into the present sea. Huge boulders, up to five metres in length, are of granite similar to that which composes the coast and the nearby islands, which are the remnants of this vanished land. However, there is no unanimity on this, because there is also evidence of glacier transport from the north, at least at times (*Campana and Wilson, 1954, 1955; Horwitz, 1960*).

The theory that land masses existed to the south of Australia in Permian and earlier times is a fascinating one. It is evident that in those ages the Tasmanian land, though its eastern part was often under the sea, extended well to the west and south. It was from this land that the greatest ice-sheets came which penetrated into the southern part of Australia. There is much in favour of the view held by many scientists that this land was, in fact, of very great extent, connecting probably with South Africa to the west and Antarctica to the south. Many geologists contend that this great southern land was, in fact, nothing less than the four southern continents and India, together comprising one huge Southland. Since the Permian these five continental masses are thought to have drifted apart again into their present positions.

An alternative theory or hypothesis is that comparatively narrow land bridges connected Australia with the other southern continents, more or less in their present positions, and that since the Permian these connecting links have sunk away into the oceans. The submarine mountain ridge 320 kilometres south of Tasmania, discovered in 1913 by the Australasian Antarctic Expedition, has been cited as evidence for this hypothesis. It rises from a depth of 3600 metres to within 160 metres of the ocean surface, and it has been claimed that it was once a land bridge to the Antarctic. However, such submarine ridges occur in all the oceans and in positions which seem to have little to do with ancient land bridges or sunken continents.

The chief evidence for the existence of "Gondwanaland" is the fact that fauna and flora of the southern continents of those times are very similar and rather different from that of the northern continents. Especially, in this regard, the southern Permian plant life—the *Glossopteris* flora, of which more will be said later in this chapter—has long been known to be of great significance. This flora developed fairly rapidly throughout the southern continents and India; even the coal seams of Antarctic mountains within 160 kilometres of the South

Pole have yielded this *Glossopteris* flora. It is difficult to conceive how this flora could have spread so widely through the southern continents, without, however, getting to Eurasia and North America, unless the former were closer to each other than to the northern continents. It is also obvious that the Permian South Pole could hardly have been located on the Antarctic Continent itself; there would certainly be no *Glossopteris*-laden coal seams in Antarctica, if that had been the case.

To conclude the story of the Permian glaciation, it may be mentioned that except for their southernmost parts the Northern Territory and Queensland show little or no evidence of the glaciation. There are tillites and varved shales in the Springsure area of the Bowen Basin, but it has recently been found that they are due to a local glacier in the Carboniferous Period (in Kuttung time). The central Australian mountains must have carried glaciers, but they carried their debris south and southeastward towards the Lake Eyre region.

TRANSGRESSIONS OF THE SEA IN EASTERN AUSTRALIA. The early Permian geography of eastern Australia, especially of southwestern Queensland and northeastern South Australia, was unknown until a few years ago. The search for petroleum and natural gas has revealed that Permian formations underlie large areas beneath the younger beds of the Great Artesian Basin from the eastern ranges right across to the Lake Eyre and Finke River region. Although there is little or no evidence of marine sedimentation throughout this area, it is nevertheless here that the Cooper Basin hydrocarbon fields have been found. Marine early Permian beds are, however, known from bores in Yorke Peninsula, west of Adelaide, as well as from the area southwest of Lake Eyre (Arckaringa Basin), and below the younger rocks of the Murray Basin. The details of the nature and subsurface extent of these early Permian marine formations in Australia's southern inland areas are still not well known, and the idea that at times a shallow marine strait could have connected a southern sea with the Queensland Bowen Basin—perhaps through the region of the Darling River—cannot be entirely discounted. The picture given on Map (17) presents a speculative extrapolation of the information currently available.

One important outcome of the new discoveries is that they limit the extent of the actual land areas making up the early Permian Gondwanaland. One could perhaps regard the marine Permian in southern South Australia as southernmost extensions of the Queensland sea, and thus do away with the large expanse of sea south of the Nullarbor Plain, but the fundamental structure of the

Australian continent does little to support such an interpretation. It is more likely that there was some kind of Southern Ocean, although it may have been somewhat smaller than the large gulf shown on our sketch. Most interesting also is the discovery of a thin formation with marine Permian fossils on top of the glacial conglomerates west of Melbourne, showing that the sea reached Port Phillip Bay via a gulf and estuary across western Bass Strait. This is of great significance to the search for oil and gas in Bass Strait, because it means that beneath the currently known oil pools of younger age there may be additional older ones.

In Tasmania, after the initial glaciation which seems to have covered the entire island (except for some then high peaks), the sea entered from the south through the middle into Bass Strait. In the west there were still high lands with glaciers, for, as on the southeastern mainland, the Permian System of Tasmania shows clear evidence of the shedding of glacial material into the sea right up into early Middle Permian times. The earliest glacial formations, however, are far more prominent and reach, in places, a thickness of nearly 3000 metres. Some of the best sections may be seen on the coast to the east of Wynyard in the northwest of the island. At Wynyard there are also some fine striated pavements which show that the ice flow was from the west and southwest.

The thickness of the marine series, which are of Lower and Middle Permian age, varies a lot, the thickest section being in the Hobart area (about 600 metres). In the great majority the rocks are siltstones and mudstones with pebble beds; good sandstones are rare and limestones, scarce in the western half of the island, increase in number and thickness east and southeastwards. A well-known locality for these marine series is Maria Island off the east coast, where the cliffs have long been a favourite collecting ground for fossils. Excellent sections may also be seen in the cliffs which form the rugged coastline in the vicinity of Eaglehawk Neck farther to the south. Here the rocks are mainly sandstones with few fossils. Some of the beds are traversed by curiously regular vertical joint cracks which cross each other at right angles, producing in one place the scenically popular Tessellated Pavement. Elsewhere the joints are on a larger scale and, when widened by marine erosion, have culminated in such picturesque features as Tasman's Arch and the Devil's Coachhouse.

In the lowest part of the marine series, the early Permian Quambi Group (*Spry and Banks, 1962*), there occurs a curious rock known as "kerosene shale". It is familiar to many Australians—a black rock, very light but of extraordinary toughness, breaking with a curved surface, and capable of producing by distillation valuable petroleum products. It occurs as seams interbedded with other rocks and is of marine origin, although it contains many plant remains, including spores. It probably accumulated in a fairly deep lagoonal backwater the bottom of which was poorly aerated, so that chemical reduction instead of oxidization of organic material took place. In other words, the chemical reactions which produced the "Tasmanite" shale were exactly the opposite of those which produced coal.

Higher up in the Permian sequence, in the Mersey Group, oil shales occur again, but this time in close association with coal measures. In fact, it seems coal was deposited in marshes and lakes close to the lagoons where oil shale was formed—different environments, but nevertheless quite close together and, at times, overlapping and interfingering each other. Very much the same picture is presented by the oil shales and coal measures in New South Wales and Queensland, which are peculiar deposits in lowland regions on the margins of marine sedimentary basins. The Tasmanian coal seams, incidentally, are of minor economic importance when compared with those of the mainland.

After the emergent phase represented by the Mersey Group Tasmania became submerged again, and this time practically the whole island went under. Only sporadic groups of rocky islands must have been left during this submergence, which lasted roughly through Middle Permian times. Rocks of this age are found in almost every part of Tasmania, though over wide areas they have been removed again by erosion or buried beneath later deposits. They cover part of Bruny Island and other areas south of Hobart, much of the Derwent and Ouse river valleys, large areas to the west and north of Lake St. Clair, the country to the west of the Tamar River below Launceston, and that near La Trobe. The rocks are mostly mud-, silt-, and fine sandstones interstratified with limestones and pebbly layers. They are known as Cascades Group (bottom), Malbina Sandstone, and Ferntree Group (top). The Ferntree Group shows the effects of increasing silting-up of the sedimentary basins towards the beginning of the Upper Permian. In that latest epoch of the period rolling sandy plains with lakes and swamps, in which peat accumulated, extended over most of Tasmania. Only the northwest and northeast hill countries rose above these plains, and much of the Permian series that had previously been laid down there was eroded away again. The peat deposits became what is now known as the Cygnet Coal Measures; thin and scarcely economic coal seams interbedded with cross-bedded and ripple-marked pebbly sand- and siltstones. The remains of the *Glossopteris* flora, including many spores, are the common fossils in these beds, marine fossils being absent.

In New South Wales the sequence of events during the Permian Period is very similar to that in Tasmania, but their magnitude and therefore the thickness of sediments which resulted from them is greatly enhanced. Not only are the marine series thicker and more richly fossiliferous; the coal measures too are incomparably thicker and therefore of great economic importance. There is only one fairly significant difference—from New South Wales' south coast northwards there is a steadily increasing intercalation of volcanic rocks in many parts of the Permian sequence until, in Queensland, major portions of the Permian rocks consist of lavas and tuffs, as we will see later.in this chapter.

The discovery of coal at Coalcliff in the Illawarra district and at Newcastle very early in the history of the colonization of eastern Australia drew attention to the associated rocks. As a result the Permian formations in New South Wales have always had their full share of geological investigation. It was a great triumph when the continuation of these economically important beds beneath Sydney, which had been deduced by C. S. Wilkinson, Government Geologist in the second half of the last century, was proved by boring and in the deep shaft of the Balmain colliery. Subsequent studies by many geologists have enabled us to establish a fairly comprehensive reconstruction of the events of that period.

We have already mentioned the early Permian glaciation in southeastern Australia. It was followed by the deposition of about 1500 metres of marine strata within a relatively small seaway, which began as a shallow embayment into the Hunter River region and farther north into the Manning River area, and later expanded from the Hunter southwest and northwest, to eventually link up past the western slopes of the New England plateau with the Queensland-South Australia seaways. Map (17) shows this extended seaway which, it may be noted, formed a little later than the glaciation geography shown on the same sketch.

Beginning in most places with early Permian marine sedimentation the whole Period falls naturally into four major epochs, each with its own characteristic series of rocks. These are in ascending order: (i) the Lower Marine Stage or Series (Dalwood Group in the north, and Lower Shoalhaven Group of formations in the south); (ii) the Lower (Greta, Clyde) Coal Measures; (iii) the Upper Marine Stage (Maitland Group in the north, and Upper Shoalhaven Group in the south); (iv) the Upper Coal Measures, which include the Newcastle, Tomago, Lithgow coal beds in the north, and the Illawarra coal beds in the south. It is noteworthy that while, as in Tasmania, marine sedimentation began—except in the southernmost part of the Sydney Basin—at the very beginning of the Permian Period, it was pushed off the continent again here well before the end of the Period, the youngest marine beds being no younger than very early Upper

Permian. The story in eastern Queensland, as well as probably throughout the problematic Queensland-South Australia seaways, is more or less the same.

The deposition areas of the Lower Marine Series have already been outlined. The sequence began with boulder beds perhaps already in the latest Carboniferous, which, as elsewhere, show the influence of glacial environments in the region. Although they are not true tillites, they represent the outwash from once glaciated valleys. At Lochinvar, near Maitland, these beds are 90 metres thick and consist of red clays with numerous boulders and pebbles which are probably derived from lands in the south and west which had existed since the Devonian, and in parts even since the Silurian Period.

The 1500 metres of marine beds which follow consist of further conglomerates, sandstones and shales, and in some places include thick beds of limestone. Subsidence of the sea floor was rather rapid and deposition was accelerated by dense showers of volcanic ash. Tuffs and tuffaceous shales constitute a considerable element in the series, and some submarine lava flows also contributed to the total thickness. Closeness of the shore, especially in the earlier, lower part of the series is indicated by such formations as that exposed in a railway cutting at Allandale, near Maitland. There the rock consists of a bouldery shoreline conglomerate, formed by breakers tearing large blocks out of a cliff face, rolling and wearing them round, and eventually dumping them among large and thick-shelled molluscs on the nearby sea bottom. Harper's Hill nearby is a classic collecting ground for marine fossils, which are beautifully preserved in a hard grey rock consisting almost entirely of fused volcanic ash. Another interesting locality in the district is the Ravensfield quarry, from which comes a fine-grained sandstone much used for building in the town of Maitland. Many fine fossils, particularly starfish, have been found in this quarry.

The Lower Marine stage came to a close because subsidence slowed down or ceased entirely, so that the whole of this area was converted for a time into a series of lagoons, backbeach swamps, and lakes. That is when the Lower Coal Measures (or Greta Coal Measures) were deposited. Although these are only from 30 to 90 metres thick at Greta, increasing to over 180 metres at Muswellbrook and to about 270 metres at Murrurundi and Willowtree, they are very important in that they contain some of the thickest and best coal seams in the State. The main Greta seam itself, in the South Greta coal mine, is no less than from four to ten metres thick.

It is time that something was said about the origin of coal. It is formed of the carbon of plants which lived in former ages, but the conditions that allowed the accumulation of such remains with only a minimum admixture of sand or mud were rather restricted. There had to be large areas of swamp or very shallow lake, with a fringe of dense vegetation to filter sand or mud from the inflowing drainage. The swamps themselves carried a heavy growth of plants—not necessarily wholly aquatic—of varying size, many being large trees. As they died and rotted these plants supplied the soil for a new growth. The whole area must have been gradually subsiding, for if static it would soon have been converted into solid ground, and the conditions for the formation of coal would have ceased. A too-rapid subsidence would have killed the plant growth and also allowed drainage to flow in unchecked, carrying silt and other impurities into the swamps. That this sometimes happened is evident from the clay and sandstone bands which do occur within the main seams of coal. When the too-rapid subsidence continued for too long so much silt was poured into the area that coal formation ceased altogether, and beds of shale and sandstone were deposited in what again had become a lake. As the subsidence slowed previous conditions might have been restored, and coal seams once again formed, above the earlier ones. Finally the pressure of the rocks above compressed the buried vegetation, and through the ages it lost much of its water and underwent chemical alteration, until at last the solid black substance called coal was produced.

If you would like to see something akin to a coal swamp at the present day, travel by launch along the Myall River in New South Wales from Port Stephens to Forster. Mile after mile the way lies through dreary tea-tree swamps, in which the water is black and the air dank with the smell of decaying vegetation. Provided that the same conditions continue for long enough, and that the area is finally buried under sand and mud, here surely is a coal seam of the future.

The third phase of the Permian history in New South Wales, represented by the Upper Marine Series, began before the end of the Lower Permian Epoch with a general subsidence which brought the sea in over a large section of the eastern part of the State. The southern coast of this sea was near the Clyde River on the South Coast, and trended from here in a northwesterly direction. As most of the rocks laid down at this time are here almost undisturbed, it is possible to reconstruct many of the details of the old coastline. It consisted for the most part of Ordovician rocks in the south and of Devonian and Silurian rocks farther north, which were folded and twisted and finally worn down into a diverse landscape of hills and valleys. As the sea advanced some hills became islands.

The township of Milton is built on a hill composed of the hard rock called quartz porphyry, while the surrounding country is composed of marine Permian rocks. Porphyry is an intrusive rock of granitic chemism which, when in a molten state, was forced under great pressure into the rocks lying above. Where such intrusions occur the surrounding and overlying rocks are invariably much altered. As the Permian rocks around the Milton porphyry are quite undisturbed and unaltered, it is apparent that this intrusion took place before Permian times, and that the intruded rocks, which were almost certainly Devonian, had been worn away also before the sea of Permian times arrived, leaving the hard quartz-porphyry as a hill or mountain in the early Permian landscape. As the land sank and became submerged at the end of the Lower Permian Epoch this hill became an island. It is shown on Map (17) which, it should be noted, indicates in this region the geography of the Upper Marine rather than of the Lower Marine (Series) time, when the sea, especially during the earliest Permian, did not reach so far south.

In other places the shore was composed of high cliffs, against which broke the waves caused by ancient storms. From these cliffs came the thick beds of pebbly conglomerates which are well exposed in the gorge of the Shoalhaven River between its junctions with Yalwal Creek and Kangaroo River. Nearby, in the flat land between the headlands, swamps and small lakes were in turn submerged beneath the advancing sea. In these were deposited thin beds of shale and a little coal at about the same time as the Greta Coal Measures were formed farther north. At the junction of the Shoalhaven with Yalwal Creek such a coal bed is seen in the face of the gorge, lying directly on tilted Devonian quartzites and overlain by sandstones containing marine fossils. Similar beds have been penetrated by bores right at the base of the Permian rocks near the coast about 50 kilometres to the south. In central and western parts of the Sydney Basin there are, however, no Lower Coal Measures, the entire sequence there being of marine origin.

Warden Head at Ulladulla, south of Nowra, is one of the most accessible and best localities to study the marine rocks. It must have been near the southern limit of the sea of the time, for at Bateman's Bay farther to the south, the rocks are again vertically upturned slates, probably of Cambrian age. The cliffs at Warden Head are upwards of 30 metres in height and are composed of shales and sandstones exceedingly rich in fossils. One bed about three metres in thickness is composed almost entirely of shells of the brachiopod *Strophalosia*, and the thin spines which covered the shells have broken off and drifted together in places to form separate layers three centimetres or so in thickness. In the cliff face, and on the rocky platform at its base, numerous boulders and pebbles protrude, varying from three centimetres or so to 1.5 metres across, which are probably glacial erratics or, at least, glacial materials dumped into the sea by a large river. These are nearly all of quartzite similar to the Devonian quartzites to the west, and the mountains which bore the glaciers were evidently not far away. In fact, some of the glaciers may have come down to sea-level. It is clear, however, that much of the land was not ice-covered. Here and there embedded in the rock are pieces of fossil wood, mostly small, though one trunk observed was over 1.5 metres in length and fifteen centimetres in diameter. This must have drifted from the neighbouring land, become waterlogged, sunk, and then been buried among shells and other marine creatures. The occurrence of such remains of trees shows that forests existed at the same time as the glaciers, and it would therefore be wrong to imagine the glaciation of that time as an ice-cap of great extent. More likely it was something similar to that of Alaska at the present day.

North of the Illawarra district the coast ran through Marulan, east of Goulburn, northward and west of the Blue Mountains towards Mudgee, whence its continuation is hidden under later rocks. Permian rocks appear beneath the later Triassic sandstones in the gullies both at Bundanoon and on the Blue Mountains. The northern shoreline can be picked up near Narrabri, followed southward to Muswellbrook and from there eastwards to the coast north of Port Stephens. In this northern area, though Upper Marine Series rocks cover much of the valley of the Hunter River, they have been considerably folded, and great masses have been removed by subsequent erosion, so that it is difficult to say where the northern shoreline really lay. The thickness of the Upper Marine is in places quite considerable. Probably more than 1500 metres of sediments must have accumulated in the Hunter River area and farther south through Sydney, where they are now buried some three kilometres beneath the surface.

The times when the Upper Marine Series were laid down must have been very disturbed. Not only was there a cold climate with glaciers probably creeping down the valleys almost to the sea, but the interval is also marked by many volcanic eruptions, some of which were on a grand scale. The columnar basalts of the "blue-metal" quarries at Bombo and Kiama are lavas which were poured out at this time, while the thick beds of red rocks which overlie them are composed of solid volcanic ash. The marine rocks at Gerringong also contain much volcanic ash as well as larger fragments of rock ejected in a molten condition by explosions in nearby volcanoes. Other lava flows and tuffs appear at different levels, but it was in the middle and the latter part of this marine interval that volcanic activity was most intense.

The existence of glacial conditions in the region is proved by large erratic blocks on many levels. A great number of these are of granite, which seem to have been carried by floating ice for considerable distances because the granite mountains from which they could in those times have been derived are well to the north and west. Erratics are common in the Branxton district. There is a large one which lies just to the west of Maitland and is embedded in shales containing marine fossils. A particularly interesting erratic found at Branxton is of quartzite and contains fossils of Devonian age of a type found at Mount Lambie, some 160 kilometres to the southwest near Bathurst. This is a clear indication of the location of the land whence came the glaciers with their burden of rock, sand, and mud.

At the end of Middle Permian time history began to repeat itself, subsidence slowed down, and in various areas elevation set in. Again lakes and swamps were formed, this time much more extensively than during the earlier part of the Permian. They occupied not only the area of the previous seaways but extended westwards towards Dubbo and Coonabarabran. In the northwest, towards the Border Rivers district of southeastern Queensland, there are also many thousand metres of Permian sediments in some areas. Most of them appear to be of early Permian age. For example, the coal seams at Ashford correspond to the older coal-forming phase like the Greta Coal Measures; but on the Queensland side of the border the presence of Middle Permian rocks has also been established. Much of this region needs to be further studied.

In the Hunter River district the Upper Coal Measures are of great thickness, in fact over 1800 metres. At the base are the Tomago or East Maitland Coal Measures containing several seams of coal; then follows about 900 metres of unproductive sandstones; and finally, and most important economically, the Newcastle Coal Measures. Both coal sequences were formed in Upper Permian times when the only marine sedimentation we know of on this continent took place in northwestern Australia. The extent of some of the swamps in Upper Permian times must have been very great. What are practically the same seams of coal extend from Newcastle to below Sydney at a depth of 900 metres, to appear again at the surface along the coast of the Illawarra district, a distance of well over 160 kilometres. To the west, on the edge of a gigantic saucer as it were, the same or closely associated seams are exposed at Lithgow and in the Blue Mountains gorges below Katoomba. The abundant plant life which formed the coal, and which is found in the associated shales, will be dealt with in the section on the life of the Permian Period.

The greatest thickness of the Permian.System in New South Wales is of the order of 6000 metres, or nearly six kilometres. The bulk of this enormous mass of material came from the highlands and mountains to the south and west which had existed since the Devonian Period. Some material, however, came from the islands to the northeast. The erosion of these high lands had been accelerated by the transport action of the glaciers, and at the close of the period most of the mountains in the region must have been reduced to rather low levels. Thus, even if the climates remained cool the lower altitude of the land helped the gradual diminution of glacial action after the termination of the main glaciation periods. Volcanic activity continued, and beds of tuff and lava are found right up to the top of the Permian formations in many places, notably at Singleton.

In Queensland, though the eventual extent of the Permian sea was greater than in the southern States, the period was ushered in rather differently—by tremendous outbursts of volcanic activity, especially in the north, but a little later also in the south where they correlate with the volcanism in New South Wales. The early volcanism in the north was centred in the vicinity of Mount Devlin and Mount Toussaint in the Bowen River district, and led to the accumulation of no less than 1500 metres of lavas and tuffs. West of Rockhampton, at Mount Britton, volcanic rocks are equally thick, while at Mount Mulligan and other localities in the north there is similar massive evidence of volcanic eruptions.

Midway through these volcanic formations there is evidence that things quietened down, the land sank, and swamps formed over the surface of the earlier lava flows for a time to allow the accumulation of 120 metres of sandstone and conglomerate containing fossil plants and one seam of coal. These are the Mount Devlin Coal Measures, and the seam, though rather poor, is of particular interest because it is the oldest in the State.

As volcanic action died down, swamps and lakes again came into existence, and even the sea found access from time to time. Some 200 metres of mixed freshwater and marine rocks, including valuable coal seams, were laid down. These are the Collinsville Coal Measures. After their formation subsidence became more general, and the sea advanced over the old swamps and lava flows over large areas. The main area of subsidence is the Bowen Basin, and the marine rocks laid down during this interval are known as the Middle Bowen Series. They are essentially of Lower and early Middle Permian age like the Upper Marine Series in New South Wales. The Bowen Basin, as far as it can be delineated on the surface, is more than 650 kilometres long in a north-south direction and from 160 to 320 kilometres across. Its northern, eastern and western margins can be traced fairly well, but in the south it disappears beneath younger formations. Beginning from about 80 kilometres southwest of Bowen, the western shore ran roughly parallel with the 147th meridian of longitude to beyond latitude 25 degrees south—that is, about the latitude of Bundaberg.

From Mantuan Downs southward it is lost and can be traced further only by deep drilling.

There is again evidence of the presence of glaciers somewhere in the Australian region during the time the Middle Bowen Series were laid down, for at various levels there are shale beds which contain numerous erratics, mostly granites, porphyries, gneisses and schists, the source areas of which are probably not in Queensland, but farther south in New South Wales and South Australia. They must have been transported northward and dumped into the sea by icebergs. In Queensland itself there is no evidence anywhere of glaciated surfaces, or of terrestrial moraines.

In the northern part of the basin some 700 metres of marine rocks were laid down. At the base is a thick bed of sandstone which stands out in prominent escarpments in various places, where it is known as "the Wall", and has proved a very useful reference bed for tracing the sequence through different localities. Much better are, of course, strata which are crowded with well identifiable fossil remains, and such are also quite common in the Bowen Basin.

Drilling for oil has now shown that shallow Permian seas spread from the southern part of the Bowen Basin southwest and westward, and around the ancient Drummond Ranges north and then east again. Probably not all of these seaways were in existence at the same time, but it is obvious now that the marine transgressions of the Permian Period were not the relatively minor and marginal events one once thought, but affected large parts of the continent and penetrated close to and even into its heart.

To the southeast of the Bowen Basin there are numerous outcrops of Permian marine rocks in the Gympie district north of Brisbane, in the Warwick area, and other places, some of which have already been mentioned in connection with the Permian of northeastern New South Wales. At Gigoomgan, in the Gympie district, is a great thickness of limestone with very early Permian fossils, and 330 metres of it is made up almost entirely of seaweeds. Just how these Permian areas were connected with those of the Bowen Basin on the one hand, and with the seaways in New South Wales on the other is not yet known. As shown in Map (17), there may well have been a complex system of islands and straits in the region between Newcastle and Rockhampton. In some of the straits very considerable thicknesses of marine sediments were deposited, in places over 3000 metres, as in parts of the Yarrol Basin. Practically everywhere, however, a substantial proportion of the series is composed of lavas and tuffs.

The last phase of the Permian history in Queensland, as in New South Wales, was marked by the silting up of the seaways and regional elevation which turned them into freshwater lakes and swamps over very extensive areas. Indeed, this terrestrial Permian is known even from regions which the sea never reached during that Period. Northwest of Cairns, in the Cooktown hinterland as well as farther inland to the southwest (Galilee and Cooper Basin), there was a large area in which coal was formed between thick layers of sandstone and shales, the material for which must have been brought down from high land in the vicinity by big rivers. The total thickness of the freshwater beds in the Bowen Basin reaches as much as 3000 metres, and since they are all laid down in shallow water there must have been very strong local subsidence of the lake-bed. This subsidence proceeded in stages because the interstratified coal beds indicate times when it ceased and the lake became a vast swamp. Some of the so formed coal seams are of very great importance. At Blair Athol, 200 kilometres west of Rockhampton, one seam is no less than 29 metres thick, probably the thickest black-coal seam in Australia. This now lies so close to the surface that it is being worked by the open-cut method—like a quarry or huge clay pit—the great thickness of coal making it profitable to remove the whole of the overburden of unproductive sands and shales.

The Permian Period is the last of the great ages grouped under the term Palaeozoic Era, meaning the era of "old life". Its conclusion saw another important folding movement in the earth's crust, called the Hunter-Bowen Orogeny. Its strongest development can be seen between the Hunter River valley in the south and Townsville in the north; pressure came mostly from the east and buckled the recently formed Permian series into folds accompanied by fractures or faults. Some of these faults are of great magnitude and have displaced the rocks on either side for hundreds of metres. Most of the folding was, however, restricted to the then extreme east of the continent. The Permian strata south of Newcastle as well as those laid down in the central and western parts of the Queensland-South Australia seaways were little or not at all affected, and in many areas they still lie in a position (not necessarily horizontal) much the same as that in which they were originally laid down. The course of the Hunter-Bowen Orogeny is demonstrated in Map (18). Especially in its Queensland areas this orogeny was also accompanied by strong volcanic activity, both intrusive and extrusive. Many granites and porphyries in northern Queensland as well as in coastal districts farther south as far as northern New South Wales penetrated into the latest layers of the earth's crust at that time. Of particular interest among these Permian intrusives are ring-shaped dyke systems of various kinds of igneous rocks, for instance those south of Kidston, about 300 kilometres inland from Townsville. In most cases, too, these volcanic activities brought valuable ores and minerals to the surface, and in all these areas there is a long history of mining for gold, silver, tin, and other ores.

Although the Bowen Orogeny was not the last mountain-building event to affect Australia, the continent was thereafter a fairly stable land mass. Elevations and subsidences of the less violent kind did, of course, still continue and influenced the distribution of land and sea repeatedly on a large scale. These are, however, not orogenetic movements; they did not produce tremendous lateral thrusts and squeezes which piled up the rocks into mountains like the Alpine ranges or the Himalayas. They are of what is known as epeirogenetic character and may be likened to a "breathing" of the earth, or to a process of gently balancing out and adjusting uncomfortable bulges and hollows of the crust, a process that goes on unceasingly and is, in fact, also responsible for most of the earthquake activity throughout the world.

LIFE OF THE PERMIAN PERIOD

Coal Forests, Insects, and Coldwater Marine Faunas

LAND PLANTS. The Permian in Australia is similar in one way to the Carboniferous in Europe and America: for the first time we get a comprehensive picture of a vigorous and extensive land flora. There the similarity ends, for the Permian plants were on the whole rather different from those which had lived before. The Carboniferous Period had seen the culmination and then the decline of the giant lycopods or scale-trees, and, when the Permian began very few of them were left—in fact, not many more than survive today.

Considering the cold climates and the nearness of huge glaciated regions it is surprising how rich the Permian flora was. In place of the lycopods, the swamps and lowlands and probably much of the glacier-free uplands were covered with forests, the chief tree of which was *Dadoxylon*. This was a tall tree, with a diameter of one metre or more, and it generally resembled a pine. It has, however, been placed in the extinct group of the cordaitales, members of which probably provided the ancestors of the true pines and the cycads. Cycads which survive today will be familiar to many Australians, as one of the order, the burrawang, is very common in coastal forests. The foliage of *Dadoxylon* has not been definitely identified, but it is thought that wedge-shaped leaves with parallel venation, found frequently as fossils, came from this tree. These leaves are known as *Noeggerathiopsis* (Plate 12).

The forest trees contributed largely to the formation of the coal seams, and their stumps are often found *in situ*, the roots penetrating the original soil which is now converted to solid rock. At Fennell Bay, Lake Macquarie, the remains of such a forest are still visible. Sections of trunks lie about in great profusion, their woody tissue converted into chalcedony and weathered out from the rocks which previously enclosed them. Fossil wood of this type is common in many localities in eastern Australia, but seems to be absent in Western Australian coal measures probably because the latter are "drift coal", that is, not grown and decayed *in situ*.

The most characteristic plant was the small and comparatively insignificant one called *Glossopteris* (Plate 12). So abundant was this, that not only in Australia, but in Antarctica, South Africa, India, and South America, as has been mentioned, the term *Glossopteris* flora is used to link the fossil records contained in the Permian rocks of all these regions. Few plant fossils have been so important stratigraphically as *Glossopteris* and its close relative *Gangamopteris*. Evidently a hardy plant, adapted to withstand a cold climate, it is closely associated with the ice age in all the countries mentioned. Its wellnigh simultaneous appearance in such widely separated lands has led to much discussion as to the way in which it could have spread so rapidly. The conception of Gondwanaland as well as the idea that continents can drift apart is partly based on the distribution of this one type of plant.

Specimens of *Glossopteris* usually consist of single leaves or fronds, typically tongue-shaped, and a few centimetres in length, though one species is over 30 centimetres. There is dense crowding of the venation into a midrib, the veins on either side being linked to form a perfect net. *Gangamopteris* is like *Glossopteris*, but the midrib is replaced by several parallel veins. Not much is known of the habit of the plant. Some species possibly rose directly from the ground, but in India the fronds have been found attached to stems, and the plant probably grew into a shrub several metres in height, or even into a tall bush. A flattened, stem-like fossil called *Vertebraria* is thought to have been the root or rhizome of *Glossopteris*.

The exact classification is still doubtful. Some think *Glossopteris* belongs to the ferns, but in the multitude of specimens examined no trace has been found of the spores which are typically found on the underside of fern fronds. Small fossil seeds, about a centimetre long, found associated with *Glossopteris* at Threemile Creek near Bowen, Queensland, would however, if proven to belong to *Glossopteris*, classify the plant as a pteridosperm related to the cycads. These seeds were described by Walkom (*1921*) under the name of *Nummulospermum*. In North Vietnam, where glossopterids are known to have survived to the end of the following period, the Triassic, cup-shaped fructification organs, called *Ottokaria*, have also been found associated with *Glossopteris*. Again direct proof is lacking—the organs could be from another plant, for they have not been found actually attached to *Glossopteris* fronds.

From the unlimited number of specimens found throughout the coal measures it appears that *Glossopteris* added a large quota of vegetable matter to the coal seams. This leads to the speculation as to whether some of the coal seams were formed on dry, or almost dry, land instead of in swamps. One may well imagine a dense mass of *Glossopteris* covering the surface like a thick green carpet, each generation growing on the decaying remains of the preceding one, until in the course of time a considerable thickness of peat-like material accumulated. When, at a later stage, the area subsided below waterlevel, layers of sand and mud would cover the peat, which would then be gradually compressed and converted into coal.

The *Glossopteris* flora first appeared in the latest Carboniferous, when it co-existed for a short while with the waning *Cardiopteris-Rhacopteris* flora *(Black et al., 1972)*. In Australia and in the other continents of the Southern Hemisphere it disappeared suddenly at the beginning of the Triassic, but it survived until the beginning of the Jurassic in Southeast Asia. It is interesting to note that the flora spread even during the Permian into Asia, where it has been found in Siberia and in the Angara region of Russia, as well as into southwestern Europe. Just why our southern *Glossopteris* flora was able to spread into the northern lands, while the northern flora was barred from reaching the southern continents, is a mystery. In fact, this northward march of southern floras continues and reaches its peak about the beginning of the Jurassic. Only from the end of the Cretaceous the separation of the present northern and southern floras begins again to take shape.

Plants other than *Glossopteris* existed in the Permian swamps and forests. Amongst them were the horse-tails, which were very common. These were bamboo-like plants with a soft pithy centre. The stem was striated vertically, and divided by nodes, from each of which sprang a circlet of slender leaves. Specimens of the flattened stems are common as fossils, but the whorls of leaves are rarer. The Permian horse-tail *Phyllotheca* had a diameter of up to ten centimetres, and formed dense thickets from 1.8 to 3 metres high. The giant tree horse-tails, *Calamites*, seem to have disappeared with the end of the Carboniferous Period, in which they were very prominent. Horse-tails still survive but they are now insignificant plants, rarely more than 30 to 60 centimetres high.

The majority of the other known Permian plants consists of ferns, both true ferns and seed-ferns. Of these it is thought that *Cladophlebis* is the frond of a tree fern, while *Sphenopteris* was a small graceful form, in habits not unlike maidenhair. Various seeds of doubtful affinity are not uncommonly found, and suggest that a still more extensive flora was living on the slopes of the higher land. This would have a much smaller chance of being preserved than that living in the swamps, and it is only from occasional fragments which have been washed to their present position that any information on the habits of such plants can be gained. On the other hand, our knowledge of the mostly microscopic seeds and spores of the various fossil floras has in recent years been mightily increased, and the specialists of this branch of science, called palynology, have contributed very greatly to the understanding and correlation of the many rock sequences which are otherwise poorly fossiliferous. Spores and later the pollens of flowering plants are, as we know now, far more widely preserved as fossils than was once thought and, very importantly, they occur in both terrestrial and marine sediments. In other words, although one does not in most cases know what the plant from which the spores came looked like, the species and genera of spores are in themselves most useful marker "floras", many of them clearly characteristic of certain ages.

INSECTS. The Permian forests reverberated to the hum of innumerable insects. The rocks of that age provide the first comprehensive information we have of this form of life in Australia, although finds described from the Wynyard tillite beds in northwestern Tasmania (*Riek, 1976*) may be as old as latest Carboniferous. It has already been stated how rarely fossil insects are preserved, and it is fortunate that in the Upper Coal Measures in the Newcastle district a great many specimens have been found. The first discovery was made by J. Mitchell, and his and subsequent finds were described by Tillyard (*1925*). The best specimens have been found at three localities, at Belmont, at Warner's Bay near Lake Macquarie, and in the cliffs facing the sea at Merewether.

Insects as a whole constitute an enormous proportion of animal life. Those of the present day have been divided into some two dozen orders, ranging from primitive, wingless types to the butterfly; the number of species is greater than that contained in the remainder of the animal kingdom, both on land and in the sea. This profusion of variety existed from the earliest times of their geological history; hence in a work of this size it is impossible to give more than the broadest of generalizations.

The first thing that strikes one about Australian Permian insects is that they were dwarfed compared with those of Europe and America, a fact probably explainable by the severity of the climate. All the species and genera have long since become extinct, as have many of the families, but most can be assigned to living orders, or to generalized groups which seem to have been the ancestors of more than one living order.

Amongst the primitive types were the plant-bugs of the order Hemiptera, insects especially adapted for sucking the juices of plants. Though it is quite impossible to work out the complicated life history of fossil insects, it may be noted that living Hemiptera do not undergo the complete metamorphosis characteristic of the higher orders.

Many species of scorpion-flies have been found at Belmont. These belong to the order Mecoptera, and they are very similar to living Australian scorpion-flies. Other species belong to the extinct order Paramecoptera, and these are interesting for they combine the characters of several living groups, the lace-wings, the caddis-flies, and the true Diptera, which include the house-flies. Even the beetles were represented, mainly by aquatic types, some of them closely related to present-day water beetles. These are but some of the forms first described by Dr. R. Tillyard who was regarded as a world authority in this special field. Other Permian insects from Belmont have been described by the former Director of the Australian Museum in Sydney, Dr. J. W. Evans, and by Dr. E. F. Riek from the C.S.I.R.O. in Canberra.

VERTEBRATES. Little is known of eastern Australian Permian fish, and finds are still awaited similar to those of Devonian and Carboniferous fish in Queensland and Victoria. The few specimens found are of freshwater type, belong to the ganoids, and are not unlike those of the Carboniferous Period. Abundant evidence of this group will, however, appear in the rocks of the next period, the Triassic. In Western Australia sets of teeth of a small shark, *Helicoprion*, have been found in several places in the Bulgadoo Shale and in the sandy siltstones of the Baker Formation of Lower to Middle Permian age in the Carnarvon Basin (*Teichert, 1940*).

Of great geological importance was the discovery of the remains of a labyrinthodont amphibian in kerosene shales from the Upper Coal Measures in the Airly coal mine near Lithgow. This is still one of the earliest vertebrates higher than fish found in Australia. There have been a few discoveries in Devonian rocks (p. 137), none from the Carboniferous, and Permian amphibians are rare in Australia, although common elsewhere.

It would appear that Australia was repeatedly, and for very long periods, badly isolated from the rest of the world. Land routes by which terrestrial animals could enter either did not then exist, or were both devious and precarious. Thus, as has already been pointed out, there is almost throughout the geological record, evidence of a time lag, the main groups of land animals reaching Australia a full period or more after their origin elsewhere. This will again be seen in later chapters, when not only the amphibians, but reptiles, birds, and finally mammals come into the picture.

The first amphibians originated in the late Devonian, for specimens have been found in Greenland in rocks which were laid down about the close of this period, but they were very abundant in Carboniferous times both in Europe and America, where over a hundred species, referable to many families and seven or eight orders, are known. As a group amphibians are highly interesting, for firstly they represent the initial stage of the invasion of the land by the larger marine animals, and secondly they are the first step on the ladder of evolution from fish to man.

There is no doubt that the amphibians developed from a fish-like ancestor, in fact, from the crossopterygians (lobe-fins) which, it will be remembered, were common in Devonian times. The first evolutionary step towards man was the ability to breathe out of water by the development of lungs, an advance probably made necessary by the fouling of the water in swamps and estuaries during times of flood. These rudimentary lungs may still be seen in the Queensland lungfish, itself a survivor of the dipnoans, one of the earliest orders of fish. The next step was the evolution of the fleshy lateral fins of the crossopterygians into horizontal paired limbs, by which the creatures could crawl over the mud of swamps or even over dry land. There are other anatomical differences, such as a considerable change in the proportions and shape of the various skull bones, but the two first-mentioned faculties alone—the abilities to breathe in the air and to crawl on land—are sufficient to separate the amphibians from the fish. It was possession of these two main faculties that opened up the entirely new environment of the land to vertebrate animals.

There was still a very long way to go. The first amphibians, like their descendants the frogs, toads, and newts, had a soft skin which needed to remain moist, so they were confined to damp localities and never moved far from water. They also underwent metamorphosis or change, and the first part of their life cycle remained aquatic. The eggs were laid in water and the young, like the tadpole of the frog, breathed by gills and had the limbs still undeveloped. This limited their distribution, and the main conquest of the land was not to take place yet.

The only Australian Permian amphibian we know of is a labyrinthodont, so called from the convolute structure of the teeth in section, and it belongs to the sub-class Stegocephalia. Animals of this group had broad flat skulls with very heavy bones; the eyes were on top of the head, and in addition to the normal pair there was a small third one in the centre. The lobe-finned fish had this third eye, and it still persists in a vestigial condition in the higher mammals, including man. The labyrinthodonts varied in size; some were very large— upwards of three metres in length. On land they were clumsy, slow-moving animals, dragging themselves laboriously over the soft mud, but they were probably much more agile in the water. They must have been able to capture their prey, probably insects or even fish, for their teeth show them to have been carnivorous.

There were vertebrates higher than amphibians in Permian times— reptiles were already firmly established in other lands—but they had, it seems, not yet appeared in Australia. Discussion of them will therefore be deferred until a later chapter.

MARINE LIFE. One might think that sea water in Permian times was cold—indeed, owing to the presence of large glaciers and floating ice, close to freezing point if not actually frozen over. For some, especially southern, regions this may well have been so, but this does not mean that the sea was deficient in life. Today, in the polar regions, the sea-bed at a depth of a few fathoms is a mass of marine life, even when the surface is covered by solid ice. Floating life is no less prolific, and the water swarms with innumerable crustaceans, fish and other creatures. The bulk of living things is as great as in tropical waters, but the forms represented are quite different. It is also important to realize that by far the richest marine life occurs in those regions of high latitude where warm-water currents from the equatorial areas meet the cold waters of the polar regions. The abundance of marine fossils in many beds of the Australian Permian may be due to such current patterns. In other words, in spite of the vicinity of glaciated lands the sea was not as cold as one might think, certainly not everywhere. Just as the Gulf Stream of today softens the climate of western and northern Europe, so, it may be imagined, the warm-water currents unfroze the climate in parts of Permian Australia.

Thus Permian marine life, even in the colder water, was abundant and varied. There was, however, a fairly marked difference between the faunas of Western Australia and those in the east. This is to be expected, for the two marine areas were separated by part of the Australian land mass as shown on Map (17). In fact, not long after the beginning of the period even the narrow seaway from western Victoria into Queensland was closed and remained so until mid-Cretaceous times. There was obviously a very effective bar to migration, particularly once the land mass extended far to the south to join up with Antarctica. Nevertheless, the faunas had certain features in common because of the similarity in climatic conditions.

There is a complete absence of the coral reefs which were such a feature in Silurian, Devonian, and Carboniferous times. Fossil corals may be found in Permian rocks, but they are mainly single, cup or horn-shaped types of small size, and comprise a very small proportion of the faunas. The commonest is a curved tapering form called *Euryphyllum*, which attained a length of about five centimetres and had survived from Silurian times. One small branching form, *Thamnopora*, is fairly common, but although it occurred in considerable masses, it was in no sense a reef builder; neither were forms such as *Verbeekiella*, *Tachylasma*, and *Cladochonus* which are common in the limestones of the Callythara Formation and other rocks in Western Australia.

Instead of the corals, the delicate colonies of the bryozoa or lace-corals were abundant. These either lived in thin, convoluted, cup-shaped colonies attached at the base, or as films on shells, rocks, or other solid objects. Some large glacial erratics show encrustations of that kind. Members of the extinct family Fenestellidae were abundant, though not confined to the period. In this family the colonies consisted of numerous thin branches connected at intervals by cross rods, giving the whole a delicate net-like appearance. The disposition and number of the minute cells which cover the branches, each cell once occupied by an individual organism, divide the family into a number of genera, which in turn are divided into many species. The chief genera were *Fenestella*, with two rows of cells on the branches; *Polypora*, with many rows (Plate 14) and *Protoretopora*, with cells on the connecting rods as well (*Laseron, 1918; Crockford, 1941*). A few bryozoans, such as *Stenopora*, were massive or branching and in appearance not unlike some corals. In some places the Fenestellidae were so abundant that their remains form thin beds of limestone; in others, mixed with mud or fine volcanic ash, they constitute the bulk of considerable strata of rock. Sometimes, as in the Branxton and Allandale railway cuttings, and on the top of Nowra Hill, south of Nowra, they are exquisitely preserved. The strata composed of them are often persistent over considerable areas, and are of great assistance in geological mapping.

Of no less interest than the bryozoans were the brachiopods or lamp-shells, which in the Permian reached their peak in numbers, size, and specialized development (Plate 13).

The family of the spirifers was particularly prominent. It was characterized by elaborate internal processes, consisting of two spirals tapering outwards from the centre of the shell. It had existed as far back as the Silurian Period, but the Permian forms were much larger and of greater variety. In some the shell had wing-like expansions on either side, and internal casts commonly found in the sandstones are not unlike rather solid butterflies in appearance. In fact, they are occasionally sent to the museums as such by correspondents. Allied to *Spirifer* is *Martiniopsis*, a particularly characteristic Permian fossil, a solid shell up to 7.5 centimetres in size, smooth, and lacking the radial ridges of the true *Spirifer*. In *Productus*, also found in earlier periods, the dorsal valve was flat, while the ventral one was large and strongly convex, with a large overhanging beak. In its close relative, *Strophalosia*, the smaller or dorsal valve is so concave that it fits right into the ventral, and there hardly seems to have been space between for the living animal. Both these genera were covered with long spines. *Derbyia*, a large form peculiar to Western Australia, had a long hinge line and a finely rugose surface, whereas its close relative *Streptorhynchus*, which is also known from Queensland, had a shorter hinge line and a more coarsely rugose surface. In the small *Dielasma* the internal processes were modified into two loops, and the animal was attached to rocks or other objects through a hole in the beak of the ventral valve. Its special interest is that it belongs to one of the few families which have survived to the present day, and it is closely related to lamp-shells now living on the Australian coast.

Molluscs were also common, and compared with those of the previous periods many were of large size. Amongst the bivalves the genus *Eurydesma* is of particular importance for stratigraphical purposes, because it is confined to the early part of the Permian Period. It is very abundant throughout the Lower Marine Series in New South Wales where, as in fact generally in the eastern Australian region, it survived longer than in the west. There it occurs only in the very early Permian Lyons Group of the Carnarvon Basin (*Dickins, 1963*). *Eurydesma* was a large, strong, smooth shell, upwards of fifteen centimetres in length and as much as 2.5 centimetres thick; with the valves closed it is in appearance perfectly heart-shaped when seen from the front or back (Plate 14). The shell lived in shallow water on the open coastline, was adapted to cool waters, and is found commonly also in coarse cobble and pebble beds, where its strong thick shell enabled it to withstand the battering of the surf. Amongst other large bivalves were *Aphanaia*, which was not unlike a giant mussel and often over 30 centimetres in length, and *Myonia*, *Pachymyonia*, and *Megadesmus* (especially its subgenus *Cleobis*), thin-shelled forms which usually lived in the mud. Another, smaller, and quite common form was *Astartila*. A beautiful little bivalve, *Nuculana*, had the hinge formed of a double row of fine teeth. Though not very common it is interesting because of its close relationship to a group of shells which today are not seldom washed up on the outer beaches. Also of great importance are *Aviculopecten* and *Deltopecten*, shells very similar to the living scallops (Plate 14). They were abundant in the Permian seas, where there is a wealth of species, some of them of large size. The above-mentioned are but a few of the many Permian bivalves, and those who would seek further information must consult the papers and monographs which have been written about them.

Univalve shells or sea-snails, though interesting, were neither so abundant nor so varied as the bivalves in Permian times. Practically all of them belonged to primitive families such as the Bellerophontidae, Pleurotomariidae, Euomphalidae, in which the shells were simple and low-spired with round mouths. The pleurotomariids had nearly always a ridge or band in the middle of the whorl, which sometimes terminated in a slit (Plate 15). Minute shells of this group, only two millimetres in size, still exist, and are to be found living on seaweed or under rocks on the Australian coast. Representatives of somewhat less primitive groups such as platyceratids, subulitids, neritaceans and a few others are known in Western Australia.

With the pteropods or winged snails are placed—for convenience, because their classification is still a matter of speculation—two curious fossil organisms which have been named *Hyolithes* and *Conularia*. The pteropods are molluscs which are found abundantly swimming in the open ocean, their shells not unlike small glass bottles. They are called winged snails because of the peculiar modification of the foot into two wing-like fins which are used for propulsion. Actually, neither *Hyolithes* nor *Conularia* resembles any known pteropod, but neither are they like any other known organisms. *Hyolithes* is a small tapering form, nearly oval in section, composed of a rather horny material, with the large end closed by a lid or operculum. *Conularia* is larger, four-sided, rectangular in section, and tapering, and has angulated ridges on the sides not unlike a sergeant's chevrons, as shown on Plate 15.

Permian cephalopods in eastern Australia are not particularly important because they are not common as fossils. They are represented by the straight-shelled orthoceratids, which still survived from the Ordovician, coiled nautiloids, and by the small coiled forms of the Goniatitidae. The zigzag sutures of their shells, formed by the intersection of the partitions with the outer shell, have already been mentioned in the previous chapter, and in the Permian their form was well on the way to becoming the complicated thing it was in subsequent ages. It is in Western Australia that most representatives of the important cephalopod group Goniatitina have been found, and there are many genera and species—the latter, as is typical for them, existing only through a short interval of time. They are thus characteristic of particular strata or horizons, and their identification in the field therefore greatly facilitates geological survey and mapping. Among the important goniatitids we may mention the large early

Permian *Metalegoceras jacksoni* and *Uraloceras irwinense*. Later we find species of *Propinacoceras, Agathiceras, Pseudogastrioceras* and others.

Amongst other marine creatures were the crinoids and the starfish. Starfish are not abundant as fossils; they disintegrated rapidly after death and are thus rarely preserved as a whole, although their tiny bits and pieces are quite common in some rocks. Nevertheless, several perfect casts of an entire large form, called *Palaeaster*, have been found in the fine-grained sandstone at Ravensfield, near Maitland, in New South Wales. The Permian crinoids, like so many other marine animals of the time, were particularly large and strong. Yet for the same reasons as with the starfish it is seldom that complete specimens are found, but single calyx plates of the large *Phialocrinus* (Plate 13) are very abundant at Mount Vincent in the Maitland district; some of them are nearly five centimetres across. In addition, many of the single-stem sections found in this and other localities are upwards of 2.5 centimetres in diameter. A peculiar form, the plates of which were once mistaken for sponges, *Calceolispongia*, is restricted to the Western Australian Permian, where it occurs with other crinoids such as members of the families Poteriocrinitidae, Allagecrinidae, and Catillocrinidae.

As with so many other groups, the number of species of crinoids so far known and described represents but a fraction of those which lived in the Permian seas. There are several forms known from Ulladulla, belonging to three entirely new genera, which still await analysis and description and are kept in the Australian Museum in Sydney. Similiar discoveries are constantly being made. At Rylstone, for instance, in rocks which were deposited in shallow water close to the shore of the Permian sea, a private collector has also discovered many hitherto unknown species, which remain to be described, and similar exciting finds must await the enthusiastic collector in many other localities.

Final mention must now be made of the trilobites, for the Late Permian marked the last appearance of this strange group of creatures on the stage of the earth's history. There are, in fact, a variety of genera and species known from various parts of the world, and one cannot help wondering why they all died out at the close of this last Period of the Palaeozoic Era. However, the fact remains that nobody has ever seen a trilobite in Triassic rocks. In Australia Permian trilobites were not known before 1944, when Teichert discovered species of *Dimitopyge* in Western Australia.

IV
THROUGH CONSOLIDATION TO MATURITY IN THE MESOZOIC ERA

11 The TRIASSIC PERIOD

WITH THE Permian ended the series of major geological periods, grouped in the Palaeozoic Era, which we have called the Era of Adolescence and Adjustment. The era which followed, the Mesozoic, meaning "middle life", includes three more great periods, the Triassic, the Jurassic, and the Cretaceous.

At the dawn of the Triassic some nine-tenths of the geological story since the beginning of the Proterozoic had already passed, but measured by human standards the Triassic was still incredibly remote. It began 230 million years ago and came to an end after 40 million years, about 190 million years ago. When it ended a large proportion of the rocks which are folded into such mountain ranges as the Alps, the Himalayas, and the Andes had not yet even been laid down beneath the sea.

The dividing line between the Palaeozoic and the Mesozoic is not an arbitrary one; nature itself presents us with the necessary clues. The division is marked by the extinction of many groups of the "old life" plants and animals, and their replacement by more highly organized and more complexly specialized types. In Australia the change is shown, for example, by the extinction of the *Glossopteris* flora on land, the trilobites, goniatites, numerous families of brachiopods, and other creatures in the sea, and their replacement by the *Thinnfeldia* (*Dicroidium*) flora, and the ammonites, new brachiopod groups and other marine animals respectively. The close of the Palaeozoic here also saw an important mountain-building movement, the Hunter-Bowen Orogeny, which produced in many areas, especially in eastern Australia, an unconformity between the latest Palaeozoic and the Mesozoic formations, thus helping the geologist to distinguish between them where fossils are absent.

Map sketches in previous chapters showed the suspected distribution of land and sea in the Australian region which involved an attempt also to reconstruct lands which may have existed over the

Map 18:

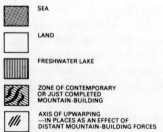

SEA

LAND

FRESHWATER LAKE

ZONE OF CONTEMPORARY
OR JUST COMPLETED
MOUNTAIN-BUILDING

AXIS OF UPWARPING
—IN PLACES AS AN EFFECT OF
DISTANT MOUNTAIN-BUILDING FORCES

The Australian region in the Early Triassic, after the completion of the Hunter-Bowen Orogeny, about 220 million years before present. Note the influence of the Hunter-Bowen orogenetic belt on the shape of the region which was to become the Great Artesian Basin.

areas of the present oceans. It is, of course, reasonable to assume that as the sea invaded the land, so the land at times extended beyond its present boundaries. For the Palaeozoic, or even earlier times, such reconstructions were of necessity very vague, though calculatedly adventuresome. Now, for Mesozoic times, the geologist can be a little more confident. There is little doubt, for instance, that there was not simply an Australia, but an Australasia in Triassic times which included New Guinea and the island area across to New Caledonia, and probably even New Zealand and the Fiji group of islands via a great arc of land around the Coral Sea. Whether this "Australian" land region was connected by a land bridge with Asia is doubtful, but it is certain that the seaway-breaks in that land bridge were not of great width and therefore not a significant obstacle to the migration of land plants and animals. However, the migration routes for animals must have been long and devious, thus producing considerable delays in the arrival of the major groups.

On Map (18) the likely land connections through New Guinea with Asia on the one hand and New Caledonia and New Zealand on the other are indicated. Evidently, the size of Australia at this time is readily comparable to its "giant" period in the Late Carboniferous, but it will be noticed that the Triassic sea makes inroads in Western Australia, which was not the case with the seas of the Late Carboniferous. Moreover, since the discovery of marine Triassic formations in New Guinea (*Skwarko, 1963*) the land connection from there to New Caledonia and New Zealand must, even in Triassic times, have been tenuous, to say the least. In eastern Australia the sea was very close to the present coastline as is shown by sediments evidently laid down in brackish waters in the Sydney Basin, and the early Triassic marine incursion near Maryborough in Queensland (*Denmead, 1964*).

The Triassic and Jurassic Periods used to be considered together, because it was thought that conditions in both of them were practically identical and that to distinguish between them was in many instances impossible. In recent years, however, much has been learnt about these matters, particularly through the efforts of palynologists, and we can now safely describe the events of these two important periods separately. Here again information received from drilling for oil and gas is of great significance in the tidying up of the records.

It remains true that, taken together, the Triassic and the Jurassic Periods can in Australia be considered largely as the age of swamps and lakes, and that some of these inland waters were of enormous extent, as large as, if not larger than any which have existed in the history of the world. For instance, Lake Walloon, in Triassic times, must have had an area of well over one million square kilometres.

Even during the Jurassic it remained at nearly that size; it was a most unusually long-persisting lake. However, as we will see later, such peculiarly long-persisting and very large inland waters are a feature of the Mesozoic and Tertiary history of Australia, quite unmatched on the other continents.

MARINE TRIASSIC. Marine formations of Triassic age occur extensively in Western Australia, where they are known from the Perth, the Carnarvon, and the Fitzroy basin areas, the Fitzroy Basin being the main area of Permian subsidence, as we have seen before; evidently, a subsidence which continued in spite of occasional reversals until the Middle Triassic. The main Triassic formation in this area is the Blina Shale, a brackish water deposit laid down probably by a large muddy river into an embayment along the western flank of the Kimberleys. Together with the plant-bearing river sand sheet, the Erskine Sandstone, the thickness of the Triassic may attain nearly 600 metres in some areas. In the basins along the far west coast the Triassic is represented by the truly marine Kockatea Shale and its correlate, the Minchin Siltstone, both of which contain fossil molluscs of early Triassic age, and are of great importance to the search for oil as possible source rocks.

Other areas where marine influences are indicated are the Joseph Bonaparte Gulf and the Sydney Basin, but they do not go beyond a brackish water environment, there being no truly marine fossils in the respective beds. The only truly marine Triassic in the east occurs in the Maryborough Basin, where a mollusc fauna was found (*Denmead, 1964*).

TERRESTRIAL TRIASSIC. As shown in Map (18) the predominant feature of the Australian scene is a pattern of inland lakes, especially in the east. The lakes and swamps in Western Australia may have been considerably more extensive too but, if so, whatever deposits they left behind have been largely removed. The Triassic age of the beds indicated in the Canning Basin is documented chiefly by plant remains and by the occurrence in them of the same little crustacean species as characterizes the Blina Shale farther north. In the Perth Basin the marine Kockatea Shale grades into terrestrial silt and sand beds with innumerable fossil spores of the same type as those in the Shale.

All Triassic beds in South Australia are terrestrial. The most important are the Leigh Creek Coal Measures; they were formed in swamps and lakes in low-lying areas of the northwestern Flinders Ranges. The lakes shown farther northwest from there on Map (18) are somewhat doubtful although—looking at the subsequent extent of Lake Walloon (Jurassic, Map 19) and considering the ancient relief of this part of the continent—it is rather likely that one or the other did exist. On the other hand, there is now no doubt that there are no Mesozoic rocks in the Ngalia Basin northwest of Alice Springs, the youngest pre-Tertiary/Quaternary formations there represent a slice of the Carboniferous history of the continent.

In Victoria, lakes existed from Gippsland across to the South Australian border, but much of the sediment laid down in them has been removed from inland areas by erosion. Remnants are visible south of Bald Hill, about 2.5 kilometres northwest of Bacchus Marsh, where there are 200 metres of sandstones and shales containing Triassic plants. The Victorian lake system may have been at times connected with that in Tasmania. Many of the rocks are very similar in composition.

The Tasmanian lakes were confined to the central and eastern region of the island, the earliest of them developing in central northern areas, from where they gradually extended east and southward. Most of the sediments were derived from mountains in the northwest, some from the east.

The main areas where Triassic sediments are still preserved lie north of Hobart to both sides of the Launceston railway line, notably near Lake Tiberias. In the northwest, on the upper reaches of the South Esk River, occur the Fingal Coal Measures, which contain fairly valuable coal seams of sub-bituminous type, but rather high content of inorganic materials such as sand and silt. The Triassic on the South Esk is about 350 metres thick, but the best coal seam, in the Cornwall Mine, is only from two to five metres thick. From the limited extent of the coal seams it may be assumed that only in certain parts of one great lake conditions were favourable for coal formation, or that there were a number of smaller lakes.

THE SYDNEY BASIN. Around Sydney lakes and swamps developed very early in the Triassic (in fact, the site of the city of Sydney lies just about at the centre of the depositional area of those times). It is possible to reconstruct the topography of those ages with a fair degree of accuracy. After the deposition of the Upper Permian coal measures the area remained low-lying long enough for a complete change to take place in both climate and vegetation. The rocks of the coal measures were in most places at too low a level to be eroded, and they lay undisturbed until the Triassic rocks were deposited upon them with practically no visible break. A very slight subsidence was sufficient to allow the resumption of lacustrine conditions. The coastline lay considerably to the east of its present position, but whether high land existed over what is now sea, or whether the Triassic (and late Permian) swamps and lakes were separated from the ocean only by one or more barriers of sand dunes and tidal mud flats, is hard to say. In Map (18) it is shown as true land, as a part of the southern extension of the mountain-building zone of the Late Permian Hunter-Bowen Orogeny, which is about to be inundated again by the sea.

"Lake Narrabeen", as it may be called, generally coincided with those of the older coal swamps of the Late Permian. Commencing off Eden, its western shore ran northwest past Goulburn, thence north through the Blue Mountains and across the Dividing Range out in the direction of Coonabarabran, where it probably linked up with that of a great lake of the inland. The eastern shore ran from Newcastle west to Muswellbrook and then north through Murrurundi and Gunnedah to Narrabri and beyond, again joining up with an inland lake shore. No doubt the deposition area of Lake Narrabeen was chiefly in the Sydney Basin, but in principle it was merely a southern extension of the great Triassic inland lake system. As the land slowly subsided it alternated between lake and swamp, and a thick series of conglomerates, sandstones and shales, containing numerous fossil plants, was laid down. In the Sydney Basin this is known as the Narrabeen Group. The thickness of the series beneath Sydney, where it has been penetrated by numerous bores, is as much as 700 metres, but it thins out toward the basin margins in the west, where it is usually less than 60 metres thick. Splendid sections form the sea cliffs north of Sydney from Long Reef to beyond Terrigal.

Almost anywhere within this area the story of their formation is written indelibly in the rocks. At one stage volcanic action was rife somewhere in the neighbourhood, for the thick beds known as the chocolate shales are composed almost entirely of volcanic ash. They contain innumerable particles of metallic copper, but this is not suitable for commercial exploitation. Apart from the numerous fossil plants, the casts of sun cracks appear in many places on the surface of the shale, showing that the original mud was not only deposited in shallow water, but also at times exposed to the air and allowed to dry. Even the rain drops of a passing shower have left pit marks on the rocks. It is wonderful to think that evidence of such a brief happening should have been preserved when so many more solid and substantial things have left no trace. Ripple marks are also abundant on the surfaces of the shale and sandstone—more evidence of shallow-water conditions—and even the tracks and burrows of worms crawling through the mud have been preserved. It is interesting to note that already in the Narrabeen Group various layers contain large numbers of a small crustacean belonging to the genus *Cyzicus* (once known as *Estheria*). This tiny creature is known to have preferred brackish waters, although being able to tolerate both fresh and salt water environments at times. We will meet with *Cyzicus* again later in the

Triassic, and it is important to realize that its presence indicates that the sea cannot have been far away. In fact, according to Conolly (*1969*), the red clays in the upper Narrabeen Group of the southern Sydney Basin are of marine origin, and so are the lowest 60 metres of the overlying Wianamatta Group (*Herbert, 1973*).

Above the Narrabeen Group, and occupying a slightly different and especially in the north much smaller area, lies the Hawkesbury Sandstone, a formation of massive sandstones, with some beds of pebbles, and some lens-shaped and rather thin beds of shales. Sydney is largely built on this sandstone, which forms the sea cliffs facing the Tasman Sea south of Long Reef and also the mighty wall-like escarpments capping the gorges of the Blue Mountains where they cut through eastward to the Cumberland plains. The western part of the Blue Mountains, including the Katoomba plateaux are carved out of the Narrabeen Group, however, which emerges westward from beneath the younger Triassic formations.

The Hawkesbury Sandstone varies in thickness from 90 to 300 metres. The exact origin of the sandstone is still somewhat doubtful because it is difficult to imagine practically identical sedimentary conditions over about 18 000 square kilometres, resulting throughout in the same types of rocks, of the same texture, and with the same interstratified peculiarities (*Osborne, 1948*). Most likely the area was a huge delta of several large confluent rivers, the largest merging from the northwest and the southwest. One of the features of the sandstone is its current-bedded texture, in which the main layers or strata are about 30 centimetres or more, horizontally laid down, but in themselves composed of fine layers which lie at an angle of about 22° to the horizontal. This kind of bedding is characteristic of shallow water fore-set deposition under the influence of strong currents, that is, a bed of sand would not be deposited evenly, but as a widening bank, the individual layers being laid at an angle on its advancing edge. The main problem is whence currents may be derived within a freshwater lake, even if it is of the size of the Sydney (Hawkesbury) Basin. Yet the fact remains that nobody has ever found marine fossils in the Hawkesbury Sandstone.

About Sydney the current-bedding dips to the northeast, but north of the Hawkesbury River mainly to the southeast, suggesting that in the southern part of the basin the delta-building rivers came from high lands to the southwest, and in the northern part from the New England mountains in the northwest. It must also be assumed that after the Narrabeen Group had been deposited there was considerable elevation of adjacent areas, which greatly increased the volume and the rate of flow of rivers draining into the lakes. The rate of erosion was also greatly increased and in both the north and the west the originally deep-seated granites which had been intruded in Carboniferous and Permian times became exposed. From these granites (but also from the Devonian quartzites) would have come the sand and other coarse material which makes up the sandstone.

There is little doubt that the Hawkesbury Sandstone is in its entirety a water-laid formation, probably deposited in one large lake, although it has been suggested that flood-plain deposition under water after heavy rains, or even wind-laid dunes, were responsible in part. Here and there in backwaters or in isolated ponds mud was deposited, which now exists as beds of shales containing the remains of plants, insects, freshwater fish, and even large amphibian animals. A peculiar feature of the shale beds is their content of smaller and larger fragments, straight and bent slabs and slivers of shales which were embedded as detritus. We imagine the origin of these fragments in the following way: we have all seen at one time or another how hard the sun-cracked bottom of a dried-out pond or billabong becomes, how the sun-baked polygons of the hardened mudcake curl up and even over at the edge of the cracks, and how they eventually break off and lie around loose. Evidently, the next river flood will wash them out, carry them some distance, and dump them into other ponds farther downstream; in other words, shaley fragments are thus deposited whole into another environment where shales are newly formed from river mud and silt.

If we are correct in this explanation of the origin of the shale fragments in the shale beds, we may just as well briefly consider what that may imply. It is rather obvious, for example, that the climate can hardly have been cool or even cold, most likely not even temperate. It takes—at least in our Australian experience—quite some time, say three to four months or more, before the billabong mud becomes deeply hardened, curled up, and broken away. In other words, there must have been long periods of drought similar to those our continent experiences over large regions today. We would then have to assume that the extensive Hawkesbury Sandstone formation was generally formed under water, but also under semi-arid tropical or subtropical conditions. Curiously enough, this conclusion is quite at variance with the results obtained from palaeomagnetic measurements which show the Sydney area for the Triassic was in a position equivalent to the central Ross Sea of the present Antarctica—namely, well over 70° south latitude. Evidently, somebody is wrong—an illustration of the extremes in contrariness of evidence, for which geologists have to seek the plausible solution. But there can be only one solution, among the many long-past situations, which will fit the records contained in the rocks. Fortunately, nature herself prevents scientific arguments becoming too "hot"—she never preserves all the records so that all arguments could be settled for all times (see p. 218).

A particularly interesting shale bed occurs in the brick pit at Beacon Hill, in Brookvale just to the north of Sydney. It has a fine history of exploration and provides an example of how a non-geologist may do work which results in outstanding scientific discoveries. The chance discovery of a fossil insect fired the enthusiasm of one of the quarry men, and of the pit foreman. One by one further discoveries were made, many of which have since been described. The collection has not only made this spot world-famous, but has added enormously to the knowledge of the life, particularly insect life, of the Triassic Period.

Above the Hawkesbury Sandstone lies the Wianamatta Group, deposited under conditions rather similar to those of the Narrabeen beds. However, the lake or lakes, though still of considerable extent, occupied a much smaller area and lay mainly in hollows on the underlying Hawkesbury Sandstone. The water surfaces, though probably not quite continuous, stretched northwards from the vicinity of Moss Vale to beyond the Hawkesbury River, a distance of about 160 kilometres, and from the eastern slopes of the Blue Mountains almost to the present coastline, about 80 kilometres. In this lake up to 200 metres of rocks were deposited, mainly carbonaceous shale with beds of ironstone and a little sandstone. Fossils are not common, but the brick pits at St. Peters, near Sydney, have yielded freshwater shells, some plants, occasional magnificent specimens of fish, and the remains of a giant amphibian. There is evidence that during the deposition of the Wianamatta Group the sea at times found access to the lake, probably breaking through a barrier of sand dunes. The fossil evidence for that event rests on the correct identification of very small uni-cellular organisms (foraminifera) and remains somewhat doubtful. Yet there can be no doubt that the sea was very close almost all the time, otherwise the small crustacean *Cyzicus* with its brackish water habits would not be so plentiful in this formation.

The Wianamatta Series was the last formation of any size and extent to be deposited in eastern and southern New South Wales. Great as its age is, its position is much the same as when it was laid down, and since the close of the Triassic Period no part of it or of any adjacent areas has been low enough to be submerged beneath water, fresh or salt.

THE CLARENCE RIVER BASIN AND SOUTHERN QUEENSLAND. At the same time as the lakes were formed around Sydney, and to a lesser extent also between Manning River and Port Macquarie, a much larger area became submerged in southern Queensland, where by the end of the Triassic Period the freshwater lakes had spread to an extraordinary extent. They began in the southeast along the north coast of New South Wales, north of Coffs Harbour, crossed the border into southeastern Queensland, northward past Brisbane and Ipswich into the Esk Trough and Maryborough Basin, and westward across today's Dividing Range into the Great Artesian Basin. The southern shoreline formed a northward protruding peninsula the backbone of which was formed by the northern extension of the New England mountains which, as shown on Map (18), had been involved in the Late Permian Hunter-Bowen Orogeny. It is not necessary, of course, to visualize the whole area as covered for long times by one large sheet of fresh water, although lakes of such size are in existence at present—for example the Caspian and Aral Sea in southern Russia—and were even larger a few thousand years ago. As a rule such large inland waters, even if unconnected with the sea, are slightly salty or brackish and contain an accordingly specialized fauna. However, it seems more reasonable, considering that a time span of say 30 to 40 million years is involved, to look upon these inland regions as having been close to sea-level, containing many larger and smaller lakes which alternated with coal swamps and even dry land. After a long time, during continuing subsidence, the positions of the lakes shifted all over the region, until eventually every part had been at some time under water, and the whole became covered with a considerable thickness of sediments, including seams of coal.

In the Clarence River Basin the rocks are probably over 1200 metres thick, and at Grafton, about the centre of the southern part of the basin, they have been penetrated by bores to that depth. The Triassic here consists for the most part of sandstones, resembling those in the Sydney Basin, and of conglomerates towards the base. Shale beds occur near the top too, and coal seams, of uneconomical type, are interstratified in the lower part. The lake in which these beds were deposited extended eastward beyond the present coastline, so that the shore at the time must have been much farther to the east than it is now. It is, however, on the whole younger than the Sydney lake, and existed mainly in the later Triassic and the Jurassic.

Across the border in Queensland the Triassic Period was ushered in with great volcanic eruptions. Near Brisbane, at Ipswich, and farther north near the Esk River, for over 130 kilometres the country was covered by dense showers of incandescent ash and scoriae from a nearby chain of volcanoes. Forests were overwhelmed and set on fire and the whole landscape was buried to a depth of many hundreds of metres—in places, with the help of ash-mud flows, to 900 metres. These ashes buried and in many places levelled out the relief of an old landscape, and vary therefore considerably in thickness. At Brisbane, where the ash overlies a relief carved out of the Brisbane Schists, it is only about 60 metres thick. In the lower part are the remnants of the destroyed forests, largely converted to charcoal, although many of the larger tree-trunks were only charred on the outside. Some of these tree-trunks were up to 50 metres in height; they were giant pines, and their timber, now changed to solid chalcedony or opaline matter, is so beautifully preserved that microscope sections show the structure of each individual cell.

At Mount Crosby, in the Ipswich district, one interesting bed lies below the ash. From it some thousands of fossil insects have been collected, many of which are still undescribed although their general classification is known. Near the top of the Triassic series, and separated from the first by a great thickness of sandstone and shale, is another famous insect bed. This was discovered in 1890 by Simmons at Denmark Hill, near Ipswich, and large collections were later quarried there under the direction of the then Government Geologist of Queensland, B. Dunstan. This most interesting bed consists of a thin layer of fine white clay-shale deposited in a pool fed by hot springs. The boiling water trapped and killed the insects, most of which were water beetles, and destroyed all but the wings. These sank to the bottom and thousands upon thousands of them may now be found on every thin layer of shale. These two occurrences, together with that at Brookvale in the Sydney Basin, constitute one of the fullest chapters in the story of former living things.

Between these two insect beds in the Ipswich district lie nearly 1200 metres of strata, both the lower and the upper portions of which contain a number of valuable coal seams. Throughout are many beds which have yielded abundant and beautifully preserved fossil plants. All these rocks are considered to have been formed in Middle Triassic times, but above them is another series of massive sandstones which are late Triassic and known as the Bundamba Group. The Triassic coal swamps of the Ipswich area were of limited extent, although other terrestrial sediments of that age are very widely distributed and in the east almost everywhere contain substantial layers of ash, tuff, and lavas. Thus, along the coast north of Brisbane there are the North Arm Volcanics and Landsborough Sandstone, in the Maryborough Basin there are at least 600 to 1000 metres chiefly of sandstones, interbedded with shales with plant remains, but no volcanics. These are present again in the Yarrol Basin towards the Rockhampton area, where there are also very important coal seams in the lower part of the sequence. These are known as the Callide Coal Measures, and the main seam is on the average about eighteen metres thick, extending over at least 130 square kilometres. In northern Queensland and

Cape York Peninsula there are a number of areas with Mesozoic freshwater beds—some of which may well include Triassic rocks; most, however, being Jurassic and Cretaceous.

Large areas of waterlaid blanket sands and shales are known in the underground of the Great Artesian Basin right across into South Australia. Although they thin out and are even absent in various areas these sands form part of the porous horizons which contain the huge underground water reserves beneath the impervious beds of Cretaceous age. Northwards they emerge to spread into the Bowen Basin where they are known under names such as Clematis Sandstone, Moolayember Shale, Carborough Sandstone, and attain in places a thickness of up to 900 metres. We shall not go into the details of the nature and subdivisions of the subsurface Triassic formations in the Great Artesian Basin except to say that they are in some areas of paramount importance to the location of commercial oil pools because of their porous nature and wide extent, an importance which they share with the similar overlying Jurassic beds.

All these inland formations were laid down in the low-lying region which for both Triassic and Jurassic times is known as Lake Walloon. In the next chapter we shall hear more about this great geographical feature of those faraway ages.

LIFE OF THE TRIASSIC PERIOD

Giant Pines and the First Reptiles

LAND PLANTS. In spite of the great wealth of fossil plants in the Triassic rocks, it is not easy to obtain an accurate picture of the vegetation of that age. Forests existed in many places, but the appearance of the trees and the nature of their foliage is often a matter of conjecture. About the smaller plants there is more information, but again much of the material is fragmentary, and it is difficult to correlate isolated specimens of seeds and spores, or fruits, with the leaves belonging to the same plant. In addition, most of the specimens are of plants that lived in swamps or on low-lying ground, and little is known of the vegetation of the highlands, most of which can have left little trace.

It can be said, however, that a great change took place in the Australian vegetation at the close of the Permian Period. In fact, this change is taken as the event which separates the last period of the Palaeozoic Era from the first of the Mesozoic. Although it is not known how long the interval of time was (if any) between the deposition of the last Permian and the first Triassic rocks, it is certain that the glacial regime of the early Permian epochs had long given way to a warmer and more congenial climate. Yet the change in the flora (*Playford, 1979*) did not come gradually during the amelioration of the climate, although one would think that such changes should be sufficient to account for the general change in the flora. The point is that at the close of the Permian there was a world-wide evolutionary step in plant life which had nothing or little to do with local or regional climatic conditions. Nobody knows why such changes take place; they are simply there, and the geologist uses them to subdivide the history of the earth into major sections.

That unaccountable change in the flora is the disappearance of the glossopterid plants and their replacement by fern thickets, after the most prominent genus of which the new vegetation type is called *Thinnfeldia* flora. This new flora, of course, included many other plants such as true ferns, horse-tails, cycads, tree-ferns, as well as various trees. Typical of many landscapes would be the swamps and lakes, or ponds, in which the Narrabeen beds were laid down. From the absence of fossil wood in this particular sequence, it would seem that in this area there were few trees, and that it may have looked rather drab and monotonous. The most common plant was the horse-tail *Phyllotheca*, which had survived from Permian times, but was larger and more luxuriously foliating (Plate 16). It probably grew in dense thickets from two to five metres high on the margins of the swamps or even in the water itself and, though lacking the large leaves, was not unlike bamboo in general appearance. Below the clumps of *Phyllotheca* were the ferns and seed-ferns such as *Thinnfeldia* and *Macrotaeniopteris*. *Thinnfeldia* (in fact its Southern Hemisphere subgenus *Dicroidium*), which was a seed-fern, had a graceful multipinnate (or much-divided) frond, like the common bracken, while *Macrotaeniopteris* rather resembled the modern bird's-nest fern. Less common in the Narrabeen Group, but prevalent elsewhere, are the remains of large tree-ferns, the trunks of which have been found in several localities. Found separately but considered to belong to these are fern fronds which have been described under the name of *Cladophlebis*.

The lone survivor of the Palaeozoic lycopodiales, *Pleuromeia*, first found in the Fitzroy Basin Triassic of the West Kimberley (*Brunnschweiler, 1954*), is now known to have been quite common in Australia (*Retallack, 1975*).

In striking contrast to the treeless landscape near Sydney were the Triassic forests in the Brisbane area. It has already been mentioned how these were overwhelmed by volcanic eruptions and their remains preserved beneath the hot ash, and also that the trees were up to 50 metres in height. Unfortunately the hot ash which charred and buried the trunks completely destroyed the foliage, and there was considerable uncertainty at first about their systematic position. As a result of examination of micro-sections of the petrified wood, however, it is now fairly certain that most of them were true conifers or pines. One of these, *Cedroxylon*, a large cedar-like tree, had features in common with the living *Araucaria*, of which it may have been an ancestral form. These forests are very nearly of the same age as the famous petrified pine forests of Arizona, which have long been a source of wonder to visitors.

Another interesting plant which was common in later Triassic and especially in Jurassic times is *Gingko*, of which a number of species have been found in various localities all over Australia. Although most of the specimens consist only of the peculiarly lobed leaves, it is thought that the plant was a small tree. The gingkos or maidenhair trees form a very special group, though they are broadly related to the pines. There is still a living species in Japan, and some trees have been cultivated in the botanic gardens in Sydney and Brisbane.

INSECTS. Our knowledge of Triassic insects comes from several remarkable deposits of fossils in which the specimens are numbered in thousands, and in which the state of preservation is remarkably good. The oldest of these is the Mount Crosby insect bed at the base of the Ipswich beds in Queensland. Thousands of insects have already been collected from this locality, only a small proportion of which have been identified in detail and described—for example, the earliest of the butterflies, *Eoses*, and cockroaches such as *Triassoblatta*. Rather better known are the insects from the famous Denmark Hill insect bed, also near Ipswich, which has already been mentioned. After its first discovery in 1890 the main fossil layer, a fifteen-centimetre stratum of fine clay shale, was carefully quarried, removed in bulk to Brisbane, where the material was carefully split into thin laminae. Thus thousands of specimens were obtained and many species described by Tillyard (*1923*). About half of them are beetles, one-seventh are cicadas, another seventh cockroaches, about one-twentieth bugs, and the rest in decreasing percentages are dragon flies, scorpion flies, praying mantises, locusts, lacewings; while other, now extinct, groups make up about one-tenth of the whole fauna.

A smaller number of insects, magnificently preserved, has been found in the brick pit at Beacon Hill, Brookvale, near Sydney. They occur in a shale bed forming the site of an old pond within the Hawkesbury Sandstone. The rocks in the brick pits at St. Peters, Sydney, which were laid down some time later in the Triassic Period, have also yielded many specimens. Thus, all in all, sufficient material has been obtained to give quite an impressive view of Triassic insect life. This is the more valuable in that no similar finds from later periods have yet been made in Australia, so that the Permian and Triassic insects are all there is to link this form of life in a remote age with that of the present day.

It is very apparent that the Triassic insect fauna had a much more luxuriant aspect than that of the Permian, possibly because of warmer climates. Individuals were larger, some of relatively gigantic size, and of a type suggestive of long days of warm sunshine—some more evidence which contradicts the results of recent palaeomagnetic studies already discussed earlier in this chapter. In the tens of millions of years elapsed since the Permian, evolution had proceeded apace, and many new orders had appeared, including some remarkable ones which have since become extinct. Amongst these was one named *Mesotitan* that measured over 30 centimetres across the wings. It belonged to an extinct order akin to that containing the modern ant-lions, and had a long spear-like proboscis with which it evidently impaled and sucked up the tissues of other insects. Still more remarkable is a single wing found at Brookvale. This is twelve centimetres in length and belonged to an insect of the same order as *Mesotitan*. Described by McKeown (*1937*), it is known as *Clatrotitan* (Plate 16), and its most striking feature is the curious grid-like structure in the centre of the wing, which must have been used for producing sound. The stridulation produced by our cicadas on a hot summer day would be but a murmur compared to the torrent of sound from myriads of these much larger creatures. Yet true cicadas also have been found at Brookvale, so beautifully preserved that even the body and head can be studied (Plate 16), and there are, as in Queensland, other insects.

TRIASSIC CRUSTACEA. Among the specimens found at Brookvale was a small shrimp-like crustacean called *Anaspides*, which has an interesting history. It belongs to a peculiar family, the Syncaridae, distinguished among other things by its lack of a carapace. It was first known from the European Carboniferous and Permian, and was thought to have then become extinct. In 1906, however, it was found living in the mountain lakes of Tasmania, and other living representatives have since been found in Europe and Malaya. The Brookvale species is the sole intermediate link in all this vast space of time.

Another small crustacean, a blind form, is found in wells at Canterbury, New Zealand, and an allied species has been located on the top of Table Mountain, South Africa. *Phreatoicus*, as it is called, is one of the isopods, a familiar member of which is the common garden slater. In 1916 several specimens of *Phreatoicus* were discovered among the fossils from the St. Peters brick pits, and these differ but little from the living forms.

The most common crustacean in the Australian Triassic is the little conchostracan genus *Cyzicus* which, because it has two hinged valves, is easily mistaken for a minute pelecypod. It occurs frequently in the Narrabeen Group, and even more in the Wianamatta shales. In the early Triassic Blina Shale in the Fitzroy Basin it is so abundant in certain layers that the rock consists almost wholly of the tightly packed tiny shells of this creature. Because the shells were chitinous, not calcareous like those of the molluscs, these *Cyzicus* layers show a considerably raised phosphate content. It is interesting to note that this tiny crustacean had the habit of living in brackish waters, that is, truly marine or truly freshwater conditions would soon drive it away. Therefore, wherever it occurs in great numbers we get an important indication as to the environment in which the sediments containing them were laid down.

FISH. Of Australian Triassic fish a fair amount is known, though most are freshwater types. The chance of fish becoming fossilized is much greater in lakes and lagoons than in the sea, where their bodies are devoured by the multitude of creatures and the skeletons dispersed or destroyed. In a lake, on the other hand, they are often preserved whole in the mud on the bottom, into which they have burrowed to escape drought, or they may have been suddenly overwhelmed and buried by excess of mud in flood time or by showers of volcanic ash.

There have been several notable finds of fish in Australian Triassic rocks. The oldest of these came from a railway ballast quarry at Gosford, from a bed of shale high in the Narrabeen Group. Here some hundreds of finely preserved specimens were collected by State Survey geologists and described by Smith-Woodward (*1890*). The brick pit at Brookvale, which lies in the Hawkesbury series, has also yielded many fine specimens (*Wade, 1935*). Still higher in the Triassic, in the shales of the Wianamatta Group, in the brick pits at St. Peters, Willoughby and other localities near Sydney, a great many forms have been and are still being found, the first fauna from there having been described by Smith-Woodward (*1908*).

In all 16 species of fish have been recovered at Gosford, 11 at St. Peters, and 29 at Brookvale. It is remarkable that though the forms from St. Peters are the youngest they more generally resemble Permian species and, if considered alone, would suggest that the enclosing rocks were thus older. Ganoids present include *Zeuchthiscus*, a salmon-shaped fish 40 to 50 centimetres long; *Myriolepis*, up to 50 centimetres long, with small scales and *Cleithrolepis*, a deep-bodied form. *Saurichthys*, a pike-like predator reached over one metre. True lung-fish (dipnoans) were also present. A complete specimen of *Gosfordia*, a deep-bodied dipnoan was found at Somersby-Gosford in

1979. Remains of a freshwater shark, *Xenacanthus* (formerly *Pleuracanthus*) come from several sites (Gosford, Bowral) but the finest specimen, 1.5 metres long, was found long ago at St. Peters.

An interesting feature of the St. Peters locality is that hard bands of ironstone contain quite different species of fish from those in the soft grey shales between them. A likely explanation of this is that the water was at times brackish, and at these periods species came into the area which at other times were absent. The sea was certainly not far distant during the latter stages of the Wianamatta epoch; in spite of the doubtfulness of the record of truly marine unicellular foraminifera by Chapman (*1909*) it can be assumed that the sea occasionally invaded the Wianamatta lowlands, much as it invades coastal lagoons and lakes today, and that it brought with it marine creatures, some of which were preserved as fossils. This also explains the occurrence of a marine shell in sandstone from the government docks at Biloela, a find that has long puzzled geologists.

Numerous shells are, however, found in the Wianamatta shales but these are true freshwater bivalves, akin to the freshwater mussel *Unio* in rivers and lakes at the present day.

AMPHIBIANS. Of the vertebrates higher than fish, the labyrinthodonts are still the predominant land animals in Australia in Triassic times, but the first reptiles had nevertheless arrived.

The record of amphibians includes a fine skeleton of the giant *Cyclotosaurus*, of which the head is 90 centimetres long and 50 centimetres wide, and the overall length about three metres. It was discovered by Dunstan in 1895 in the brick pits at St. Peters, and is in the British Museum. The preservation was so excellent that it has been possible to make a cast of the brain cavity showing the roots of the principal nerves. Another complete skeleton has since been found in the Brookvale brick pit. It is at present in England for study by specialists but will be returned to the museum in Sydney thereafter. Other amphibians from the Sydney Basin are members of the genera *Mastodonsaurus*, *Capitosaurus*, and *Platyceps*. Skull bones, jaws and teeth of capitosaurians and especially of the trematosaurians *Erythrobatrachus*, *Deltasaurus*, and *Blinasaurus* are also found in the Blina Shale of the Fitzroy Basin and in the Kockatea Shale of the Perth Basin, both formations being also notable for an abundance of *Cyzicus* and of ganoid fish scales.

Though the specimens are not numerous, it does not follow that amphibians were uncommon at the time; in fact, the few genera that are known and many others were probably quite common, dragging their cumbersome bodies through the swamps and lagoons round Sydney and in Queensland, or along the margins of the great lakes in the interior, and the delta lagoons in Western Australia. Like tadpoles, they spent the early stages of their lives in the water, and were never far from it; the real conquest of the land was left to other creatures. It is significant in this context that the *Platyceps* from the fish beds at Gosford is, in fact, a larval or tadpole form of which the true, adult generic classification is not really known.

REPTILES. Compared with the rest of the world there was in Australia during the Triassic Period once again a peculiar delay in the appearance and spreading of higher vertebrates. The Permian elsewhere had been marked by great evolutionary steps, but very few indeed of the new animals seemed to have succeeded in finding a migratory route to Australia. Even the first amphibians in Australia did not appear before Middle Permian times (*Bothriceps* from the Airly coal measures at Lithgow and the Newcastle coal measures; doubtful *Microsaurus* from the Mersey coal measures near Hobart), and now we see that the elsewhere already dominant reptiles are extremely rare on our continent during the Triassic. This is the more perplexing inasmuch as, since Devonian times, Australian plant life not only had kept abreast of that in neighbouring countries, but had, in fact, spread its *Glossopteris* flora northwards into Asia. Why more reptiles did not find their way into Australia, in spite of the fact that other Gondwanaland regions such as South Africa had many remarkable Permian and Triassic forms, remains one of the great problems in palaeogeography. The same problem is encountered with the mammals which began to appear in other lands towards the end of the Triassic, but of which we know nothing in Australian rocks until the Tertiary. However, evidence of the presence of birds (feathers), which had evolved from reptiles about mid-Mesozoic times, and for whose migration no land bridges were necessary, recently turned up in Australia in Late Jurassic to Early Cretaceous beds in South Gippsland (Koonwarra). This is very shortly after the first appearance of the birds (p. 225).

Until recently there were rather few records of reptile remains from the Australian Triassic. One is a *Thecodontosaurus* from a bone bed on the North Queensland coast of which the locality is somewhat uncertain; the second is a single femur or thigh-bone found in sandstone at Government House in Hobart, Tasmania. Once believed to be from an amphibian it is now known to be from a captorhinomorph reptilian. In appearance these animals—the most primitive of the reptilians, such as *Cotylosaurus*—were not unlike the amphibians from which they had evolved. They were clumsy and slow-moving animals with short legs, incapable of lifting their heavy bodies above the ground. There was, however, one important difference which definitely separated them from the amphibians. No longer was there any metamorphosis between the immature and the adult stages. The eggs were laid on land, and the young reptiles emerged as small replicas of the adult, and though some species retained a semi-aquatic habit, others became fitted for an entirely terrestrial life.

However, a more varied story is beginning to unfold. Lepidosaurians and thecodonts—together with amphibians—are reported from the Lower Triassic Rewan Formation in central Queensland (*Bartholomai and Howie, 1970*) and have been described also from Tasmania (*Cosgriff, 1974; Camp and Banks, 1978*). Footprints of reptilian origin have been discovered both in the Sydney Basin (Berowra flagstone quarry) and Queensland (*Hills, 1958*). The slabs from the Berowra quarry have been extracted and are on exhibition in the Australian Museum. It is, of course, not possible to reconstruct the animal which made the footprints; all that can be said is that it was about 90 to 120 centimetres in length and that the foot had two prominent claws, with smaller ones at the sides, and a pad. The tracks are not unlike those of crocodiles, but the tail did not drag. In one place, though, the animal paused and its tail touched the ground and left an impression which shows that it was short and pointed. It is wonderful to think that while no trace of the animal itself remains, its tracks made on soft sand have been preserved for about 200 million years, and give at least a vague idea of the kind of animal it may have been.

LIFE IN THE TRIASSIC SEA. Until a few years ago nothing was known about the local marine life in Triassic times; now we know that substantial parts of the west coast and at least one small portion of the east coast country were inundated. The molluscan faunas found in both these regions are of early Triassic age and show that by that time the isolated provincialism which characterized the earlier Permian marine faunas had disappeared. The forms present are those common to other Triassic seas which spread through the East Indies and the Himalayas to Europe. Thus there are ammonites, bivalves, brachiopods, and worms described from Western Australia (*Dickins and McTavish, 1963*) which link this region closely with Southeast Asia, and the same may be conjectured of the species found in Queensland (*Denmead, 1964*).

Early Triassic faunas the world over are quite different from those of the Permian Period, although some Permian species lingered on. The chief feature was the decline of the brachiopods and the steep rise of the molluscs, particularly the cephalopods. Throughout the Mesozoic Era the seas were dominated by the cephalopods, as the land was by reptiles. They were predatory creatures, and already in the Triassic increased vastly in numbers, variety, and size. There may have been shell-less forms like the octopus, but if so they have left no trace. The majority had shells, which were generally involute like the nautilus, but which had the complicated sutures of the group known as ammonites. In these molluscs the shell partitions or septa had their edges so convoluted that the suture line was bent into an intricate pattern of lobes and saddles, and these are very useful in classification because they reflect, to some extent, internal organization of the animals. In the characteristic Triassic type, *Ceratites*, the suture consisted of a number of plain saddles, while the lobes between them were serrated. This group became extinct at the end of the Triassic and was replaced by the true ammonites, in which both saddles and lobes are serrated.

The Australian Triassic ammonites are not true *Ceratites*, but belong to a closely related group, the Otoceratidae, which is typical of the early part of the period. Four genera are known to occur here, *Ophiceras*, *Subinyoites*, *Glyptophiceras* in the west, *Protophiceras* in Queensland, but more will certainly be found sooner or later.

The pelecypods are represented by taxodonts like *Trigonucula*, and *Nuculopsis*, which have a double-row of many fine teeth along the hinge. The scallops with *Claraia*, the mussels with *Bakevellia*, and other bivalves with various species make a strong appearance both in the east and west. Of the lamp-shells, however, only the primitive and long-surviving genus *Lingula* has been found. The annelid or worm is *Serpula*, tiny spiral tubes usually found attached to shells of molluscs.

12 The JURASSIC PERIOD

THE ORIGIN of the names of the great geological periods goes back to the early part of the 19th century when geology began to emerge as a rational natural science. Some names, like Carboniferous or Cretaceous, refer to important characteristics of the formations laid down during the period, especially in the Northern Hemisphere; other names simply indicate where the respective rock system was first studied as a whole and shown to differ markedly from other systems. Since subsequent students naturally referred to the earlier work the name, if at all convenient, became firmly established and perpetuated. The name Jurassic belongs to the latter category. It is derived from the Jura Mountains in western Switzerland, where the formations of that system are beautifully exposed in deep gorges which rivers have cut across the strike of a series of mighty folds, in a manner similar to that by which the Torrens and other rivers have cut through the folds of the Adelaide Hills.

The duration of the Jurassic is roughly of the same order as that of the Permian; in other words, as Periods go, it is one of the longer ones. It began 191 million years ago and ended 55 million years later—that is 136 million years before our time.

In Australia the Jurassic story, like that of the Triassic, is to a large extent one of terrestrial sedimentation in great lakes and swamps, and of continuing erosion of the great mountain ranges formed during the Palaeozoic. Yet the sea keeps on nibbling at the continent, and by the end of the middle epoch of the period it manages to inundate quite substantial areas in Western Australia. For the first time since the Proterozoic it finds access to northwestern Queensland by way of a new feature in the north, the Gulf of Carpentaria. In the east truly marine beds are not known yet, but the sea was, as in Triassic times, very close to the present coastline and, in fact, began to transgress it in a few places, especially in Queensland, just about the end of the period.

MARINE FORMATIONS OF WESTERN AUSTRALIA.

The two regions in the west to which the sea found access are the Canning Basin and the coastal strip between Perth in the south and Onslow in the north, that is, the Perth and the Carnarvon Basins. The earliest inundation took place in the area of Northwest Cape where a probably narrow trough began to subside fairly rapidly even during the Liassic, the oldest epoch of the Jurassic. By Middle Jurassic or Dogger time the sea spread into the Perth and much of the Carnarvon Basin, and towards the end of that epoch it had also penetrated deeply into the Canning Basin. It is true that, except in the Northwest Cape area, these inundations were shallow and in places short-lived, but together with the newly shaping Gulf of Carpentaria and the persistent nibbling along the east and south coasts they herald the pattern of the great marine transgression that was to take place in the next period.

The best known marine Jurassic beds are those in the Geraldton district, 480 kilometres north of Perth. They consist largely of sand- and siltstones, some of them non-marine, and include a few tens of metres of a richly fossiliferous limestone, the Newmarracarra Limestone (*Arkell and Playford, 1954*), which has yielded a rich harvest of ammonites and other molluscs. The few metres of shale and sandstone below the limestone and the nine metres of sandstone immediately above it also contain many molluscs, but ammonites are absent. The entire Jurassic sequence in this area is only 100–150 metres thick, most of it laid down during the early part of the Middle Jurassic (or Dogger) Epoch. The Geraldton sequence extends southward through the Hill River district to Gingin near Perth; thereafter it disappears beneath younger beds.

The most complete Jurassic series known in the eastern part of the Southern Hemisphere was encountered in the Cape Range No. 2 oil bore in Western Australia. The drill remained in marine rocks of that age from 1112 to 4551 metres below the surface, when operations terminated, but only the upper part of the Lower Jurassic or Liassic beds had been reached. The top of the series is of Upper Jurassic age, but the very latest stage of the period is missing. The entire sequence consists of silt- and clay-stones, a most monotonous affair, but the very fact that such a colossal marine series exists on the fringe of our continent gave heart to the oil explorers—it provided the incentive to carry on with the search and they finally struck oil not far to the northeast of this trough, on Barrow Island.

The only surface occurrence of marine Jurassic rocks in the Carnarvon Basin is a tiny outcrop of sandstone with some molluscs and algae halfway up the Minilya River, near Curdamuda Well.

Map 19:

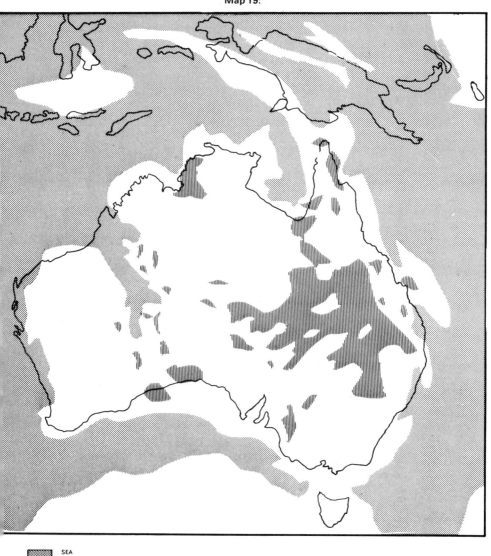

SEA

LAND

FRESHWATER LAKE

The increasing isolation of the Australian (and Antarctic) region during the Mesozoic Era becomes evident. This sketch shows Lake Walloon and the other freshwater expanses which foreshadow the large depressions into which the sea will eventually enter during the early Cretaceous. The extent of the marine embayments is that which existed at about the turn from Middle to Late Jurassic times, some 150 million years ago, except for eastern Australia where the situation shown is that at the end of the Jurassic Period.

Much more extensive are marine Jurassic beds in the Canning Basin south of the Fitzroy River, where they form a huge thin sheet of sand- and siltstones which, though patchy, reaches far southward towards the Nullarbor Plain. However, obviously marine beds with telling fossils have not yet been found south of latitude 22°.

Map (19) shows the extent to which Western Australia was invaded by the sea during the Jurassic, the widest inundation taking place during the earlier part of Upper Jurassic times. Since the sea also advanced into it from the west, the Gulf of Carpentaria may be mentioned here. From a single find of a little mollusc near the Queensland-Northern Territory border one must conclude that towards the end of the Jurassic this gulf had been established.

NON-MARINE JURASSIC FORMATIONS. River, lake, and swamp deposits of Jurassic age are most widely spread, though largely hidden beneath younger formations in eastern Australia. As indicated on Map (19), such deposits are known also in Western Australia and the Northern Territory, but they are of lesser importance there. Perhaps the most intriguing non-marine event—if we may call it that—in the west during the later Jurassic is the extrusion of basaltic lavas in the Bunbury district down south, and of peculiar potassium-rich lavas, called carbonatites, which blew up—perhaps, according to Wellman (*1973*), as late as mid-Tertiary times—through narrow necks and pipes near the northern margin of the Fitzroy Basin and in adjacent regions in the Proterozoic Kimberley Block to the north and east. These explosive volcanics came from great depths, and in recent years one has found they contain diamonds. The most famous place for such valuable volcanics is the Kimberley region in South Africa, and it now seems rather prophetic that the northernmost part of Australia was long ago also called Kimberley—at a time when nobody had the faintest notion that many decades later diamonds would be found there too. It should be noted, by the way, that the Bunbury Basalt and the Kimberley Carbonatites are the only manifestation of volcanic activity throughout the vast Australian west and northwest since at least mid-Palaeozoic times. In eastern Australia strong volcanism persisted in many areas until only a few thousand years ago.

LAKE WALLOON. The largest of the eastern Australian lake areas, Lake Walloon, has already been mentioned in connection with the Triassic period. This lake belongs to the most extensive bodies of freshwater which have ever existed in the history of the world. As we have seen, it lasted a very long time although its shape did not remain the same. What may be called the cumulative area, which at one time or another was part of the lake, is shown in Map (19). From the Brisbane district, skirting the Darling Downs, its eastern shore ran northward, following the trend of the main divide, raised in the Late Palaeozoic, for over 800 kilometres almost to the 20th parallel of south latitude. Turning west-southwest in the Charters Towers region it crossed through central and southwestern Queensland into central Australia; thence back, forming several bays and peninsulas, toward the east into central northern New South Wales; and from there northward again to link up around the New England mountains with the Ipswich-Brisbane embayment, our starting point.

Along the southeast to northeast perimeter lay the then mighty mountain ranges which in their still more worn-down form are known as the Great Dividing Range today. In the north the waters lapped on the Proterozoic mountain remnants about Mount Isa and Cloncurry, farther to the west on the flat-lying Cambrian beds of the Georgina Basin. In central Australia the water found its way into the valleys between the worn-down folds of Cambrian and Ordovician rocks in the eastern part of the Amadeus Basin. In the south the Flinders Ranges formed a prominent peninsula to the west of a great gulf in the Dubbo district and the Darling River.

Many rivers must have flowed into Lake Walloon, particularly from the high mountains in the south, and from the breaking down of these mountains came nearly 1500 metres of shales and sandstones, which were laid down during the Triassic and the Jurassic, most of it in the latter period. These rocks are only visible now around the fringes of the area, but they have been penetrated by numerous artesian bores, and their extent and thickness at depth is well established. At their base are massive sandstones covering an enormous area. In the Gulf country in the north they are known as the Inoruni Sandstone, towards Tambo as Hutton Sandstone, and still farther south as Marburg and Precipice Sandstone. Then follows a widespread sequence of sands and shales with coal seams, for instance those of the Cornwall and Alcurah districts. All these rocks are of great economic importance, not because of the coal, but because they are an essential element in the structure of the Great Artesian Basin, which will be discussed in the next chapter.

For the greater part of its existence the water of Lake Walloon was fresh, showing that it must have had an outlet to the ocean. Bodies of water which are isolated for a sufficient period invariably become salt, owing to gradual concentration by evaporation. Although the course of the large river draining the lake has not been traced, it probably lay in the north of New South Wales or in southern Queensland, and flowed through one or more of the group of lakes in that area. Very likely it was not always in the same place, varying its position with the geographical changes which took place in the region during long intervals of time. It is possible, for example that at times the lake was drained northward into the Carpentaria Basin, or even southward along the western foot of the Flinders Ranges into Spencer Gulf and the Southern Ocean.

In the latest shrinking stages of Lake Walloon at the end of mid-Jurassic times the sluggish river ceased to flow. The swampy ponded waters salted up, all the more as conditions seem to have been semi-arid or even arid. That is when the Walloon Coal Measures with their calcsilt interbeds were laid down (*Whitehouse, 1953*). These geological records again appear at variance with palaeomagnetic data because the latter place Australia into antarctic 60°–80° South latitudes during the Mesozoic Era. However, much milder polar climates did not allow of ice-caps then. In fact, absence of polar ice-caps appears to be the normal state in the earth's history. Their presence is exceptional and short-lived.

As shown on Map (19) the southeastern extension of Lake Walloon reached through the Ipswich district, where we had the great Triassic coal swamps, to the New South Wales border north of Lismore. In this southern embayment developed the coal swamps in which now lies the Walloon coal field. The Walloon Coal Measures are underlain by the early Jurassic Marburg Sandstone, which in turn lies upon the late Triassic Bundamba sandstone beds. Conspicuous amongst the sandstones, shales, and coal seams of the Walloon sequence are some thin red beds containing beautifully preserved Jurassic plants which, being white, stand out against the red background. These red beds are very useful for geological mapping, notable localities being Beaudesert, Boonah, Purga, Kalbar, and Harrisville.

To the west and northwest the Jurassic lakes and coal swamps merged with the main area of Lake Walloon in the interior. Thus we find the respective rocks covering large areas around Clifton and Toowoomba, extending northwestward to Dalby and Chinchilla and over the whole of the southern slopes and summits of the Great Dividing Range to the headwaters of the Warrego River. Along the margins of Lake Walloon in this area there formed a number of coal seams.

Lakes and coal swamps also existed in various parts of the coastal areas from north of Ipswich to the vicinity of Maryborough. The coal measures here, known as the Tiaro Coal Measures after the town of that name a few kilometres south of Maryborough, unfortunately have largely lost their value because of the intrusion of igneous rocks which in places, notably at Mount Bopple, has converted them into solid graphite.

It will be noticed that on Map (19) some of these coastal areas are shown as invaded by the sea. These inundations did not, in fact, take place until very latest Jurassic times. During the formation of the Walloon series—that is, in the earlier part of the period—the sea stood some distance off the present coast. The embayments of Grafton, Maryborough, Mackay, and Laura in the far north were either much shallower, or did not exist, in spite of the fact that subsidence in some of these then still non-marine troughs led to the accumulation of several thousand metres of sands and shales—for example, as much as 4000 metres in the Maryborough Basin—during Triassic and Jurassic times. It is this subsidence which must inevitably have led to unbearable stresses in the earth's crust, so that eventually, in the latter part of the Jurassic, the hot viscous magmas of the deep broke through. Thus we now have a belt of plutonic rocks, ranging from granitic to dioritic chemism, associated with and intruding the lower Jurassic Tiaro Coal Measures through a broad sweep of country from the Gympie district in the south to 30 to 40 kilometres west and northwest of Bundaberg.

Lakes existed also in the extreme north of Queensland, where they covered much of the northern extremity of Cape York Peninsula and down the eastern side as far south as Cooktown. Later they shrank, and seem to have divided in two, one near the Pascoe River in the north of the peninsula and a larger one near Cooktown. In both the northern and the southern area coal seams occur, but the coal is not of good quality. The age of the coal in the southern Laura Basin is early Jurassic, like the Walloon Coal Measures.

The northern freshwater basin was once famous for the accumulations of gold in the form of placer or deep lead deposits (fossil river gravels) in its sandstones, one of which, the rich Batavia Deep Lead, forcefully demonstrated the vicissitudes and vagaries of such gold deposits by suddenly ending against a steep wall of older rocks which formed the shore of a lake or the bank of a large river pool.

THE LAKE OF TALBRAGAR. Amongst the innumerable smaller lakes which remained around the shrinking margin of Lake Walloon there is one of particular interest. It was probably little more than a large pond a few hundred metres across, situated 30 kilometres northeast of Gulgong and some 290 kilometres northwest of Sydney. It lay in a hollow of sandstones belonging to the early Triassic Narrabeen Group, and was swarming with small fish. Apparently these were killed off suddenly, either by a drought or more likely by a shower of hot volcanic ash. A layer of hard brown shale, only five to ten centimetres thick, is packed with their remains, and a small slab frequently shows a dozen or more specimens, together with ferns and other plants. The fossils stand out perfectly in white against the brown background, and are so exquisitely preserved that the shape of each vertebra and small bone can be clearly seen with a lens. This fossil bed is world famous because of the light it has thrown on the anatomy of Jurassic fish.

THE VICTORIAN LAKES. In Victoria there are three main outcrop areas of Jurassic rocks of freshwater origin. The lakes, in which they were deposited, probably formed at times one big body of water. The most westerly of these outcrops is in the Glenelg River district on the South Australian border, the central one forms the Otway Ranges and the Barrabool Hills west of Geelong, and in the east there is the large area of hill country to the south of the Pacific Highway—that is, the South Gippsland hills east of Western Port Bay—and a small area a little to the north of it between Yallourn and Toongabbie. The maximum thickness of these eastern Jurassic and Early Cretaceous sequences reaches several thousand metres in places; in fact, drilling in search for oil has nowhere yet reached their bottom, although up to 1800 metres have been cut along Ninety Mile Beach. The rocks consist of massive sandstones—containing tuffaceous material—and shales, and fossil plants as well as freshwater mussels have been found in a number of places. Near Wonthaggi a few coal seams have been worked, the only black coal deposits in Victoria. The coal is of fair quality but the seams are thin and erratically discontinuous because of complex minor faulting.

THE TASMANIAN VOLCANICS. The Jurassic Period in Tasmania is characterized by mighty extrusions of lava. No definite Jurassic sediments are known although some palaeontologists suspect that the youngest parts of the sequences regarded as Triassic may in fact be of early Jurassic age (*Dettman, 1961*).

The volcanic events had a pronounced effect on the existing Tasmanian topography. From deep within the earth liquefied rock forced its way upwards and sideways beneath Triassic strata, lifting them up so that they often floated like islands on the molten mass. These intrusives cooled and solidified into the black rock dolerite, which is like basalt in appearance but more coarsely crystalline in texture. Sheets of dolerite now cover much of the eastern half of the island, and the highest mountains in this area are composed of them. Ben Lomond in the east and Mount Wellington in the south are capped by dolerite, and in the centre the picturesque Cradle Mountain and Barn Bluff are composed of the same material. On the southeastern coast the dolerite sheets come down to sea-level and the rock forms the high cliffs on the rugged coast at Cape Raoul and

Bruny Island. A feature of the dolerites is their columnar structure, the rock when cooling having contracted to form gigantic vertical and hexagonal columns closely packed together.

LIFE OF THE JURASSIC PERIOD

The Rise of the True Ammonites and the First Bony Fish

LAND PLANTS. As the Triassic Period passed into the Jurassic, the main change in the flora was the rise of the cycads. These were palm-like plants, but of much lower organization. A living example familiar to many is the burrawang or *Macrozamia* common in Australian forests. A common cycad was *Podozamites*, which had small leaves pointed at each end and with a parallel venation; it is extremely abundant in the fish beds at Talbragar. The genus *Taeniopteris* is generally placed in this group, although its classification is not definitely settled. This plant had narrow lanceolate leaves with strong midribs, from which the smaller veins branched at right angles, dividing once or twice before they reached the margins. One species, *Taeniopteris spatulata*, is very important because it is confined to the Jurassic Period, and its presence as a fossil is therefore considered decisive when the age of the enclosing rocks is in doubt. Another doubtful cycad is *Otozamites*, which in Australia is also confined to Jurassic rocks. The leaves of this form are narrow, and alternate on either side of the stem; there is no midrib, and numerous fine veins spread out and branch once or twice before reaching the margin. So far *Otozamites* has not been found in New South Wales, but it has been discovered in a number of localities in Western Australia, the Northern Territory, and Queensland.

The seed-fern *Thinnfeldia* barely survived into the very earliest Jurassic; it is found in Victoria in a few places and in the Evergreen Shale, which overlies the Precipice Sandstone, in the Great Artesian Basin. By the time the Walloon Coal Measures were deposited *Thinnfeldia* had disappeared. The most prominent fern form was *Cladophlebis*, a large tree-fern, which was so common that it has been termed the "Jurassic weed". In the forests the conifers or pines were well established, and one of them was a true *Araucaria* closely related to the bunya-bunya or "monkey puzzle" still growing on the Queensland coast.

In recent years a great deal has been learnt about what is known as the microflora of rocks of various periods. By this is meant the pollen and spores of numerous plants the true nature and macrostructure of which may, or may not, be known—which does not really matter because these microscopic things are quite distinct in themselves, change their form through the ages, and are therefore extremely useful species and genera of a special kind for the palaeontologist-specialist, in this case the palynologist. Delicate special techniques have been developed to extract these minute plant remains from various types of rocks.

FISH AND CRUSTACEANS. There are several localities in Australia where Jurassic fish have been found, but only one large find has been made—that at Talbragar already mentioned. This is remarkable in that it comprises innumerable specimens, magnificently preserved in a deposit of small extent. The rock is fine and compact and splits readily into very thin slates, and in one or two layers the fish remains are so abundant that a dozen or more may be visible on the one slab. Although somewhat deformed by the weight of the overlying rocks, the shape of each small bone is visible under a lens. With the fish are uncountable specimens of finely preserved plants. The fish are of several species, and all of them are small. The most abundant is rather like a sardine in appearance and is called *Leptolepis*. It is classed with the Teleostei or true bony fish, but in many ways it is a connecting link with the cartilaginous ganoids, which began to decline in the Jurassic. For instance, the tail is neither truly heterocercal nor truly homocercal; the backbone is bent upwards near the extremity, but does not extend right into the upper lobe of the tail. From specimens at Talbragar much has been learnt of the anatomy of the early bony fish. Moreover, the change from the side-by-side ganoid to the imbricated, rounded, cycloid scales was already well advanced in the Talbragar fauna which, apart from the true teleosteid *Leptolepis*, includes a coelacanthid lobe-fin, a palaeoniscid *Coccolepis*, and some holosteid species, such as *Uabryichthys*, *Aetheolepis*, *Aphnelepis*, *Archaeomaene*, and *Madariscus*. In Victoria *Leptolepis* has also been found near Carapook, and teeth and scales of the lung-fish genus *Ceratodus* elsewhere.

While freshwater crustaceans are unknown from Australia's Jurassic, marine species of the order Ostracoda are common in the Western Australian Middle Jurassic, especially in the Newmarracarra Limestone of the Geraldton district.

AMPHIBIANS AND REPTILES. The reptiles had overrun the earth in the Triassic—however scarce they may have been in Australia—but their development reached its zenith in the Jurassic and the Cretaceous Periods. Apart from innumerable comparatively small forms such as crocodiles, the predominant group were the dinosaurs, remains of which have been found in every part of the world. They included the largest terrestrial animals which have ever lived: the gigantic *Brontosaurus* or "thunder lizard", for instance, was 22 metres long and weighed upwards of 40 tonnes. However, not all were large, the smallest being about the size of a dog.

The dinosaurs are divided into a number of groups, and include harmless plant-eaters as well as terrible carnivorous creatures over nine metres in length. Some were amphibious, others were purely terrestrial. In some the skin was comparatively smooth, while others were armoured with rows of huge vertical plates upon their backs. Their thick legs supported their bodies above the ground, but in some—such as *Iguanodon*—the forelegs were so reduced that their posture was not unlike that of the kangaroo. Volumes popular and scientific are available on the dinosaurs, but as the present account is essentially of Australia, space does not permit more than this short generalization.

The dinosaurs did reach Australia in Jurassic times but, so far, discoveries of their remains have been scarce. One interesting find was made in the Walloon Coal Measures at Rosewood, Queensland, where at the top of the coal seam in the Lanefield Extended Colliery were impressed the three-toed footprints of one of these creatures. The middle toe was 45 centimetres long, and measurements showed that it had a stride of two metres. Similar footprints are known from Donnybrook, Western Australia. The largest footprints were found in the Walloon Coal Measures at Balgowan Colliery, Oakey, on the Darling Downs; with 70 centimetres toe length and a stride of almost three metres the animal was a three-toed, bipedal, beast of the *Iguanodon* type.

Most important of all was the discovery at Durham Downs, 72 kilometres northeast of Roma in Queensland, of a mass of colossal bones belonging to one individual. Less than half the complete skeleton was found, but the bones recovered weighed more than a tonne. From these it has been possible to make an approximate reconstruction of the creature, which has been called *Rhaetosaurus* (Fig. 5).

Rhaetosaurus was typical of many of the herbivorous dinosaurs. It was about twelve metres in length, of which one-third was a thick massive body. This was about 3.5 metres high and was supported by two pairs of equally massive legs, nearly equal in size, so that it had a true quadrupedal habit. The neck was long and thick, tapering to a ridiculously small head with a brain little larger than a man's fist. This brain was so inadequate that a secondary nerve centre was situated near the pelvis to control movements of the rear parts. The long tail, which dragged upon the ground, was enormously thick near the body, but tapered abruptly and was thin and whip-like near its extremity.

Rhaetosaurus belonged to the sauropods, a group mainly aquatic in habit, and very clumsy on the land, to which they came only at night to feed on the vegetation. The enemies of the sauropods were the terrible theropods, the large carnivorous dinosaurs whose presence in Australia remains conjectural, although the claw of a carnivorous dinosaur of smaller size has been found in early Jurassic rocks at Cape Paterson in Victoria (Plate 18).

Fig. 5
Reconstruction of the Australian Jurassic dinosaur, *Rhaetosaurus*, which was 12 metres long and 4.5 metres high.

Durham Downs is also the locality of a find of a small quadruped dinosaur's footprints.

Other curious reptiles which have not so far been found in Australia are the pterodactyls or pterosaurs, which had true powers of flight and are the forerunners of the birds. The fourth finger of their forelegs was lengthened to support a wing-like membrane not unlike that of a bat, and the wing-spread varied from a ten centimetres to seven metres. Two other groups of reptiles are the ichthyosaurs, or fish lizards, and the long-necked pliosaurs. They are not known from the Australian Jurassic, but their remains occur in Cretaceous beds, and they will be mentioned in the next chapter.

It is necessary now to mention two developments outside Australia which were destined to have far-reaching effects. These were the steady growth of mammals and the first appearance of true birds.

Mammals, of course, suckle their young, but one of the main differences between them and the reptiles is that they are warm-blooded; that is, they are able to maintain a constant body temperature, and are thus better able to resist climatic changes than are reptiles. The first mammals had many reptilian characteristics; they were small, rat-like creatures and many lived in trees. The time had not come when they were to challenge the reptiles for possession of the earth.

The first true birds were also reptilian in many characteristics, but they too were warm-blooded, and in this differed from the still quite reptilian pterosaurs—apart, of course, also from having feathers instead of scales, an evolution which gave them greatly improved powers of flight. The earliest known birds were found in Upper Jurassic rocks in Bavaria. Two specimens are so beautifully preserved in an extremely fine-grained lithographic limestone that even the details of the feathers may be seen. This bird, *Archaeopterix*, is famous as one of the proverbial "missing links" in the chain of evolution. Many details of its skeleton were still reptilian; the bones were not hollow, the head, though bird-like, was armed with reptilian teeth, and the backbone was prolonged into a tail from which the feathers grew out on either side.

Mammals have not yet been found in Australia's Mesozoic but bird feathers occur in South Gippsland (p. 210). The story of the vertebrates on this continent can be reconstructed from only the few individual specimens that we know of here. Besides those already mentioned, the only other find to date is that of skull fragments of a fairly large labyrinthodont amphibian, which has been named *Austropelor*, from the lower part of the Marburg Sandstone at Lowood in Queensland. It shows that this group of larger amphibians survived into the very earliest Jurassic (the skull was about 60 centimetres long). Other discoveries will yet be made. Unfortunately, the great masses of Jurassic lake and swamp rocks which underlie the artesian basin are nowhere intersected by valleys and gorges to make them accessible, but no doubt the bones and other remains of many creatures must there lie buried, and from the coal mines and outcrops nearer the coast many may yet come to light.

LIFE OF THE SEA. Since the only marine Jurassic rocks in Australia are in Western Australia, all our knowledge of the local sea life of the period comes from there. The most easily accessible collecting locality for Jurassic fossils is in and around Bringo railway cutting, 30 kilometres east of Geraldton. Shales, sandstones, and limestones in that area contain a rich fauna of univalve and bivalve molluscs, tiny ostracod crustaceans, and unicellular foraminifers which have found the interest of palaeontologists since the middle of last century. Of special importance are the many true ammonites in the limestone horizon known as Newmarracarra Limestone, because they allow an exact determination of the age of these beds with reference to stages and zones in other parts of the world, especially southern Asia and the circum-Pacific countries. The most significant result of the study of the mid-Jurassic ammonites of Western Australia by the late Dr. W. J. Arkell (*Arkell and Playford, 1954*) is that peculiar genera such as *Pseudotoites*, once believed to be confined to Western Australia, occur in the Moluccas, British Columbia, Alaska, and Argentina; whereas *Zemistephanus*, previously known only from North America, occurs also in Western Australia. This indicates that migration ways through the Pacific to the Moluccas and Australia's west coast were open at that time, as indicated on our Map (19), but that ways to Europe were more devious.

The ammonite fauna of Geraldton represents only a very minor interval of the earliest epoch of the middle Jurassic, the Bajocian, but even so some 25 species of seven genera are now known and described from there. The ammonites during the Jurassic Period attained a peak (there is another one in mid-Cretaceous time) and occurred in extraordinary variety and abundance. The evolution of this great group of molluscs was of phenomenal rapidity; new species and genera, some of bizarre form, originated, spread far and wide, and as rapidly became extinct, to be replaced by others. The result is that within the one formation of rocks one can distinguish many zones, in each of which occur species of ammonites found in no other zone, each thus representing a specific interval of Jurassic time (of which we usually do not know the duration in terms of years). The evolution of the ammonites has enabled the period to be subdivided by its fossils as has perhaps no other in the geological record. It was the study of the Jurassic rocks of England that led surveyor William Smith, the father of historical geology or stratigraphy in its modern meaning, to discover that geological formations may be recognized and correlated by their fossil contents, and the ammonites were the main basis of his studies.

In the Jurassic of the northern Carnarvon Basin and the Canning Basin ammonites are of great importance because they show that the sea remained in these areas much longer than in the Perth Basin. Genera such as the fat *Macrocephalites*, or *Perisphinctes*, and *Kossmatia*, indicate the presence of formations the age of which is late Middle, early Upper, and middle to late Upper Jurassic. Most of these ammonites are small to moderately sized fossils, that is, no more than fifteen centimetres in diameter, but in the early Upper Jurassic Jarlemai Siltstone of the Canning Basin a species of *Perisphinctes* was found of which the diameter is just over 60 centimetres. The largest ammonites known reach a diameter of about two metres, but appeared only later in the Mesozoic.

Another interesting group of cephalopods, strongly represented in the Western Australian formations are the belemnites. These creatures were something like the modern squid in appearance, and impressions of the soft parts of their bodies have been found, especially in the fine lithographic limestones of Bavaria, whence also came the fossil bird mentioned earlier. They had no external shell like the ammonites, but an internal bone like the squid and the cuttlefish. Unlike the cuttlebone, however, this was slender, cone- or finger-shaped (in Germany they are popularly known as "devil's fingers"), hollow, and divided like the external type of cephalopod shell by numerous partitions, through which passed a tube or siphuncle.

Most of the other Western Australian fossils are bivalve shells, some of which belong to genera still living, such as *Ostrea* or oysters, *Pecten* or scallops, and *Trigonia*, which is probably the most interesting. It had a more or less triangular shell outline—hence the name—with a very intricate interlocking hinge. It is difficult to separate the valves without breaking this hinge (Plate 18). *Trigonia*, which evolved from *Schizodus* in the Permian through *Myophoria* in the Triassic, was once considered confined to Jurassic and Cretaceous rocks. But then it was found alive on the Australian coasts—a discovery which, like that of the coelacanth fish *Latimeria*, caused a sensation in scientific circles. The handsome pearly shell will be familiar, as it is used in brooches and necklaces, particularly in Tasmania.

With some other bivalves *Trigonia* has the power of progressing through the water by sharply opening and closing its valves. The first live specimens dredged in Sydney Harbour by Samuel Stutchbury were placed on the seat of a rowing boat, and to his chagrin actually leapt overboard. The Jurassic species of *Trigonia* (the modern forms are actually known as *Neotrigonia*) were larger, thicker, and had more ornate sculpture than those living today. In fact, the coarse sculpture became more and more attenuated during the Cretaceous, and the few *Neotrigonia* species known from Tertiary rocks are already much like the present-day form.

In northwestern Australia, and in New Zealand, the East Indies, and the Himalayas at about the same time—the turn from Middle to Upper Jurassic—appears also the bivalve genus *Inoceramus*, which was to become the most characteristic and common bivalve of Cretaceous times. The lower part of the Upper Jurassic beds in the Canning Desert have also yielded starfish and a number of well-preserved *Ophiura* or brittle-stars (*Brunnschweiler, 1954*) of two or three species with a small central disc and fine slender arms up to five centimetres long. Whole specimens of these delicate creatures, like those from Western Australia, are rare finds; one finds minute bits and pieces of the compound discs and arms individually as grains in the rocks, and it is not possible to reconstruct the animal.

A very important fossil of microscopic size, less than 0.05 millimetre in diameter, is the infusorian unicellular genus *Calpionella*. It is confined to the very latest part of the Upper Jurassic and occurs in a richly fossiliferous small outcrop near Langey Crossing on the lower Fitzroy River, 50 kilometres from Derby, together with the cephalopod *Kossmatia*, and belemnites of the group of *Belemnopsis* and other molluscs. Among the molluscs is the little ear-shaped bivalve *Buchia*, which is also found further inland in the Canning Basin, in bores at depth along the coast near Broome and, last but not least, in northwestern Queensland near the Northern Territory border, showing that the sea had advanced through the Carpentaria region to this point at the end of the Jurassic.

13 The GREAT FLOOD in the CRETACEOUS PERIOD

THE WHITE CLIFFS OF DOVER are composed of chalk, a kind of rock deposited over large areas of the Old World in the period which followed the Jurassic. The Latin word for chalk is "creta", hence this period is known as the Cretaceous. It is characterized in many regions of the world by unusually strong and far-reaching transgressions of the sea. The heart of the Gobi Desert in Asia was sea in those times as were many other large areas on other continents which are now dry land. In Australia also it was a time when the sea managed to inundate very large regions of the continent, so as to break it up into several groups of islands.

The later part of the Jurassic Period saw the disappearance of most of the lakes which had been such a feature for long ages. Of Lake Walloon only small lakes and swamps remained; it must have shrunk to a fraction of its former size. At this stage the interior of the continent must have looked much as it is now, a region without prominent features, much of it endless, monotonous plain, and parts of it perhaps desert. Even the high mountains which had existed, particularly around the southern margin of the basin, must have been markedly reduced in altitude, for their erosion had provided much of the mass of sediment which filled Lake Walloon. But this desolate and monotonous picture did not last very long.

After an elevation of the land, which not only pushed the Jurassic sea off the continent just before the end of that period, to leave only small embayments in the Kimberleys and in northwestern Queensland, but also probably drained the Lake Walloon area dry, events reversed themselves.

The first effect of this reversal in the Australian interior was the reappearance of extensive lakes in the old basin of Lake Walloon and other, smaller, inland basins along the coast of Queensland. At the same time the sea began to inundate various areas along the Queensland coast, for instance the Stanwell district near

Map 20:

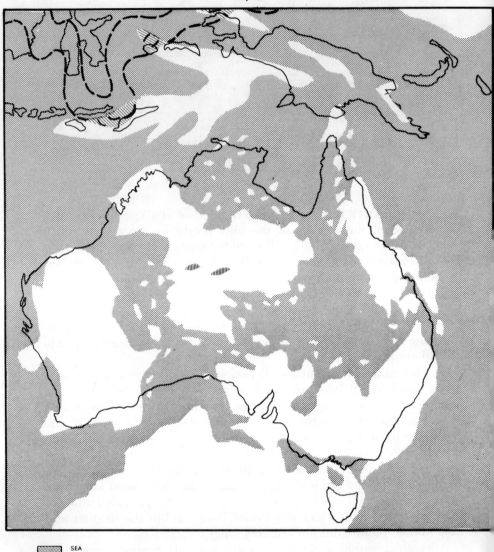

The great Australian deluge of early Cretaceous times, about 130 to 120 million years before present. Note the disastrous narrowing of the living-space for whatever flora and fauna was in existence at that time, both in Australia and Papua New Guinea.

Rockhampton. In Western Australia it moved into the Northwest Cape region and Dampier Peninsula between Broome and Derby, as well as some tens of kilometres across the present coastline between Broome and Port Hedland. The deepest penetration into the continent's interior, however, took place from the Timor Sea into the Northern Territory in a southeasterly direction right down to the Queensland border—the first marine transgression on this large tract of land since the Cambrian Period ended some 500 million years ago.

These first incursions of the sea in Cretaceous times took place right at the beginning of the period, about 130 million years ago, and they did not last more than perhaps a few million years. The marine sediments left behind by this transgression are rather thin, and in many areas they have since been largely removed by erosion.

The main marine invasion came, after an interval of elevation of the land, somewhat later at the beginning of Middle Cretaceous times. It flooded the continent to such an extent from the south, west, and north, that for a time there just wasn't a visible Australian continent; only about four major islands—or five, if one also counts the Timor-Moluccas island to the north—and probably innumerable smaller ones, of which we do not know the exact location. True, the immersion of Australia was a shallow one and did not last long, relatively speaking, but while it lasted it changed the geography of the region radically, and literally wiped Australia from the face of the earth. In fact, until recently geologists had not realized just how extensive this early mid-Cretaceous, or Aptian, inundation was. In Map (20) the island groups in the Australian region during that time are shown diagrammatically; in other words, this is not an accurate geographical map. It simply shows the fundamental pattern of land and sea during this great flood, which took place around 110 million years ago and may have lasted for something like 10 or 20 million years.

During the Aptian Epoch the sea thus entered the Perth and the Carnarvon Basins, and formed a broad strait across the interior of Western Australia from the Timor Sea to the Southern Ocean. The southern part of this strait, beneath the Nullarbor Plain, is known as the Eucla Basin. From the Eucla Basin there was a shallow strait eastwards to the Great Artesian Basin, and this in turn was connected with the open sea eastwards by a narrow channel north of Brisbane, and northwards via the Carpentaria Basin and the shallow and island-studded Northern Territory sea.

The southeastern corner of Australia remained land through the first half of the Cretaceous; the larger part of it, in fact, through the whole of the period, of which the duration was 71 million years. As over the Yilgarn-Pilbara Shield in the west, the Kimberley-Central

Map 21:

- SEA
- LAND
- FRESHWATER LAKE
- ZONE OF CONTEMPORARY OR JUST COMPLETED MOUNTAIN-BUILDING

The re-emergence of the Australian continent towards the end of the Middle Cretaceous (Albian Epoch) about 110–100 million years ago.

THE GREAT FLOOD IN THE CRETACEOUS PERIOD 233

Australian region, and the eastern range country of Queensland, there are few records of its passing, or none at all. This applies also to Tasmania which was high and dry land in the Cretaceous. The lakes in southern Victoria did, however, persist from the Jurassic into early Cretaceous times and in the sandstones and shales of that age in South Gippsland (Strzelecki and Otway Group) there are indications that the sea was very close by at times, for there are microscopic marine creatures, called hystrichospherids, in some of the beds. In western Victoria there are similar indications and one wonders whether the marine influences came from the north via the Murray Gulf of the Artesian Basin, or from the south. In any event, it seems the land connection between Tasmania and the mainland became increasingly tenuous during the Cretaceous. It is also thought that the great, theoretical Gondwanaland either shrank greatly or was broken up in Cretaceous times, although precarious land bridges may from time to time have existed from Asia to the southern continents, including Australia.

The great flood of Aptian times receded considerably towards the end of that epoch, so that in the next, the Albian, only the great inland sea of eastern Australia (the Great Artesian Basin) and some marginal lands in western Victoria, Western Australia, and the Northern Territory remained within the marine realm. As shown on Map (21) the Victorian lakes shrank very much too and had already disappeared before the middle part of the Albian Epoch. On the other hand, there seems to have been a narrow marine strait connecting the Southern Ocean northwards across the Murray Basin with the sea of the Great Artesian Basin, while the previous connection from the Eucla Basin across the northern part of South Australia was closed for ever.

Towards the end of the Albian Epoch, the later part of the Middle Cretaceous, the Murray-Darling as well as the Carpentaria connection between the inland sea and the open oceans were closed too, and the region of the Great Artesian Basin once more became a land of lakes and swamps—the area of Lake Winton, as it has been called, Map (22). During the later Cretaceous Epochs the sea withdrew in the north from the Carpentaria embayment as well as from the Darwin district and nearby offshore islands. In southern coastal areas, e.g. in western Victoria, marine influences managed to hold on, with interruptions, a little longer, but half way through the Upper Cretaceous (towards the end of the Turonian Epoch) they disappeared too. In eastern Victoria the sea remained well out in the offshore region of the Gippsland Basin, and the Bass Basin between Victoria and Tasmania was all land. Only in the marginal depressions along the far west coast of the continent the inundation continued,

Map 22:

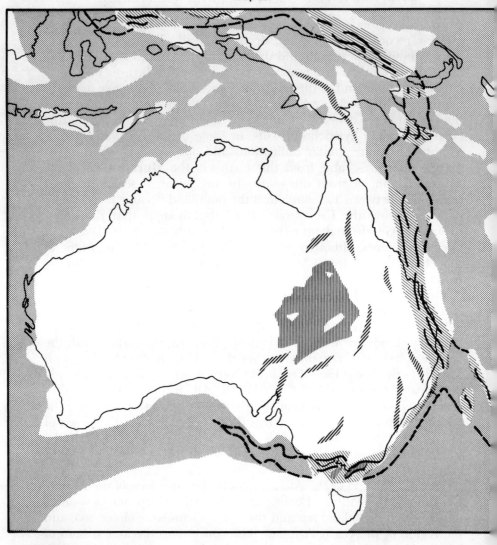

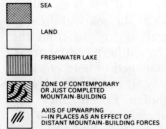

SEA

LAND

FRESHWATER LAKE

ZONE OF CONTEMPORARY
OR JUST COMPLETED
MOUNTAIN-BUILDING

AXIS OF UPWARPING
—IN PLACES AS AN EFFECT OF
DISTANT MOUNTAIN-BUILDING FORCES

The Australian region at the end of the Maryburian Orogeny in the Turonian and Santonian Epochs of the Late Cretaceous (ca. 85–75 million years before present). Of the many inland lakes only the largest, Lake Winton, is shown. Note also that the Antarctic Continent is still close to Australia.

albeit with occasional interruptions. It is interesting to note that a regression of the sea during the middle and the early upper part of the Cretaceous Period affected also the islands of Timor and the Moluccas, Celebes, and others to the north of Australia. It is believed that these widespread movements of the earth's crust at this time indicate the beginning of the circum-global mountain-building event known as the Alpine Orogeny, which reached its greatest intensity in the last of the great geological periods, the Tertiary. But let us now see briefly what the rocks and records of the passing of the long Cretaceous Period are in the various Australian regions.

EASTERN QUEENSLAND. The mountain ranges of eastern Queensland, which had been formed in the Late Palaeozoic, were a formidable obstacle for any transgression of the sea from the Pacific, and it is not surprising that where the sea eventually conquered them the breach was narrow and of short duration. In this context it is important to realize that the first Cretaceous transgression along the Queensland coast, like that of the Northern Territory and the northwest of the continent, was so early in the Period as to appear as a hangover from the Jurassic rather than the beginning of the great Cretaceous story, and it certainly shows that through much of the Mesozoic Era this coastline was never far from where it is now.

In the early part of the first epoch of the Cretaceous, the Neocomian, there were marine embayments near Cooktown (Laura Basin), in the Stanwell area near Rockhampton, and in the Maryborough district.

Subsidence in the Maryborough Basin was particularly marked. At first, and probably still in latest Jurassic times, thick series of tuffs and lavas, together with carbonaceous shales and sandstones, were laid down in a freshwater environment, and this went on into the Neocomian Epoch until 1000 to 2000 metres of this material had accumulated. This series is known as the Graham Creek Formation. As subsidence continued, the sea broke into the depression and up to 1800 metres of silts and sandstones, often with an admixture of tuffaceous material, were deposited before the sea withdrew again towards the end of the Aptian Epoch. This marine sequence is known as the Maryborough Formation, and it contains abundant fossils, of which few have yet been studied. The retreat of the sea was followed by another freshwater environment, in which a further 1550 metres of sediments accumulated, mostly silts and sandstones with iron-carbonate nodules and bands, but also some shale and several coal seams in the middle part of the series. These are the Burrum Coal Measures which are mined near Howard and Torbanlea and in recent years still produced about five per cent of Queensland's output of coal, although they have been mined for over a hundred years. Good specimens of fossil plants as well as freshwater shells such as *Corbicula* and *Rocellaria*, the latter a boring mussel, have been found at many levels.

Another major embayment, like the Maryborough Basin connected at times probably with the Great Artesian Basin sea during the Aptian, is the Laura Basin near Cooktown. Here too the sea entered early in the Neocomian Epoch, as has recently been proved by the presence of ammonites of that age, and stayed on into the Aptian, but the thickness of these beds is considerably less than that in the Maryborough Basin. Coal seams underlying the early Cretaceous beds here are of Jurassic age and are of poor quality.

Between these two large embayments there was a small gulf near Rockhampton, the Stanwell coal basin. Here too sandstones and coal seams of Jurassic age are followed by marine, partly calcareous sandstones with a molluscan fauna of early Neocomian age. During the great inundation of the Aptian Epoch this area was probably land again. Map (20) shows the situation before Aptian times. A short distance farther north are the Styx River Coal Measures, formed towards the end of the Middle Cretaceous, which contain several seams of good bituminous coal, as well as evidence of a brief marine incursion in Albian times.

THE GREAT ARTESIAN BASIN. Although the comparatively small Cretaceous embayments just discussed are by no means unimportant, it is the vast expanse of the central seaways, covering just about half of the whole of Australia, that really captures the imagination. Their main communication with the open ocean was through the Gulf of Carpentaria, and they also overlapped much of Cape York Peninsula across to the Laura Basin. Of the brief existence of oceanic connections eastwards to the Maryborough and southwestward to the Eucla Basin we have already heard. Both existed only during the early stages of the great transgression, as one can see that in the critical areas in South Australia the early marine beds reach much farther southwestward than the later ones. In fact, in many areas around the Artesian Basin it can be seen that the level of that great inland sea, although fluctuating to some extent, was gradually lowered after the first major transgression. This allowed rocks already formed to be exposed on the surface and suffer erosion before the waters of a fluctuation cycle returned. For this reason the sequences of sediments, especially in marginal areas of the basin, are nowhere really continuous; there are more or less noticeable "breaks" in them which may be recognized by the absence of certain species of fossils.

One of the major breaks along the eastern basin perimeter is that between the Aptian Roma Formation and the Tambo Formation of Albian age. However, the more we move away from the margin of the basin towards its centre the less distinct is this break. Evidently, in these central parts the marine regime was uninterrupted from Aptian into Albian time, and the change from one epoch into the other is simply indicated by the changing fossil content of the rocks. Yet the picture is not quite so simple in that the effects of the fluctuations in the level of the sea are noticeable to some extent also in the central parts of the basin, that is, within the continuous marine sequence itself. Many marine organisms are very sensitive to changes in the depth of waters in which they live, the degree of salinity of these waters, their mud content, or the amount of light penetrating to the depth in which they prefer to live, and other conditions affecting the marine environment. Thus, sea-level fluctuations, which are indicated clearly by breaks or disconformities in marginal areas of the basin, are recorded in more central regions as a special type of formation, containing a peculiar marine fauna. For example, in the greensand-limestone bed at the junction between Aptian and Albian in South Australian areas of the Great Artesian Basin the only fossils known are lamp-shells and scales of fish, whereas the beds below and above are teeming with the remains of many species of molluscs and other animals.

Another point to be remembered is that the floor of this huge basin was not as simple as the surface of a saucer; there were prominent ridges which subdivided it into several individual depressions. In the north there is the Carpentaria depression, and this is separated by the Euroka Ridge, which crosses from Cloncurry to the Georgetown area, from the central Eromanga (or Coopers Creek) depression which reaches into South Australia. Into this Eromanga depression, or sub-basin, advance several spurs; one, for instance, points south through the Boulia area, another juts out from the east and is known as the Longreach spur, and still others point northeastward and are sub-surface ridges extending from the old ranges in South Australia. The Eromanga sub-basin is separated by the Nebine Ridge from the Surat depression in the southeast which itself is underlain by the hidden southern extension of the Permian Bowen Basin. Little or nothing of all these subdivisions can be seen on the surface, we know of them only from drilling for water and oil.

The whole of the Cretaceous sequence in the Great Artesian Basin has long been known as the Rolling Downs Group, but with our ever-increasing knowledge about it this generalizing name is no longer deemed appropriate. From the late Jurassic into the early Cretaceous the entire region was still a freshwater area in which large sheets of sandstones accumulated. These transitional beds are known as the Blythesdale Group and are in some areas as much as 450 metres thick. Then follow the mostly silty and shaley beds of the marine Roma Formation, from 300 to 500 metres thick, and these in turn are overlain by another 300–350 metres of siltstones and shales, known as the Tambo Formation. This latter formation contains a markedly different set of marine fossils from that of the Roma beds. The Roma Formation is early, the Tambo later Middle Cretaceous in age, representing the Aptian and the Albian Epochs respectively. In all, these two marine formations cover an immense area, especially the Aptian, which also extends over the northern part of the Northern

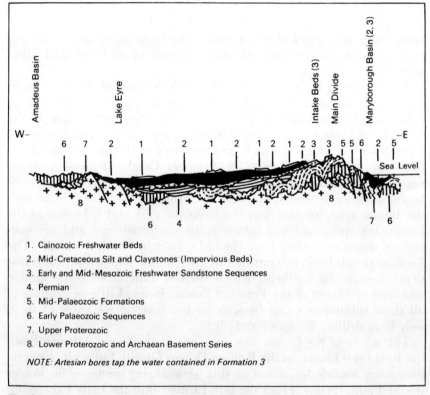

Fig. 6
Diagrammatic section across the Great Artesian Basin from the southern part of the Northern Territory to the Pacific shore, showing its general structure. (Altitudes greatly exaggerated!)

Territory, into areas south of Alice Springs, across into the Nullarbor Plain and from there northward into the Canning Basin.

The sea was always shallow, but in the central region of the Artesian Basin there were probably no islands, and even those around the margins were not of large extent. Subsidence was rather slow, keeping pace, approximately, with the deposition of rocks. Although there was an abundance of marine life, it was of rather limited variety because of the peculiar shallow-water conditions. These conditions kept the open-ocean fauna out, and most creatures living in these restricted waters were mud-dwellers and crawlers, some species being local forms not found anywhere else in the world.

The complete absence of corals, together with what appeared to be ice-borne erratics in the formation in the opal fields at White Cliffs in New South Wales and the Stuart Range in South Australia's north, was long believed to indicate that the Cretaceous waters were cold, and that glaciers existed in the southeast of the region. However, there is little doubt that the so-called erratics are derived by wave action and submarine mudflows from glacial beds of Permian age which in places formed the coastline during the mid-Cretaceous inundation. The absence of corals simply means that the waters were on the whole outside the tropics; which can, in turn, be taken to indicate that Australia was just above where it is now with reference to the equator. Even today the greater part of the Artesian Basin area lies outside the tropics, that is, south of the Tropic of Capricorn. It is interesting to note that the odd beds which contain "erratics" enclose also numerous fragments of fossil pine wood, again indicating nearness of land and rivers.

An important factor in the withdrawal of the mid-Cretaceous sea from almost the entire continent was the Maryburian Orogeny, the last of the true mountain-building events in Australia. It affected directly only the eastern perimeter, but indirectly it brought about the re-emergence of the continent as a whole. The name of the orogeny is taken from the Maryborough district north of Brisbane, where the mid-Cretaceous formations are strongly folded. Folding also took place at about that time—the Upper Cretaceous lies unconformably on the Lower—along an east-west trend in southern Victoria, and it is possible that these movements are connected with the main orogeny in the northeast, as shown on Map (22). Still another effect are a number of intrusions of granite along the eastern margin of the Bowen Basin and probably farther north in Queensland, as well as to the south along the coast of central New South Wales and even in northern Tasmania. None of these intrusions are of great extent, but they show one more aspect of the extensive repercussions created by the Maryburian Orogeny throughout the Australian region—including New Guinea and New Zealand.

The Cretaceous formations as a whole, as well as those of the Jurassic and Triassic which precede them, are of immense importance to the economy of eastern Australia, for together they comprise what is known as the Great Artesian Basin of hydrological fame.

This enormous depression began to form in Triassic times, and it still remains as a depressed area although it is now dry land, even desert in parts. Although its floor is by no means evenly curved it may be likened to a gigantic saucer, the upturned edges of which are composed of the older rocks, while the centre is covered by those which were last deposited. The eastern edge of the saucer is the highest and generally conforms to the main divide in southern Queensland and northern New South Wales. Along the western slope of the divide, and sometimes on its actual summit, are exposed the Trias-Jura sandstones which are the intake beds of the basin. These sandstones are highly porous and absorb a portion of the rain which falls upon them. This moisture seeps by gravity downwards into the centre of the basin, where it is imprisoned by the overlying impervious shales laid down in the Cretaceous Period. It thus forms a vast subterranean reservoir, fed continually by the rain falling on high lands many hundreds of kilometres away. It is also under pressure; there is a considerable hydrostatic head, caused by the greater elevation of the intake beds, by the weight of the rainwater that is constantly being absorbed, and by the weight of the rock above. Thus, when tapped by bores in the artesian country—and there are thousands of them—the water rises in the bores and in many places even spouts above the surface. It is generally hot, and often bitter from salts it has dissolved in its passage through the rocks, but it has brought life and fertility to many regions otherwise arid and almost desert-like. In Fig. 6 the principle of this artesian water storage is illustrated schematically.

OPAL. Connected closely with the Cretaceous rocks of the Great Artesian Basin are the famous opal fields of Australia. Opal, which chemically is silica containing from 4 to 9 per cent water in extremely minute capillary spaces, is a common enough mineral, but only rarely and in a few places does it display the colours that make it one of the most beautiful gemstones in the world. Opals have been prized by mankind since the days of Pliny, but the finest stones, which then came chiefly from Hungary, were of poor quality compared with those now found in Australia. For many years the chief field was White Cliffs in western New South Wales. The White Cliffs opals are the "white" opals, beautiful indeed, but now less popular than the darker-coloured "black" opals from Lightning Ridge and Andamooka.

At White Cliffs the opal is found as thin veins, pipes, and patches in Cretaceous sand and siltstones, but is also found in many curious shapes, having "petrified" or, more correctly speaking, replaced the original substance of fossil shells, wood, and even the bones of extinct reptiles. The opals of Coober Pedy in the north of South Australia are also of the "white" variety.

Lightning Ridge is in northern New South Wales, Andamooka not far from the northern end of Lake Torrens in South Australia. The other area where "black" opals are found in places is an extensive belt of country in Queensland, some 400 kilometres wide, and extending 900 kilometres northwards from Hungerford on the New South Wales border to the Kynuna district. Such fields as Cunnamulla, Eromanga, Jundah, and Opalton have all produced first-class gems, but are not now worked. Most of the current production comes from the South Australian fields.

In all these places the opal is found in rocks of Cretaceous age, but it is not itself a deposit of the Cretaceous Period, for it was deposited from wandering solutions long afterwards, perhaps as late as Upper Tertiary times, a few million years ago. The reason why the Cretaceous rocks are so especially suitable as hosts for the opal formation is not known. Opaline material is found also in rocks of other ages, but not so commonly. One such deposit is in New South Wales, at Tintenbar on the lower Richmond River near Ballina. This opal originated as a secondary mineral filling steam holes in an ancient basaltic lava flow. Some of the stones, though small, are of fine quality, but their capillary water content is rather high and after cutting they show a tendency to develop cracks. Similar stones of lesser quality have also been found in the vesicles of basaltic lavas in Victoria.

CRETACEOUS ROCKS IN WESTERN AUSTRALIA. In the area north of the Great Australian Bight it is known that the sea penetrated inland for a considerable distance both in Aptian and Albian times. Cretaceous rocks in this basin, known as the Eucla Basin, are buried and we know of their presence only from bores drilled in search of water through the Tertiary limestones of the Nullarbor Plain. Such a bore on the Trans-Continental line reached Cretaceous rocks at a depth of 180 metres, and penetrated them for a similar depth before reaching a bedrock of granite. At Madura, near the coast, over 300 metres of Cretaceous beds were penetrated by the drill, while other bores between the railway and the coast show thicknesses of 200–240 metres, mostly made up of shales and, towards the bottom, of sands and conglomerates.

North of the Nullarbor Plain, towards Leonora and Laverton and beyond, isolated patches of marine beds of Aptian age occur over a wide area, showing that the Eucla Basin on the Bight and the great Canning Basin of the Northwest were for a time connected by a shallow strait. It is also known now that these Cretaceous beds lie upon thin boulder beds of Permian age in many places. The boulders often show ice-scratches and—to confuse the geologist—many of them have been rolled around by the waves of the Cretaceous sea, and thus became incorporated in the Cretaceous shales and sandstones, simulating an environment of glaciers in the Aptian Epoch of the Cretaceous.

On the west coast of Western Australia the mid-Cretaceous sea penetrated inland for upwards of 160 kilometres in the Carnarvon Basin, and for several hundred kilometres in the Canning Basin. The basal sandstones of the series in the Carnarvon Basin, together with some beds in the immediately underlying late Jurassic sequence, are of great importance because they contain petroleum (Rough Range, Barrow Island).

Unlike Eastern Australia, the Western Australian sequence contains formations of Upper Cretaceous age along the west coast (but not in the Canning Basin), where they are exposed—for instance, in the Gingin district, just north of Perth, then on the lower Murchison River, and along the eastern margin and the central parts of the Carnarvon Basin (Cardabia-Giralia Range). The thickness of these later Cretaceous beds increases northwards from 500–1500 metres near Gingin in the Carnarvon Basin. In many places they are richly fossiliferous and tell much of the story of the changes in marine life during the period.

An interesting feature of the Western Australian Cretaceous is the presence among its limestone and marl series of thick beds of chalk, a white, friable rock composed largely of the minute shells of simple unicellular foraminifera. This chalk is reminiscent of the chalk cliffs of southern England, with which it is approximately contemporaneous. As the Toolonga Chalk, it is well exposed on the northern bank of the Murchison River, where it forms cliffs for about twenty kilometres above the mouth, and also for over thirty kilometres along the coast to the north. Farther north, in the Carnarvon Basin, the chalky character disappears and the rocks of that age are simply limestones and marls.

In the northern part of the Carnarvon Basin the sea remained over parts of the land almost to the very end of the Cretaceous Period, and the very latest of these beds, known as the Miria Marl, is particularly rich in fossils, especially ammonites and gastropods. Because of the abundance of marine life in the warm and shallow shelf seas such marl beds contain in places accumulations of the mineral apatite—more commonly known as phosphorite, the raw material for the preparation of superphosphate. In some areas a large proportion of the limestone consists of larger and smaller fragments of a big species of the bivalve *Inoceramus* which must have lived in extensive colonies on the muddy bottom of the shelf waters, in fact, almost to the exclusion of other marine molluscs.

Another interesting kind of rock, laid down during Aptian times, is an extremely fine-grained silica formation, of which a large portion consists of the minute skeletons of the unicellular organisms known as radiolaria. This rock, when dry, is very light in weight, and where exposed to the weather it is often turned into a material resembling light china-ware. In this form it is known as porcellanite and occurs also in some other areas in Australia, for instance around Darwin and in some parts of the Great Artesian Basin. It is a strange, and as yet unexplained, puzzle as to why this particular type of marine rock is, in Australia, limited to exposures of Cretaceous formations. Although similarly fine-grained rocks occur in series of other ages none of them seems to be capable of being turned into such fine porcellanite upon weathering.

In the Northern Territory, where there was a wide but shallow transgression of the sea in the earliest part of the Cretaceous (up to the early Aptian), a small area about Darwin and the nearby islands was again inundated in Albian time. This transgression lasted until the Cenomanian Epoch, the first of the epochs of the Upper Cretaceous— thereafter, and until this very day, the sea never again managed to spread over the northern part of the continent. Many of the Albian and Cenomanian fossils collected in this area, especially on Bathurst Island, are beautifully preserved ammonites on which even the colours of the nacreous shells are still recognizable (*Wright, 1963*).

LIFE OF THE CRETACEOUS PERIOD

Giant Reptiles and the First Flowering Plants

PLANT LIFE. Cretaceous fossil plants, like those of the two preceding periods, are abundant but fragmentary, and the exact classification of many remains uncertain. Nevertheless, through the mists of antiquity a broad picture of steady if unspectacular progress can be discerned. Many of the earlier Mesozoic plants still survived— the tree-fern *Cladophlebis*, other ferns such as *Sphenopteris*, the cycads

Podozamites and possibly also *Otozamites*, the unclassified *Taeniopteris* and others. In addition, many new ferns and other lowly plants had made their appearance; but perhaps the main feature was the prominence of the cycads, the gingkos, and above all the pines.

Forests of pines evidently existed in many places around the borders of the great inland lakes and sea straits, as well as in the coal swamps of eastern Queensland. Logs are found embedded in the marine rocks of the Aptian Roma Formation in the positions in which they became waterlogged after drifting far from shore. An interesting feature about these petrified logs is that they have sometimes been bored by marine organisms similar to the so-called ship-worms or teredo of the present day. These, by the way, are not worms at all, but true bivalves in which the shell part has become much reduced as the animal adapted itself to its special environment.

Among the pines were two of particular interest, *Araucarites*, similar to and possibly the ancestor of the Queensland hoop pine, and *Protophyllocladus*, the forerunner of the celery-top pine of southern Tasmania. Many of the fossil woods have not been identified, but some may well belong to the angiosperms or higher flowering plants. In fact it is these which give a distinctive character to the later Cretaceous flora, and which herald the close of the great Mesozoic Era. Apart from fossil wood, the leaves of angiosperms have been found in many localities, notably at Winton in central Queensland and in the coal measures on the Styx River near the coast. Although most of the leaves cannot be identified in the absence of flowers and fruits, they generally resemble plants now living in the Australian coastal rain forest, and future research may well determine their proper classification. Another increasingly useful line of research is the study of the microscopic plant spores and pollens which are found mainly in shales and coal measures, but with these, too, there are great difficulties in assigning them to definite genera and species which are otherwise known from stems, leaves, and flowers.

Difficult though it is to identify the remains of the Cretaceous flowering plants, their mere presence is of great significance. The final severing of the tenuous land connections with Asia at the end of the period meant that henceforth Australian plants and animals were left to develop their own characteristics in isolation from the rest of the world. The Cretaceous flowering plants are no doubt the ancestors of the present Australian bush, and from them evolved the eucalypts, the tea-trees, the acacias, the casuarinas, and other unique elements of our flora. A more detailed story of this has been told in two chapters in *The Face of Australia*, where it is described how the bush came in the first place to Australia, and then in some 60 to 70 million years became truly Australian.

FISH, REPTILES, AND MAMMALS. The story of the development of vertebrate life in the Cretaceous follows logically on that in the Jurassic Period and there is no evidence of the arrival of new and strange immigrants, certainly not among the land animals.

The remains of marine fish found in Queensland and Western Australia are mainly of sharks and true bony fish, but the dipnoan *Epiceratodus* or lung-fish occurs in the Upper Cretaceous freshwater deposits of western New South Wales and thus affords an interesting link between early Mesozoic and present-day representatives of the order.

The reptiles were still predominant and, if anything, larger and more grotesque than before. Amongst them was the dinosaur *Tyrannosaurus*, the largest carnivorous animal which ever lived. This fearful monster had very small forelegs and walked nearly erect on its massive hind legs. It was close to fifteen metres long and stood 6 metres high; its head was 1-2 metres in length and armed with sharp teeth 15 centimetres long.

Dinosaurs undoubtedly roamed the Australian land in Cretaceous times, but so far very little information is available. Our sparse knowledge is based on a find of several large vertebrae on Clutha Station in northwest Queensland. These are upwards of 40 centimetres across, and have been identified as belonging to the genus *Austrosaurus* (*Longman, 1933*). There is no means of judging what this creature looked like, but it was probably a plant-eater, and it has been estimated that it was about 15 metres in length.

More is known about Australian ichthyosaurs and pliosaurids, for their remains have been found in many places. In the opal fields at White Cliffs vertebrae and other bone fragments are often found converted into common and even precious opal. In Western Australia their remains have been found in the chalk near Gingin (*Teichert, 1944*), in South Australia in shales near Oodnadatta (*Freytag, 1964*), and in New South Wales and Queensland in various places in both the Roma and the Tambo Formations.

The ichthyosaurs were dolphin-shaped reptiles, entirely aquatic in habit, with the fore-limbs modified into swimming paddles (Plate 24). They were good swimmers and divers, and their food, like that of the sperm whale, consisted of the cephalopods so common in the Cretaceous seas, such as the squid-like ammonites and the belemnites. An ichthyosaur found on the Rolling Downs was 5.5 metres in length, and a great swimming reptile, the pliosaurid *Kronosaurus*, was still larger (Plate 25). Remains of other pliosaurids have been discovered in the same and other localities. These were extraordinary creatures, the body shaped like that of a turtle, the limbs modified into four strong swimming paddles, the neck extremely long and snake-like, the head small and the mouth armed with sharp teeth. True turtles were also common and some were of gigantic size, notably the two forms *Cratochelone* and *Notochelone* found in the Cretaceous beds of the Rolling Downs. One Cretaceous turtle, *Archelone*, not found in Australia, was 3.3 metres long and 3.6 metres across the flippers.

Of the smaller life which lived in the Cretaceous woods and forests there is scarcely any knowledge. There must have been a great variety of insects, smaller reptiles, and even birds and primitive mammals which were already plagued by fleas (*Riek, 1970*). If it was in the later part of the period that the land connection with Asia became finally severed, then the original stock of the present unique fauna of Australia must in the main have been already present, even though no trace of it has yet been found.

Marsupials are amongst the earliest mammals known, and their remains have been found in the Northern Hemisphere in deposits laid down in Mesozoic periods prior to the Cretaceous. Some of these must have found their way to Australia before the general development of mammals accelerated elsewhere. Henceforth evolution in Australia and the rest of the world took diverging paths. Australia was isolated, and in the story of the growth of the higher mammals—of the evolution of the horse and of the large carnivores, of the appearance and disappearance of mammoths and cave bears and sabre-toothed tigers and a thousand others, of monkeys, apes, and finally of man—that country played no part. Instead it was to develop its own unique assemblage of animals, comparatively primitive, but in its own way no less varied and interesting, and as specialized for different roles in the general balance of life as were the animals developing elsewhere.

MARINE LIFE. Cretaceous seas were still dominated by the cephalopods, especially the ammonites (Plates 22 and 23), though this group had passed the peak of its evolution. In this, the last period of their existence, species were still short-lived, most being continually replaced by others, so that once again the subdivision of the fossil record can be made very fine and exact over the whole world wherever fossil ammonites are found, or where the evolutionary sequence of other marine animals can be tied in with that of the ammonites. Study of the ammonites is a specialist's field, and their classification into numerous families, genera and species depends on anatomical differences which cannot be discussed here. For more detailed information the student is referred to textbooks on palaeontology, and for the Australian Cretaceous forms to the writings of *Etheridge (1902, 1903), Whitehouse (1926, 1927, 1928), Brunnschweiler (1959, 1966), Spath (1926, 1940), Woods (1962), Wright (1963), Reyment (1964)*, and others.

Vast numbers of ammonites have been found in the Middle Cretaceous of the Great Artesian Basin from Central Queensland and the Gulf Country through New South Wales into South Australia, as well as in the southeastern part of the Northern Territory and in the Carnarvon Basin in the west. In size they varied from less than 2.5 centimetres in *Falciferella* to giants such as the *Tropaeum imperator* found in Arckaringa Creek west of Oodnadatta and over 75 centimetres in diameter. Their form also varied considerably: some were flat and disc-like (*Aconeceras, Sanmartinoceras*), the last whorl of the coiled shell completely covering earlier whorls, while others were openly coiled with all or most of the inner whorls visible (*Beudanticeras, Boliteceras*), or even, like *Crioceras* and *Australiceras*, so loosely coiled that especially the later whorls of the shell were quite separate from each other. The even more curious groups of *Hamites* and *Baculites*, of *Turrilites* and *Nostoceras*, which had open hook-shaped and straight stick-like, or conical spiral and open cork-screw type shells respectively, are rare or absent in eastern Australia but common in the latest Middle and Upper Cretaceous rocks of the Darwin area and Western Australia.

The Upper Cretaceous ammonite fauna of Western Australia is a very rich one and shows close relationship to the faunas of Southern India, Madagascar, Antarctica, Chile, and New Zealand; whereas the relationship of the Middle Cretaceous ammonites is less with the southern than with the northern continents. The western and northern Australian forms of the Upper Cretaceous are of great variety and many are beautifully preserved. On Bathurst Island off Darwin one finds the very strongly ribbed and often almost spinous *Acanthoceras* and *Euomphaloceras*, and the fat little *Scaphites* with its partly uncoiled whorls. In Western Australia the fat *Pachydiscus*, which is often as much as 60 centimetres in diameter—overseas specimens up to two metres are known—is common around Gingin and together with finely and ornately ribbed *Phylloceras, Maorites, Gunnarites*, smooth discs of *Phyllopachyceras, Anagaudryceras, Saghalinites, Pseudophyllites, Hauericeras*, squat little spinose *Brahmaites*, and many others abundant in the Carnarvon Basin. Even more abundant there are the sword-sheath shaped sticks of *Eubaculites* and the last descendants of the Hamitidae, *Glyptoxoceras* and *Neohamites*.

The ammonites were to become extinct at the end of the Cretaceous. In fact, it seems remarkable that they had been able to survive and evolve for so long. It is true that many had strong shells which may have given protection against many predatory creatures but, like the dinosaurs on land, most were clumsy animals, relatively small-bodied in comparison to the size and weight of their shells, even though these were partly filled with air for buoyancy. Many were not fitted for the active pursuit of larger prey, and probably fed crawling along the bottom of the sea or by floating or slowly swimming among the clouds of plankton, which is the collective name for the minute life of the sea. If, for any reason, the supply of food became diminished, they seem to have been less fitted to survive than less specialized types, and they also probably fell victim themselves to the many new types of predatory fish already mutiplying within the sea. Whatever the reasons for survival or extinction may be—and we know precious little about this central problem of evolution—one can only marvel at the phenomenal effectiveness of all the adjustment mechanisms with which nature has been endowed.

The other cephalopod form, *Belemnites* and its related genera, common in the Jurassic, remained so in Cretaceous times, but disappeared from the world's oceans together with the ammonites. In Australia they are very common in Middle Cretaceous rocks and very rare in the Upper Cretaceous although abundant elsewhere in the world. Like that of the ammonites, their study is a specialized one, and the differences between various forms, though important both zoologically and geologically, are not sufficiently striking to be given in detail here.

Although the Cretaceous seas in Australia were not cold there is, as previously mentioned, a puzzling absence of corals of the reef-building type. Molluscs were, however, abundant, particularly bivalves. From the Upper Cretaceous of Western Australia comes one giant, a species of *Inoceramus*, which was 50 centimetres long and looked somewhat like a cockle. Fragments of similar shells from Queensland show that this species grew even larger. True oysters were abundant and large, with many other genera, some of which are still living. The remarkable *Trigonia*, common already in the Jurassic, remained so in the Cretaceous and developed into various new related genera and many species (*Skwarko, 1963*). Univalves or gastropods, while small and only moderately common in Lower and Middle Cretaceous rocks, grew large and in great abundance in the late Upper Cretaceous sea in Western Australia. The same applies to the brachiopods, among which the early Albian *Australiarcula* from the greensand at the base of the Tambo Formation is of particular interest because it is the ancestral form of the distinctively Southern Hemisphere family Terebratellidae of the present day, a family which was, during the Upper Cretaceous, very widespread also in the Northern Hemisphere, but has since almost completely disappeared there (*Elliot, 1960*).

There were also many crinoids, heart-shaped sea-urchins, and brittle stars. The latter, allied to true starfish, have a small disc-like body with long slender arms or tentacles which are very brittle. These have been found as far back as the Silurian and are still common beneath rocks on the seashore today.

It has already been mentioned that a number of calcareous and especially chalky deposits, for instance in Western Australia, are composed largely of the minute shells of foraminifera such as the globular *Globigerina* or the similar, but more conically coiled *Globotruncana*, and others. Though very minute creatures, they have long been numerous enough to play a very large part in the general balance of life or ecology of the sea. Unlike the radiolaria, which have shells formed of silica, the shells of the foraminifera are composed of carbonate of lime. We have already seen how radiolarian cherts of various periods—in Australia mainly Palaeozoic—as well as Cretaceous chalks may be made up of the remains of such tiny creatures, and both types of organisms are even more abundant at the present day. It is, however, only in the depths of the ocean or in areas far removed from land that they are able to quietly accumulate with little or no admixture of foreign matter. For such deposits the term radiolarian, or foraminiferal, ooze is used.

The study of foraminifera, like that of many other micro-organisms, has become of very great importance, especially in the search for petroleum. In drilling cores these microfossils are, of course, far more abundantly found than the larger molluscs, brachiopods, or even vertebrates, and the oil geologist must largely rely on zoning by means of those minute organisms with only an odd larger fossil giving a helping hand. However, through the diligent work of hundreds of palaeontological specialists throughout the world the exact dove-tailing of macro- with micro-fossil zoning of the geological record has for most periods been quite well established. This was a much-needed development simply brought about by the means and needs of one of man's greatest industries. It could not operate with the traditional fossils used for zoning and, while employing the main framework of epochs and stages established in the 19th century, it substituted the evolutionary sequence of macro-fossils by a sequence of micro-fossils which, though having a different rate of evolution which did not quite fit that of the traditional zoning, was quite exact enough to serve the same purpose. Especially for the Cretaceous, but also for some Palaeozoic Periods, this type of substitute zoning has by now been so perfected that there is scarcely any discrepancy left.

V
THE LAST 65 MILLION YEARS—SERENE ISOLATION IN THE CAINOZOIC ERA

14 From the PALAEOCENE to the PRESENT

BECAUSE MORE is preserved of and known about them, the later chapters of the geological record tend to become more voluminous though covering shorter periods of time. Thus the whole of the Cainozoic Era is about the equivalent of the length of an average Mesozoic or Palaeozoic Period. The Cainozoic is divided into two periods: the Tertiary, covering most of the Era, and the Quaternary, the opening stages of which lead to the geological happenings of the present day. Of the 60 to 65 million years of the Cainozoic Era, the Tertiary Period occupied all but about the last two; in other words, we are at present living in, and the sediments laid down now will contribute to, the formations which are going to be grouped into Stages and Series representing and recording the events of Ages and Epochs of a new period, that will stretch tens of millions of years into the future, and which we have chosen to call the Quaternary. In fact, we even assume tentatively that the first Epoch, represented by the Great Ice Age or Pleistocene, has already passed and that our time is part of the second Epoch of the new Period. Let us hope that we are right. Who would wish the Ice Age upon us again!

The Tertiary Period is subdivided into five Epochs which are, beginning with the oldest, the Palaeocene, the Eocene, the Oligocene, the Miocene, and the Pliocene. It is interesting to know what these terms mean—they are so strikingly different from say Carboniferous, Jurassic, or Artinskian, Neocomian, Danian, which are derived from rock types or from localities.

The epochs of the Tertiary arose from the idea of a countback applied to the genera and species which populated the world through various intervals of the past 65 million years, the question being: How many of the forms which live, or have lived, in our period (the Quaternary) already existed 60, or 30, or 10 million years ago? That is, how many forms are recorded already in the earliest, early, middle, or later stages of those formations which are older than the Great Ice Age, but younger than the youngest beds of the Cretaceous (with ammonites and belemnites, or with dinosaurs)? We need not go into the details of the matter, nor explain why that well-meant system is no longer seriously applied, even though its terminology has survived; but the outcome was that latest formations of the Tertiary showed a very large percentage of modern species, somewhat earlier ones rather less, the next earlier ones even fewer, and right down at the bottom of the Tertiary sequences of the world one found only the odd few modern forms. Thus were created the words Palaeocene from the Greek "palaios", meaning ancient, combined with "eos", meaning dawn or daybreak, and "kainos", meaning recent or new; Palaeocene then denotes "the ancient part of the time of the dawning of the new forms". Consequently, Eocene means "the dawning of the new". The Greek "oligos" means little, thus Oligocene denotes the epoch when there was still "a little new". Miocene includes "meion", meaning less; thus in the Miocene there were less of the new forms than in the Pliocene, when there were more ("pleion" means more). Entering our Quaternary Period, we come to the Pleistocene or Great Ice Age, with "pleistos" meaning most; and finally we arrive at our own time, or Holocene, in which "holos" means the whole, all, entire—the recent and present as man knows it, and in which happen the geological events that we can see and study ourselves.

In some countries there is an orderly and in places complete sequence of Cainozoic rock formations in which all five epochs of the Tertiary and the Quaternary ones are well represented. But in Australia there is no one rock sequence which would show the records of all epochs in full. This, however, is no problem to the geologist nowadays. There are many characteristic fossils, from the minute pollens and spores, foraminifers, and ostracods to the larger molluscs, brachiopods, echinoderms, and even the large vertebrates, including sharks and whales, which enable us to correlate and fix geological events in time—even if the relevant time interval represents only a minor portion of an Epoch. Of course, the fewer or shorter the fossil-documented time intervals are in a sequence, the more numerous or longer are the non-documented intervals between them, and the less precise will be our knowledge of the timing of events which took place during the gaps. Nevertheless, though the precise age of some of the happenings cannot be determined, it is possible to compose a fairly accurate picture of what took place.

THE PALAEOCENE AND EOCENE EPOCHS

In most parts of the continent conditions in the Palaeocene seem to have been rather similar to those in the latest Cretaceous. Lake

Winton, that great body of inland water in which the early Upper Cretaceous freshwater beds had been deposited, had largely dried up, but there still were a number of lakes in various parts of the area. In these were deposited sandstones, clays, and limestones which contain algae, ostracods, and gastropods, all of freshwater character. It has long been suspected that these beds, which are in places over 100 metres thick, cover large areas in Queensland, northwestern New South Wales and northern South Australia. Some years ago it has indeed been proven that these beds, although they arose from a patchwork of lake, swamp, and river sediments formed more or less independently from each other through early Tertiary times, had coalesced eventually into what we may regard as a single contiguous sheet of deposits, called the Eyre Formation (*Wopfner et al., 1974*). Later this vast sheet of sediments became dissected again, and its remains are now found either in the form of isolated flat-topped hills rising above the surrounding plains, or as infillings of ancient depressions and channels below the alluvials of the later Tertiary covering the plains. It is not always easy to distinguish the Eyre Formation from the Cretaceous Winton Formation, but very laborious studies of the respective characteristics of the sediments as well as diligent search for fossils, helped by the analysis of many wells drilled for oil and gas, have succeeded in keeping the two series apart. It is interesting to note that these early Tertiary beds contain much fossil leaves and wood, among which remains of *Eucalyptus* are prominent.

Also referred to earliest epochs of the Tertiary are several lake deposits in eastern Queensland. In the Brisbane area, overlying the Triassic Ipswich series, are sandstones and shales 100–200 metres thick. They are exposed at Redbank, Oxley, Darra, and many other localities, and in places are rich in fossil fish, insects, and innumerable leaves of flowering plants. Similar deposits occur at Lowmead between Bundaberg and Gladstone. Their main significance is that they contain colossal reserves of oil shales which can be treated to yield petroleum.

The drying up of the "Eyrian" lakes in the interior was a slow process and never really ended, mainly because slow and very gentle warping of the earth's crust created continually changing patterns of low- and high-lying areas, with lakes or swamps forming in the depressions, while the raised areas were attacked by erosion and the detritus from them brought into the low-lying areas. Even so it is obvious that vast areas on this continent had now been reduced to an unending plain or to use the geomorphological term, peneplain, which means that the land has been worn to such a low level that there is scarcely anything left to be eroded from it. Never since the dawn of geological history had the country been so featureless. The mountain ranges which had been successively upheaved during the

Palaeozoic Era had been gradually worn away to provide the enormous mass of sediments deposited in the intervening ages. The Mesozoic rocks of the Great Artesian Basin are over vast areas upwards of 1500 metres in thickness. Here alone is bulk sufficient to form many lofty mountain ranges. To this must be added perhaps an equal amount of material carried by coastal rivers from the same high land, and distributed afar over the bed of the open ocean.

Plains, endless plains, made up most of the Australian Palaeocene and Eocene landscape. They reached from the far east to the far west, from the far north to the far south, broken only here and there by low ranges and isolated low hills, the residuals of the hardest and most resistant rocks. There was little deposition of rock at this stage, but there was redistribution of materials by wind or by rivers in flood, especially in the hill ranges in the east.

Marine Palaeocene rocks were laid down in two marginal regions of the continent, the northern part of the Carnarvon Basin in Western Australia and the coastal parts of western Victoria. In both cases the sea remained there only for part of the epoch, and was rather shallow. In Western Australia these beds contain a rich fauna of lamp-shells, oysters, sea-urchins, and small single corals, all more similar to forms of Cretaceous than of Tertiary age; but the micro-fossils, especially the foraminifera, show an entirely Tertiary aspect. This discrepancy in the character of various parts of a fauna is typical of the Palaeocene Epoch and remains the cause of much argument among palaeontologists; the larger fossils imply that the Palaeocene is an epoch of the Cretaceous, the micro-fossils claim to be Tertiary. In western Victoria, though, this discrepancy is not so evident because properly identifiable macro-fossils are few and of long-ranging type.

Once the Eocene is reached there are so few survivors from the Cretaceous that all arguments cease. Even in earliest parts of this epoch there is no doubt where one stands, and in Australia we do not even have marine formations of early Eocene age. The first significant transgression of the sea over marginal parts of the continent after the Palaeocene happened about the middle of the Eocene Epoch, and by then the character of life on earth had completely lost its Cretaceous face.

The mid-Eocene transgression is a very important one, especially in Western Australia, where it penetrated as far inland as Norseman in the south and the Kennedy Range, some 160 kilometres inland from Carnarvon, in the northwest. The Perth Basin was also inundated, together with the Eucla Basin under the Nullarbor Plain (Wilson Bluff Limestone). In South Australia the sea entered St. Vincent Gulf along the Mount Lofty Ranges as well as western parts of the Murray Basin. Along the Victorian coast line the new marine transgressions did not take place until some time into the Oligocene. There was no Bass Strait before the end of the Eocene, only a gulf extending from the Tasman Sea westward into the offshore area of the Gippsland Basin. At the turn from the Eocene to the Oligocene the continent must therefore have been rather larger than it is now, with Tasmania linked to the mainland as a large peninsula.

Of Eocene age are the Moorlands Coal Measures on the northwestern rim of the Murray Basin in South Australia, and so are much of the Eastern View Coal Measures in the eastern Otways and the Bass Basin in Victoria, as well as the offshore coal measures in the Gippsland Basin.

The Moorlands Coal Measures show very clearly the sequence of events leading to the formation of coals—in this case lignites, or brown coals—in the earlier part of the Tertiary. Beneath them is a hard pavement of very ancient rocks, mainly older Palaeozoic, which had been eroded to a low level long before the coal beds were laid down. When the great subsidence which originated the Murray Basin began, lakes and swamps formed in depressions on the old land surface and became gradually filled with vegetable matter, partly from plants living in the swamps, and partly from leaves and branches washed in by drainage from surrounding higher terrain. The thickness of this vegetable matter differed in adjoining depressions; sometimes the beds of lignite are but a few tens of centimetres thick, in other places they are over 10 metres. As the land sank low enough the sea gradually inundated this area, and above the coal there lie therefore marine sands, clays, and bryozoan limestone lenses containing many fossils, including the teeth of extinct sharks.

THE OLIGOCENE EPOCH

The widespread withdrawal of the sea from the continent at the end of the Eocene was one of the effects of the circumglobal crustal movements which were soon to lead to the emergence of the mighty mountain ranges of the Alps, the Himalayas, the Rockies, and the Andes. This Alpine Orogeny culminated at different times in the various regions of the world, its youngest major manifestations being particularly noticeable in Southeast Asia and Australasia. These mighty movements had also stirred the pent-up forces in the semi-molten masses of rock beneath the outer crust. Moreover, along the eastern seaboard erosion of the ancient mountains had lessened the pressure on the subcrustal magmas. From innumerable cracks and fissures this now erupted and there sprang up a chain of volcanoes which was nearly continuous from Victoria (already in the Eocene) to northern Queensland. Vast floods of basaltic lava poured over the country; they filled the shallow valleys and spread far over the plains. In Victoria the Oligocene lava near Bacchus Marsh is even now over 300 metres thick; at Western Port 600 metres. In New South Wales it covered parts of what is now the Monaro Tableland; remnants of it form the higher summits of the Blue Mountains; it forms the cap of the high tableland at Guyra in New England; it covers extensive areas at Deepwater and also in the Richmond River district on the coast. In Queensland there were also many great lava flows, that at Oxley being 450 metres thick.

All these eruptions were not simultaneous, but occurred over a considerable length of time from the early Eocene-Oligocene into the early Miocene. The first eruptions were accompanied or closely followed by an elevation of the land along the coastal belt, which, by the end of the Miocene, raised the Palaeozoic Tasman Fold Belt region of eastern Australia by at least 1000–1500 metres above its level during the later Mesozoic (*Ollier, 1978; Wellman, 1979*). This led to a renewal of erosion, and the later lava flows were not poured out before the new high land had been well worn down. For these reasons, geologists divide the Palaeogene volcanic rocks into two series, the older basalts, including in New South Wales those at Kiandra, the Bald Hills near Bathurst, and the flows near Gulgong, and the younger basalts, which include among others the great flows of New England and the North Coast.

Closely associated with basaltic lavas are the deep leads. These are the gravels and alluvium of the old river beds which were buried and preserved beneath the lava which flowed down the valleys. The effect of this was the diversion of the drainage and the formation of new river beds along quite different courses from the old. In the course of time the new rivers have cut their beds to even lower levels than the old, and the hard resistant basalt now forms the summit of the hills, with the ancient river gravels still buried beneath them. Near Kiandra the bed of one of these former streams has been traced from hill to hill for over 30 kilometres. To the miner the location of the deep leads is of the greatest importance, for they are often rich in alluvial gold ·or tin, and sometimes contain precious stones— sapphires, zircons, garnets, topaz and even diamonds.

Apart from volcanic rocks, some lake deposits in eastern Queensland are considered to be of Oligocene age. One of these is at Petrie to the north of Brisbane, and from here many fossil fish, freshwater shells, insects, and leaves have been obtained. A similar formation containing many fossil fish is exposed at the Narrows, between Gladstone and Rockhampton; another is at Duaringa in the Dawson Valley and farther north at Waterpark and Plevna. A particularly interesting series is the Silkstone Formation which in the Ipswich district overlies the Eocene Redbank Plains Formation. It consists of an alternation of basaltic flows and lake sediments, all together about 270 metres thick in places. Most of the lake sediments are shales, but there are also substantial layers of magnesian limestone. These were evidently chemically deposited in the lakes, probably fed by hot springs on the surface of the old lava flows. In spite of the volcanic environment the lakes were well populated with freshwater molluscs such as *Planorbis*.

The bulk of the enormous lignite deposits of southeastern Victoria are also of Oligocene age. They were laid down in coastal lakes and swamps mainly in the area of the La Trobe Valley, 150–200 kilometres east of Melbourne. The closeness of the coast is indicated by the fact that a new transgression of the sea was already inundating the Sale and Lakes Entrance districts immediately to the east and southeast in the Gippsland Basin.

Drilling in the search for oil and gas has also revealed that the coastal lowland environments favourable to the formation of coal measures had prevailed farther out in the Gippsland Basin ever since latest Cretaceous times although the lignite reserves out under Bass Strait are not nearly as big as those in the La Trobe Valley. The entire more or less coal-bearing sequence in the region, called Latrobe Group, is therefore of Late Cretaceous and Eocene age offshore, but Eocene and Oligocene onshore in the La Trobe Valley. The large Bass Strait oil and gas fields—capped commonly by Oligo-Miocene marine beds—are found almost exclusively in the Eocene upper portion of the offshore La Trobe Group, and the porous reservoir rocks are mostly sands of coastal stream channels and deltas of that time (*Leslie et al., 1976*).

THE MIOCENE EPOCH

In southern and western marginal parts of the continent the story of the Tertiary is one of repeated transgressions and regressions of the sea. We have already heard of a Palaeocene and a mid-Eocene transgression, and now, after a regression lasting through much of the Oligocene, we see the sea advance again on to the continent shortly before the beginning of the Miocene Epoch. This transgression, as far as southern Australia is concerned, was the most extensive during the Tertiary but again, about mid-Miocene time, the sea began to withdraw. Thereafter we have evidence of two more major transgressions, one lasting through part of the later Miocene, the other through the early Pliocene, but neither of these two extended as far inland as

Map 23:

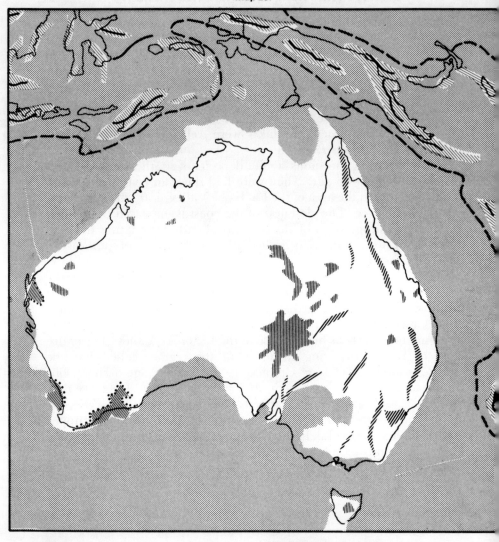

	SEA		ZONE OF CONTEMPORARY OR JUST COMPLETED MOUNTAIN-BUILDING
	LAND		AXIS OF UPWARPING —IN PLACES AS AN EFFECT OF DISTANT MOUNTAIN-BUILDING FORCES
	FRESHWATER LAKE		EOCENE MARINE TRANSGRESSION IN WESTERN AUSTRALIA Zone between dashed lines: INTENSE ALPINE FOLDING

The Australian continent approaching its present shape was due finally to events during the Tertiary Period. The main outline on this sketch shows the situation during the early part of the Miocene Epoch some 21 million years ago. Note that the major Tertiary inundation in Western Australia is of an earlier, Eocene, age and that except for the Murray Gulf, Nullarbor and Carpentaria, the shoreline is coming close to that of today. Observe also how the Indonesian region and Papua New Guinea are being strongly reduced in their land areas by the crustal movements of the Alpine Orogeny at that time. The broad upwarping especially along Australia's east coast took place mainly in Eocene and Oligocene time. Moreover, Australia has finally and decisively moved north and separated itself from the Antarctic continent since the latest Cretaceous. The point-dash line indicates the southern coast line towards the end of the Pliocene Epoch.

the Oligo-Miocene transgression which brought the sea in over a large part of the La Trobe Valley coal measures. It is interesting to note that these younger transgressions affected in the west only the Carnarvon, but not the Perth Basin, and even in the Carnarvon Basin they did not penetrate as far inland as the great transgressions of the middle and upper Eocene times.

On Map (23) we can see the extent of the Miocene inundations, and it will be noted too that the sea now inundates the area of the Barrier Reef extensively for the first time since the late Palaeozoic—Australia begins to take on its present-day shape. The presence of marine Miocene in the Barrier Reef region is known from a bore drilled on Wreck Island off the coast near Gladstone.

In southern Australia there were three main areas over which the sea transgressed: the area of Bass Strait including parts of Tasmania and Victoria, then western Victoria together with the Murray Basin and the Gulfs of St. Vincent and Spencer, and finally the Nullarbor Plain or Eucla Basin. Map (23) also shows the Tertiary pattern of lakes on the continent and Tasmania, but this pattern must not be taken to represent the situation during Miocene time alone—it shows the position of all localities from which there are records of Tertiary lake deposits of importance, whether they are of Eocene or Pliocene age.

The most easterly of the marine inundation areas is that of Bass Strait. The subsidence here was in the form of an east-west depression, and the encroaching sea not only converted Tasmania into an island, but covered parts of northern Tasmania and southern Victoria. The Miocene rocks are still visible in many places around the margin of the transgression area in the coastal regions of Victoria and Tasmania. Their thickness is quite considerable, especially in the offshore parts of the Basins; in Gippsland it increases from 1000 metres maximum onshore to 2000 metres out in Bass Strait (Seaspray Group). The corresponding Torquay Group in the Bass Basin also increases from a few hundred metres onshore to 1800 metres offshore, and the Heytesbury Group in the Otway Basin has a similar onshore-offshore variation in its thickness. The same applies to all other sedimentary basins which are marginal to the continent and extend seaward onto the continental shelf.

The rocks consist of clays, marls, and limestones and are rich in fossils. There are many collecting localities. One of these is Bird Rock Bluff at Springs Creek, near Torquay, where there is a splendid section on the coast. Another is Balcombe Bay near Mornington, Port Phillip Bay, where, from clay beds exposed right on the beach, over 300 species of extinct shells have been described, many of them in an exquisite state of preservation. Another noted locality is Muddy Creek near Hamilton in western Victoria, where an even greater variety has been found. In Tasmania great numbers of fossils have been found at Table Cape on the northwest coast, and also in some localities on King and Flinders Islands in Bass Strait itself.

West of Melbourne the Miocene sea made an island out of the Otway Ranges and covered a broad strip along the coast past Warrnambool, then turned northward to join up with the great Murray Gulf. This large embayment extended from Spencer Gulf and St. Vincent Gulf across southern portions of the Mount Lofty Ranges, over the mallee country of northwestern Victoria and the whole of the valley of the lower Murray to 160 kilometres above its junction with the combined Lachlan and Murrumbidgee. The ancient shoreline can still be traced with fair accuracy. South of the mallee country in central western Victoria a long arm of the land ran out from the east and has been termed the "Dundas Peninsula". Murray Gulf was shallow, certainly near the margins, even if it reached in places in the centre moderate depths, and therefore there is considerable difference in the sediments deposited. Those in the centre of the gulf are about 180 metres thick, and this thickness increases slowly toward the southwest. The rocks consist of limestones, clays, sandy clays, and shales, and are very rich in marine fossils. Excellent sections may be seen in low cliffs flanking the river banks on the Lower Darling and Murray rivers, and much accurate information has been gleaned from the numerous artesian bores in the mallee country.

Miocene marine rocks underlie a large part of southern South Australia and their outcrops can be studied on the western side of the Mount Lofty Ranges and in many places along the eastern shores of St. Vincent Gulf. At Blanche Point near Aldinga they overlie marls with some limestone of upper Eocene age, and these in turn rest on early Tertiary freshwater beds which overlie either glacial boulder beds of Permian age or rest on a pavement of early Palaeozoic rocks. Here too the Tertiary formations are in many places rich in marine fossils. Gulf St. Vincent is itself a great rift valley which has sunk down since the Miocene marine rocks were elevated into dry land. In places the Miocene strata, which are elsewhere lying horizontally, are pitched down at a high angle on the margin of the great fault which borders the sunken area. Such a locality is Sellicks Beach where the beds are almost vertically tilted, recalling the folded and contorted strata of the Palaeozoic Era.

The great inundation across the area north of the Great Australian Bight extended westwards from Fowler's Bay in South Australia to Point Culver in Western Australia, a distance of 800 kilometres, and the sea penetrated inland to cover the whole of what is now the Nullarbor Plain; in fact, as is indicated by isolated outliers in the

Victoria Desert north of the Trans-Australian Railway, it must at times have reached as far northward as the 28th South Parallel.

The Nullarbor Plain is one of the most remarkable physical features in Australia. Composed nearly entirely of the Miocene Nullarbor Limestone, it is so flat that the Trans-Continental railway runs in a perfectly straight line for 530 kilometres, the longest straight stretch of railway in the world. The plain has a low but regular rainfall, yet there is no trace of a stream or watercourse. Every drop of water either evaporates, or sinks into the numerous cracks and fissures of the limestone, which is honeycombed with innumerable caves, many of them still unexplored. On its southern edge the plain is cut off suddenly by a vertical escarpment, 75 metres high, against which pound the mighty waves of the Southern Ocean. In the central part of the Bight, however, this scarp is set back up to 50 kilometres from the coast for a distance of about 240 kilometres and is known as the Hampton Range. To the south, in front of the range, extends a low-lying area, covered with sand dunes, which is one of the largest marine erosion platforms in the world, carved out of the upper part of an up to 300-metre-thick Tertiary limestone sequence by the pounding seas of the Pliocene and Pleistocene Epochs.

Marine fossils are abundant in the limestones almost anywhere on the Nullarbor Plain, but they are not as a rule well preserved. There is one occurrence, however, which is of considerable interest, for it throws some light on the Miocene climate. This is the presence of reef-building corals, first found at Forrest, on the railway in the heart of the plain. This is clear evidence of the Miocene seas having been warmer than the Eocene and Cretaceous ones, as well as that of the present day. Early Miocene faunas in the Murray Gulf also indicate warm water environment, as do those of western Victoria and Gippsland, but reef-building corals are known only from the Eucla Basin. Later in the Miocene the warm water aspect of the fossil faunas disappears gradually.

The only region with Miocene marine formations in Western Australia is the Carnarvon Basin north of latitude 25°50′South, where there are three limestone sequences, the oldest of late Oligocene age, which cover areas up to 50 kilometres from the coast, and are richly fossiliferous, very large foraminifera and echinoids being especially common. From bores one knows that the thickness of these mid-Tertiary limestones is in some areas as much as 450 metres, and they are underlain by 100-200 metres of Palaeocene and Eocene beds, mostly limestones again. Here too the Miocene faunas indicate warm water environment and contain corals, but the contrast between them and the earlier faunas of the Cretaceous and Eocene is not as marked as in southern Australia. We must, of course, again remember that the

Carnarvon Basin, like other basins on the margins of the continent, extend well out onto the shelf beneath the sea. It is out there, in fact, where the major hydrocarbon accumulations have been found. One of the main reasons for this is that the thickness of most formations, particularly those of marine origin, increases considerably seaward. We have seen that in Bass Strait, and it is the same here on the Northwest Shelf of Australia.

THE PLIOCENE EPOCH

The Pliocene, the last of the epochs of the Tertiary, witnessed many changes which had a large effect in moulding the present topography of Australia. The first of these was elevation in the south and west and therefore the recession of the sea from these parts. Simultaneously came a renewal of intense volcanicity in eastern Australia, producing many ranges of volcanoes, jagged outlines of which are still visible. Another feature of the period was a system of great rivers in the interior, the courses of which are in many places clearly traceable.

Let us take these events in turn. First there was the general recession of the sea. Probably the first area to emerge as dry land was the Nullarbor Gulf, for there is no clear evidence of Pliocene marine fossils there. In the Carnarvon Basin only a narrow coastal zone shows evidence of marine Pliocene. The Murray Gulf was in part still occupied by the sea, though shallowed and rather restricted. Many of the fossils found in bores in the mallee country are indicative of estuarine and brackish environments, probably because of the large volumes of fresh water fed into these areas by the great rivers from the north. Bass Strait also remained open, and marine Pliocene rocks are known from various parts of Victoria, northern Tasmania, and the islands in the Strait.

The rivers, the Darling, Murrumbidgee and Murray, much larger than they are now, poured their waters by separate mouths into the Murray Gulf, and built up considerable deltas, through which they were eventually to unite to form one stream. The climate in the interior of Australia at this time was warm, and the country had an abundant rainfall. The courses of other great rivers which have long ceased to run have also been traced. One of these came from the MacDonnell Ranges and flowed south through Lake Eyre and Lake Torrens to enter the sea at the head of Spencer Gulf. Another, which has been called the Frome, lay farther to the east, and was fed by the ancestral streams of the Diamantina River and Coopers Creek. This flowed south through what are now the salt lake beds of Blanche, Gregory, Callabonna, and Frome, and entered the sea near the head of St. Vincent Gulf.

The volcanic eruptions of the Pliocene differed on the whole little from those poured forth during earlier periods of the Tertiary. Most of the lavas are of basaltic composition but, as in the Eocene and Oligocene, in some areas more alkaline types—types rich in potash for example—were extruded. Most of the larger volcanoes consisting of such light coloured lavas are, in fact, of older Tertiary age, especially those in southeastern Queensland; probably also Mount Canobolas near Orange in New South Wales, which still rises 450 metres above the plateau on which it is based, and the jagged Warrumbungles and Nandewars farther north and west. These mountains present a striking appearance as they rise sheer from the surrounding plain. Even more picturesque are the Glasshouse Mountains, visible from the train between Brisbane and Gympie, which were originally isolated cones from which outside portions were worn away so that only the solid plugs of once molten rock which filled the central vents, are left. These now tower in great monoliths hundreds of metres above the surrounding country.

Interstratified with the lava in some volcanic ranges are deposits of the interesting rock called diatomaceous earth. This is a white and friable rock, very light in weight, composed entirely of the microscopic skeletons of unicellular plants called diatoms. These minute plants are ubiquitous throughout the waters of the world, from polar seas to the tropics, but here they evidently lived in hot springs and small lakes on the surface of lava flows. Some of the deposits are of considerable size and are valuable economically for a number of purposes.

Until recently it was believed that the 240-400-kilometre-wide mountainous zone along Australia's east coast was due to a very young and mighty regional uplift event—called Kosciusko Uplift by Andrews (*1938*)—which happened within a very short time right at the close of the Pliocene Epoch. However, according to more recent research (*Ollier, 1978; Wellman, 1979*) the raising of this eastern rim zone, that is, the remnants of the Palaeozoic Tasman Fold Belt, had been going on slowly and steadily all through the Tertiary, and its initial impulses may even have something to do with the Maryburian Orogeny at the beginning of the Upper Cretaceous. In fact, it looks more like the close of the Pliocene was the time when the uplift movements were coming to an end. They were, in any event, not evenly distributed over the entire zone. The Bass and Gippsland Basin areas remained relatively depressed in spite of the uplift going on around them in Tasmania, Victoria and southeastern New South Wales. The Bass Basin emerged as low-lying land only at the time of the greatest glaciation (Würm Glaciation) in the late Pleistocene. After that it was quickly submerged again.

Because of the comparatively strong total elevation in the southern Australian Alps one has long assumed that erosion in the short time of the last 2–3 million years did lower the raised ancient surface—which need not at all have been a peneplain—by at least 1000–1500 metres there. However, a peculiar feature of the southeastern highlands of Australia, revealed in the past 20 years through the engineering works of the Snowy Mountains Authority, is the great depth of weathering in the rocks on the one hand, and on the other the colossal amount of deeply broken-up weathering material which has simply been left lying there on the valley slopes. The power of the rivers in undercutting and removing these softened-up materials was insufficient, substantial lowering of the highland surfaces was therefore unlikely. Thus, although there was steady uplift throughout the Tertiary (not, however, a short single "Kosciusko Uplift"!), its magnitude has probably been over-estimated, especially if we care to remember that some plateau surfaces, such as the one Canberra stands on, may well be as ancient as mid-Palaeozoic. They would hardly be preserved if erosion in the region had removed over 1000 metres of material close by.

Naturally, it would be surprising if the effects of such a general elevation were over a great distance everywhere the same. There is in eastern Australia a great diversity of rocks and rock structures, and in different places the rising land yielded differently to the tremendous stress and strain. As a whole the uplift shaped the region in the form of a broad arch with the steeper face upon the eastern border. In some areas this eastern side sloped evenly to the sea, in others it was fractured by great north-south trending faults. Large masses of country reacted independently and either rose above the general level or sank below it. The highest area in Queensland, the Bellenden Ker Range, is a huge block about 50 kilometres long and 16 kilometres wide, bounded by faults, and pushed up 1000–2000 metres to some 600 metres above the neighbouring plateau. There are many similar blocks forming such ranges in Queensland, all running nearly north-northwest and south-southeast, parallel to the main plateau, and separated by valleys or corridors which either remained at a low elevation or sank again later. Some of these corridors are actually below sea-level—the famous Hinchinbrook and Whitsunday passages, for instance. The mountainous islands on the seaward side of these shipping passages are themselves part of the general elevation. In New South Wales are similar valleys on the eastern side of the main plateau, all running approximately north and south, but they are far less well defined.

Nearly the whole of Tasmania rose at the same time in the form of a huge table tilted towards the north and sloping gently to the south. The northern edge of the table stands like a great east-west wall, the Western Tiers, an erosional feature crossing the major northwest-southeast trends of the faults which cut Tasmania up into several raised blocks or horsts and sunken corridors or grabens.

The Tertiary Uplift determined the main topography of Australia as it is now, though it has been somewhat modified by erosion and by lesser movements in the earth's crust. An important result of the uplift was the shortening of the coastal streams, for their head-waters were gradually cut off and flowed westwards towards the interior instead of eastwards to the Tasman Sea. The warping and faulting of the eastern edge of the plateau also affected the drainage; the coastal streams were diverted and flowed for considerable distances parallel to the coast before they finally broke through to the sea. Much of the latest Tertiary and Quaternary geological history comprises the vicissitudes of these rivers for ever seeking the shortest journey to the ocean.

THE PLEISTOCENE—THE FIRST EPOCH OF THE QUATERNARY PERIOD

The time that has elapsed since the close of the Tertiary Period, that is, since the end of the Pliocene Epoch, is generally referred to as the Quaternary Period, and as we have seen it is divided into one complete Epoch, the Pleistocene or Diluvium, and the beginning of another, the Holocene, in which we are living at the present day. Together they have lasted about two million years, of which the Holocene or Recent Epoch accounts for about the last 10 000 years. Though the events of the later Tertiary largely moulded the present topography of Australia, those of the Pleistocene modified its details in many ways.

Chief of these modifications was the erosion of the elevated highlands, shown not so much by a reduction of their maximum elevation than by deep dissection of the tablelands, so that only a portion of them remains. This can be seen perfectly on the Blue Mountains just to the west of Sydney. These are capped by the horizontally bedded Hawkesbury Sandstone with above it the remnants of the early Tertiary basalt. Prior to the Tertiary elevation, the area seems to have been an almost flat plain 100–200 metres above sea-level. Now the plateau, over 900 metres high, looked at from a distance still appears solid, with an even skyline, above which are a few slightly higher ·peaks composed of basalt. A view from the air or an examination of a contour map shows, however, that the plateau is but a skeleton, and its contiguity a sham. It is not only dissected by a labyrinth of deep gorges, but these occupy by far the greater part of the map. The original plateau surface is confined to long, narrow, sandstone ridges forming the divides between the innumerable watercourses. Probably over one-half of the total mass of mountain has already disappeared, and as sand and mud been carried by the streams far below away to the sea.

The Blue Mountains are typical of many other dissected plateaux along the eastern border of Australia, all showing the effect of at least 20 million years of erosion. It will be but a short time, geologically speaking, before they lose the appearance of tablelands. All that will then remain will be a number of rounded hills and gently flowing streams meandering down broad flat valleys. Although the eastern coastal regions of the continent show these young erosion features most spectacularly, they are not the only regions where the land surface was elevated and freshly dissected since mid-Tertiary times. The rivers on the southern half of the coast of Western Australia have also cut young gorges in many places, but the elevation of the plateau there was only of the order of 100–200 metres, not 1000–2000 metres.

Volcanic activity, which had been so important in the Pliocene, continued in some areas right through the Pleistocene and even almost into recent times. The areas affected were northern Queensland, western Victoria, and a small part of South Australia. In western Victoria, from Melbourne nearly to the South Australian border, and from the coast northward to the Grampian Mountains, volcanism was rife. There were many centres of eruption, and from these basaltic lava flows and showers of incandescent ash covered the whole country to a depth of hundreds of metres. It can in some instances be no more than a few thousand years since the volcanoes were active. Not only is the surface of the lava in places as fresh as if poured out yesterday, but the actual cones are intact and almost unweathered. Many of the craters are now filled with small lakes, sometimes fresh and sometimes salt. Mount Noorat, Mount Laura, Mount Elephant, and Tower Hill near Warrnambool are but some of the many extinct volcanoes in this region.

Across the border in South Australia a belt of hills 50 kilometres long in the vicinity of Mount Gambier is composed of extinct volcanoes. The best preserved is Mount Schank. This is a perfect cone consisting mainly of ash and other material thrown out during the eruptions, of which the latest took place only 4700 years ago, as shown by radiocarbon dating (*Gill, 1955*). Much of the equally young crater of Mount Gambier itself is preserved and is occupied by several small lakes. There was relatively little lava poured out from the volcanoes in this area. The eruptions were of the explosive type and the material ejected consisted mainly of ash.

In Queensland the Atherton Tableland with its average elevation of 900 metres above sea-level is composed of basaltic lava flows, and many craters, now converted into lakes, are to be found, notably Lake Eacham and Lake Barrine. Mount Le Brun, near Biggenden, was also a volcano, and its twin craters now form the Coalstoun Lakes. West of Charters Towers is another plateau composed of Pleistocene lava, and in Mount Emu the crater, which has been breached by the outflow of lava, is still well preserved. The volcanoes of Murray Island, in north Queensland, were active during the early stages of the Barrier Reef, for large fragments of coral are found included in the tuff which was then ejected. There are similar occurrences at Darnley Island and at Bramble Cay.

Remarkable formations which may have originated in this epoch have been revealed by deep-sea sounding. It had been known for a considerable time that a ridge rises from the sea-bed about 160 kilometres east of Southport in southern Queensland. This has since been shown to be seven separate and gigantic mountain peaks rising 4200 metres from the bottom of the sea to within a short distance of the surface. A similar mountain range rises from the floor of the Tasman Sea off the New South Wales coast in the latitudes of Sydney and Grafton. These could well be extinct submarine volcanoes sitting on top of folded mountain ranges of Tertiary and Cretaceous age, the volcanic peaks themselves having been formed during the Pliocene and Pleistocene, analogous to those on the nearby land.

It is curious that throughout the long ages since the earlier Palaeozoic times, volcanic activity in Australia has been almost entirely confined to the eastern regions. With the exception of a few minor eruptions, western and central Australia have been practically free, comprising one of the most stable portions of the earth's crust. Nevertheless, in central Australia some striking geographical changes did take place during the Pleistocene.

At the beginning of the Epoch the great rivers of the Pliocene were still flowing southwards to the sea, draining a vast area of country stretching nearly to the Gulf of Carpentaria. Now the flow of these rivers was interrupted by a large-scale earth movement which gave rise to an east-west upwarp and an elevation of the land just north of Adelaide. Not great in itself, the elevation was sufficient to make a barrier across the rivers and to impound their waters into another large inland lake with an area of about 100 000 square kilometres. This lake, called Lake Dieri by T. W. E. David (*1932*), lay over the present site of Lake Eyre, but its borders reached far beyond, particularly to the east, where it extended almost to the New South Wales border. Lake Frome, Lake Blanche, and the other numerous semi-permanent salt lakes in this area are but remnants of this stretch of water. Lake Dieri must have been filled in fairly rapidly by the great rivers flowing in from the north, and eventually overflowed, possibly by more than one channel—one of them very likely through Lake Torrens into Spencer Gulf.

Central Australia must then have been a wonderfully rich country. The climate was generally a cool one, though with warmer intervals, and for much of the time there was abundant rainfall, rich soils, and the land was intersected by permanent rivers and many lakes. There was food in abundance and the whole country teemed with life.

It was at this stage that the great alluvial plains of the interior began to accumulate. Some of the great mass of material produced by the erosion of the eastern highlands had been carried to the sea by eastward flowing streams, but more was carried to the west. The inland rivers, particularly in flood times, were laden with silt, which they spread far and wide over the plains, filling the small lakes and depositing the residue in Lake Dieri in South Australia. In the featureless country the rivers must have changed their courses many times, and the alluvium was distributed over the whole area. In many areas it is of considerable thickness—on the Darling Downs in Queensland 30 metres, while at Peak Downs in Central Queensland the black soil with underlying sand and gravel is 37 metres. About one-third of Queensland is so covered, as also are large areas in New South Wales and South Australia, and some lesser areas in Victoria.

It is in the lake and alluvial deposits that most of the fossils of large Pleistocene animals have been found. The discoveries have mostly been scattered bones in the eroded banks of rivers and creeks, but in a few places, notably at Maryvale in northern Queensland, the bones are in such quantity that they form distinct deposits known as bone breccia. In Lake Callabonna in South Australia large herds of giant marsupials died around the last muddy pools of water when the great inland lakes finally dried up. Bone breccias have also been found in limestone caves, as at Marmor and Gore in Queensland, and more notably in the Wellington Caves in New South Wales. Perhaps the greatest number of fossils so far known have come from the Darling Downs in Queensland and from the interior of South Australia. Finds made at Clifton, Drayton, Eton Vale, Warwick, Westbrook in Queensland and in a number of localities around the shores of Lake Eyre and the smaller lakes in that area have thrown much light on the terrestrial life of the times.

More is known of the Pleistocene climates throughout the world than of those of most other periods. The Great Ice Age, with its interglacial periods of warm climates, has left a clear story of the sequence of events in the Northern Hemisphere, events which had their repercussions throughout the world. One of the effects of the alternate withdrawal of water from the sea in the form of ice and its return to the sea in the warm interludes, has been a variation of the general sea-level from about 80 metres below its present level to over 30 metres above it. The latter height was attained about the middle of the epoch, and at this stage there was practically no ice at all on the surface of the earth. In Australia throughout the fluctuations of temperature the climate seems to have remained humid in most regions, and there was no diminution of life in the vast interior—certainly not for as long as there were permanent watercourses along which animals could migrate from drier into more humid regions of the continent, if and when the climate in one or the other part of the land deteriorated.

THE GREAT ICE AGE. The Pleistocene as a whole covers the last period of repeated intense glaciations the world has experienced. The chilling of the climate was worldwide, but it was in the Northern Hemisphere that the action of glaciers was most widespread. Four major glacial stages, separated by warm or inter-glacial interludes, can be distinguished. During the cold periods vast ice-sheets, similar to those of Greenland and Antarctica at the present day, covered much of northern America and Europe to a depth of 1000–2000 metres and more. At its height the ice-sheet overlay not only the whole of Canada but parts of the United States as far south as latitude 30° North. The estimated area was ten million square kilometres and the thickness of the ice over three kilometres. A similar ice-sheet covered northern Europe as far south as Germany and included the whole of the British Isles. The Alps too sent colossal valley glaciers out into the lowlands surrounding them and they carried mighty erratics on their back to dump them eventually scores of kilometres away from where they originally fell onto the ice stream.

In Australia glaciation was on nothing like this scale. The main regions directly affected by ice action were a considerable part of Tasmania and a much smaller area of the mountains on the New South Wales-Victoria border around Mount Kosciusko. Although Lewis (*1945*) thought there was evidence of three glaciations in Tasmania, thus suggesting some parallelism to the events in the Northern Hemisphere, more recent studies (*Jennings and Banks, 1958*) have shown that Browne (*1945*) was correct in saying that there was only one glaciation phase. It is now believed that this phase coincided approximately with the last of the major glaciations in the Northern Hemisphere, the "Würm Stage". On Map (24) the position of the glaciated areas in the Australian region are shown, including those suspected to have been formed on the high mountains of New Guinea. It should be noted, however, that the areas are shown rather larger than they actually were, and that the glaciation on the Darling Range near Perth is highly doubtful, though not entirely impossible.

Map 24:

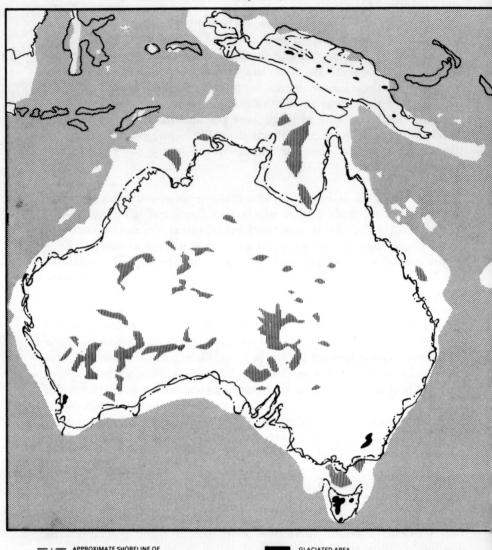

The Australian region during the Pleistocene Epoch. The main outline shows the low sea-level of the Würm Glacial Stage, some 20 000 to 15 000 years ago, when flora and fauna of Papua New Guinea were replenished by immigrants from Australia. The ancestors of our Aborigines had arrived during earlier glacial low sea-level stages, at least 40 000 years ago, from the north if, that is, they really needed such lower sea-levels to migrate here. The point-dash line shows the coast line during the higher sea-level stages of the Epoch.

On the Tasmanian tableland and in the wild country of the west and south there are many areas where one can see the effects the ice had on the landscape. Most spectacular are the mountain cirques formed in later stages of the glaciation when the glaciers had retreated to the higher mountain slopes. Rising from the edge of the tableland in the scenic reserve, Cradle Mountain is a perfect example of a cirque mountain. Its jagged skyline, composed of vertical columns of dolerite, lies in an arc within which are deep U-shaped valleys containing a whole nest of small glacial lakes. All the peaks within this area, picturesque and rugged mountains rising from 450–600 metres above the general level of the plateau, have been carved out by the action of ice. The rough and almost inaccessible Arthur Range in southern Tasmania as well as many peaks south of Queenstown in the west show similar evidence of the former ice age.

As in northern and central Europe and America the presence of innumerable lakes is evidence of the melting of the ice. Many of the small lakes lie in hollows scooped out by the ice; others, such as the beautiful Lake St Clair, lie in glacial valleys and have been dammed back by extensive moraines left behind by the retreating glacier. Elsewhere there are thick deposits of glacial clay with varved layering, as well as wide areas of rock scratched, smoothed, and polished by the overriding glaciers. Port Davey is one of the localities where the ice came down close to sea-level, or may even have reached the sea. This inlet looks like a typical fiord, similar to those of Norway and the west coast of the South Island of New Zealand.

On the mainland of Australia the story of the Pleistocene glaciation is much the same as in Tasmania, but on a smaller scale, the ice-affected area being only of the order of 800 to 1000 square kilometres and the ice sheet probably only about 30 metres thick. On retreating it left a number of moraines behind which dammed back small lakes or tarns.

The Australian glaciers, while of great local interest, had little topographic effect outside their immediate areas, but the northern ice-sheets, in one way, had a profound influence on the oceans throughout the world. The repeated withdrawal of so much water in the form of frozen rain resulted in lowerings of the level of the oceans by 100 metres or more, and this meant the recession of the sea from many shallow coastal regions. On the other hand, the melting of the ice in the inter-glacial periods raised the sea-level and re-flooded the land. It has been estimated that at the beginning of the Great Ice Age, when there were few ice-sheets in the world, and when even Antarctica was almost ice-free, the sea-level was 42 metres higher than it is now. Evidence of this is given in many places by the presence of raised beaches and terraces which were formerly at the level of the sea. As the ice-sheets formed, sea-level fell eventually almost 120 metres below this, or to about 80 metres below the present level.

After having been separated from the mainland since the Eocene, Tasmania was again annexed for a short time during the greatest of the Pleistocene glaciations. The submarine topography of the Bass Basin area suggests that there was a large lake between the King Island and Flinders Island land bridges, as shown on Map (24).

The final melting of the northern and the reduction of the Antarctic ice-sheets raised the sea 70 to 80 metres back to its present level. That it has not returned to its pre-Pleistocene maximum is no doubt due to the extensive ice-sheets still remaining in Greenland and Antarctica, sufficient probably, if they melted, to raise the sea-level another 30 metres or more. Evidence of a continuing rising of the sea has been noticed all over the world, although in many places one cannot be quite certain whether it is the level of the sea that rises because of an increase in the volume of the ocean waters, or whether the slow encroaching of the sea is due to local downward movements of the earth's crust.

In Australia the melting of the ice again made Tasmania an island, produced Torres Strait, and separated many small islands from the mainland. It flooded many lower reaches of river valleys such as that of the Derwent in Tasmania. Port Jackson, which was a small valley with a little stream, became a mighty harbour, while low-lying parts of the coast were flooded and turned into embayments such as Botany Bay and Jervis Bay. The glacial valleys of New Zealand were also flooded and became fiords. On the other hand there is no doubt that considerable crustal movements raised the eastern Indonesian islands and New Guinea from 200–1500 metres higher out of the sea than they were during the early part of the Pleistocene Epoch. On the island of Timor, for example, early Pleistocene beach deposits are now found as high as 1700 metres above sea-level—quite obviously, that kind of "raised beach" has nothing to do with the freezing and melting of the glaciers of the Great Ice Age.

The end of the Great Ice Age was about 10 000 years ago, and that was also about the time when Port Jackson was finally flooded; in other words, at the beginning of the Holocene or Recent Epoch. The elapsed time of the Recent Epoch is too short for great geographical changes to have taken place, for such changes, though seemingly rapid in geological retrospect, are imperceptibly slow by human standards. Yet sedimentation of eroded material is for ever going on. Already Sydney Harbour has been shallowed by the deposition of some 25 metres of mud and sand, and it is well known what problems are created by the constant shallowing of the port at Newcastle from material brought in by the Hunter River.

THE RECENT EPOCH

Here are crowded the events of the last 10 000 years. To mankind these events are of momentous importance, yet they represent but tiny pieces in the vast mosaic that had been built up for an infinity of time.

With the conclusion of the Ice Age the climate again became warmer, but this time, and in fact already towards the end of the Pleistocene, also drier. No longer did moist winds regularly bring masses of rain-laden clouds across the country. The rainfall became less and less; the large rivers became intermittent in their flow and many dried up altogether. The remnants of Lake Dieri became smaller and smaller, their content of salt and gypsum higher and higher, until the muddy and mineralized waters were useless to animal life. Conditions became much as we know them now. The centre of Australia and many large regions stretching right out to the coast are still very dry and parts of these areas practically desert. Animals, and especially birds, are still present, but in only a fraction of their former abundance, and most of the larger forms have become extinct.

Although the level of the sea did not remain steady since the beginning of the Holocene, the minor fluctuations there were did not involve very noticeable climatic changes, even if they were due to changes in the volume of the polar ice masses. In the last 4000 to 5000 years, for example, the sea-level has fallen over all by about six metres. This is why we now see beaches raised in many places along the continent's coast. Previously submerged platforms, cut by the breakers maybe 5000 to 7000 years ago, became narrow coastal plains, and many a small island became joined to the coast. Just north of Sydney, in the vicinity of Narrabeen, what were formerly channels between Pittwater and the sea, became land.

On the other hand, there have also been times when the sea-level was somewhat lower than it is now. The coastal features—beach terraces, platforms—of such periods lie now drowned a little off the present coast.

One of the important geological features created during an arid stage with a considerably lowered sea-level was the formation of the sand-ridge deserts which now cover vast areas of the continent. The original arid phase, which caused such deserts, was probably at the end of the final or Würm stage of the Great Ice Age rather than any of the low sea-level stages of the Holocene. Nevertheless, the point is that the country has never recovered from the sandy devastation then created, even though the original high aridity has since been somewhat reduced and the sand-ridges almost stopped from wandering by some sparse plant growth in many areas.

The largest expanses of such sand-ridge desert are found in a boomerang-shaped, broad belt stretching from the coast on the Timor Sea in the northwest to the region north of the Nullarbor Plain in the south, thence eastward across the Centralian Railway, and through the large Lake Eyre Basin, ending up against the foot of the eastern MacDonnell Ranges. The northwestern part of this desert zone is known as the Canning or Great Western Desert, in the south as the Victoria Desert, and in the northeast as the Simpson Desert. Most of these regions are so inhospitable that until a few years ago they had seldom been penetrated on the ground, even by Aborigines. Modern desert vehicles and helicopters have changed this situation completely in recent years, and many of these desertic areas are now in many ways better known than much of the thickly timbered range country in the more densely populated regions of eastern Australia.

The sand-ridges of these areas may rise to a height of 30 to 40 metres, but the majority of them are broad-crested and no higher than nine to twelve metres. Their length may reach dozens of kilometres but they also form a kind of interlocking network in many areas (*Veevers and Wells, 1961*). Throughout the Canning and the Victoria Deserts the predominant alignment of the dunes is approximately west-east and conforms to present-day directions of the strongest winds. In the Simpson Desert, however, the alignment is prominently north-northwest to south-southeast and turning to north-south in some areas in the southeast.

Sand dunes of this type are typical of deserts in many parts of the world—for instance, the Great Sand Sea in the northern part of the Libyan Desert in Africa. They consist of material transported and concentrated by the wind, but the reasons why parallel ridges should form in some localities and meshworks or isolated crescent-shaped dunes in others are not yet quite clear. Most dunes, though formed of loose sand on the surface, are commonly cemented into fairly solid sandstone in the centre, and thus constitute real geological formations of considerable extent. Such formations were formed also in earlier periods (*Burn and Crowe, 1977*), but are not always easy to distinguish from those formed by sedimentation in water.

The Coastal Limestone forming much of the coast of Western Australia and portions of that in western Victoria is partly of Holocene and partly of Pleistocene age. It is a large geological formation, in places 100–200 metres thick. It consists of dunes that for the most part have been consolidated into granular limestone by percolating water which dissolved and redeposited the lime of the innumerable calcareous fragments of sea shells contained in the sand. In the Carnarvon Basin region, close to the coast, there is an area of dunes which run north-south and may also have originated early in recent times.

Belonging also in the main to the Quaternary Period are two other natural phenomena, the Great Barrier Reef and the numerous caves which honeycomb the various limestone formations throughout Australia. Another common feature, which may still be forming at present, is the hard siliceous crust or "Duricrust" which is present over large expanses of fine-grained rock outcrops in the interior (the softer weathering crust, known as laterite and bauxite, was formed in Tertiary times), and the pebbly desert country known as "gibber" plains. All these are, however, together with the Coastal Limestone, so much part of the present landscape formation that they have been dealt with in *The Face of Australia*, and there is no need to recapitulate the story here. One thing, however, may be said—coral reefs and limestone caves both represent the balancing forces of nature, those of construction and destruction, outlined in the first chapter of this book. The Barrier Reef is still growing, building up what may be new lands of the future, while caves are ephemeral things, features of the gradual dissolution of all the high lands of the present.

LIFE IN THE TERTIARY AND THE QUATERNARY PERIODS

No cataclysm of nature separates the Cainozoic from the Mesozoic, and local geographical changes are no greater than those occurring between epochs or stages. Nevertheless, the contrast in life between the two eras is rather abrupt, and in many ways spectacular. This is true for the life of both the land and the sea, but more particularly for the larger vertebrate animals. If the Mesozoic was the age of reptiles, the Cainozoic is the age of mammals and birds. It is true that the mammals are found as far back as the Triassic, that birds arose in the Jurassic, and that reptiles still survive in great numbers, but these facts do not obscure the fundamental picture of abrupt worldwide change.

The large and unwieldy reptiles which had dominated the earth for long ages—the dinosaurs, the fish-lizards, the bat-like pterodactyls—disappeared as it were almost overnight, and of all the vast hordes only a few groups, such as the crocodiles, lizards, and turtles survived. In their place, and with comparative rapidity, came order after order of mammalian creatures developing, most likely, at a pace related to the evolution of the grasses, which was also largely an event of the

Tertiary Period. The first mammals were primitive types which have in turn become extinct, rhinoceros-like animals called *Uintatherium*, and creatures like bears which lived in the Gobi Desert and had skulls 90 centimetres long. Then came dogs, true bears, cats, elephants, true rhinoceroses, giant sloths, horses, oxen, deer, and many others, including, of course, such sea-living mammals as the whales and seals. In the air, the toothed birds of the Mesozoic had likewise disappeared, and in their place came all the families of modern birds. Modern plant life had appeared a little earlier and, although grasses were scarce, the higher types of flowering plants were abundant in the Cretaceous, since when, however, they have become far more varied and highly specialized.

In the sea the most striking change was the disappearance of the ammonites and belemnites, which had been so conspicuous throughout the whole of the Mesozoic Era. Their extinction was as sudden as that of the dinosaurs on land, though it is interesting to note how simple, more generalized types such as the nautilus, which had persisted practically unchanged from the earliest Ordovician, still survived.

Such is the general picture of the worldwide changes in life during Cainozoic times, but it has already been pointed out that, with the exception of the birds in part and some plants, Australia had little share in this; developing in isolation its own unique flora and fauna.

PLANT LIFE. In spite of the great wealth of fossil specimens it is surprising how little is really known of plant life during the later stages of its evolution. The reason is that the majority of fossils consist of wood or impressions of single leaves. Flowers, which are the basis of classification of higher plants, are the most fragile and transient of structures, and the chance of their preservation is infinitesimally small. Most fruits also are soft and decay quickly, and even when they are hard and woody and preserved as fossils it is difficult to link them with leaf impressions, although these may be found in close proximity. Even beautifully preserved fossil wood, which shows in microscopic sections each individual cell, is rarely differentiated enough to do more than indicate that it has general affinity to some natural order; the true genus or species can hardly ever be determined. The same difficulties, of course, arise with the classification of such minute remains as spores and pollen. Leaves by themselves may even be completely misleading. Both the shape and the veining may be similar in quite different orders, or there may be considerable variation in leaves from different parts of the one tree, as we know from the eucalypts. This may give rise to much confusion and has at one time or another misled even great botanic authorities. From the superficial resemblance of various fossil leaves the theory once arose that in Tertiary times a similar flora lived in every part of the world and many Australian plants, for example, were mistakenly identified as oaks and other European trees which have never existed in Australia.

A few generalizations about the identification of leaves may be made. *Eucalyptus* leaves are distinctive and may be definitely recognized. These have been identified from many localities and from all periods from the latest Cretaceous to the present day. It is impossible to say, however, if they were gums, stringybarks, or ironbarks, or if they belonged to any of the 500-odd species at present forming the bulk of the Australian forests; but, following the theory which demonstrates that there is a relation between the venation of *Eucalyptus* leaves and the chemistry of the contained oil, it may be said that some had a peppermint oil while others had an oil rich in eucalyptol.

Leaves of some other genera may also be identified with living Australian plants; *Banksia* has been found in the Eocene in Queensland and South Australia, and the tea-tree *Melaleuca*, the bottle-brush *Callistemon*, and an *Acacia* are known from Oligocene beds. Waratahs and some other indigenous plants have also been identified with some certainty. Others, such as *Cinnamomum*, common in Eocene beds, no longer occur on this continent. Usually it is possible to say that the leaves resemble many shrubs and trees still living in the coastal rain forests or in the open forests of the tablelands and western slopes. One may conclude, then, from the general distribution of these leaves, and in view of the moister climates of the Pleistocene, that the rain forests or jungles now confined to the coastal belt previously extended much farther into the interior.

THE MONOTREMES. The order of the monotremes, confined to Australia, includes the platypus and the echidna, solitary members of two separate families. These are enigmas of the animal world. If they had been found as fossils in Jurassic or even Triassic formations they would have created less astonishment than they do as living creatures. They are the most primitive of mammals. Though being furry animals which suckle their young, they lay eggs and have many other characteristics in common with the reptiles. Bones of an extinct platypoid species, *Obdurodon*, occur in Miocene beds in the Lake Eyre-Lake Frome region. Extinct types of both platypus and echidna are common in Pleistocene beds in Queensland, but of the pre-Miocene development of the monotremes there is no trace anywhere in the world. This is extraordinary, for both their primitiveness and their extreme specialization suggest that they are of great antiquity, and it would be reasonable to expect that somewhere fossils ought to have been found to link them with a distant past. Such a discovery may yet be made. Primitive mammals occur as early as Late Triassic in England, Germany, Switzerland, South Africa, and in the Gobi Desert. They are small rat-like creatures in no way resembling or related to the Australian monotremes, the origin of which must be left at present as another one of the unsolved problems of nature.

THE MARSUPIALS. The unique Australian fauna of marsupials presents a very similar problem to that posed by the monotremes, although their fossil record goes well back into the Tertiary Period— the finds are not many yet. The great number of marsupials, their variety and high specialization today was in certain directions exceeded during the Pleistocene but before that epoch of marsupial abundance the record is very meagre. Members of the order still live in North and South America, and they have there been found also as fossils as far back as the Cretaceous. The same applies to Europe except that they became extinct at the end of the Miocene. None so far have been found in Asia, either living or fossil. All of these fossils are more or less related to the American opossum and bear no resemblance to the numerous distinctive families peculiar to Australia. One must therefore assume that the original entry of marsupials into the Australian region took place before it was cut off from the outside world late in Cretaceous times, and that the immigrants were simple, generalized types from which all others have since sprung. This evolution took place locally in isolation, and it is in Australia alone that evidence will be found to trace backwards the ancestral tree of the many families.

One general feature of the marsupials is at once striking. In their development in isolation, there has been a strong convergence towards the extreme characters of the other orders. In the course of time they have adapted themselves to fill all the roles in the natural environments that are elsewhere filled by animals of other orders. The result is a general resemblance to these other animals in form and habits, though no real relationship exists. Thus, representing the Carnivora, we have dog-like forms in the Tasmanian wolf and cat-like forms in the native cats or dasyures. Insectivorous animals are represented by the bandicoot, rodents by marsupial mice, the European squirrel by the possum, the beaver by the wombat, and the mole by the marsupial mole, while the larger grazing animals, though dissimilar in appearance, are represented by the kangaroos and wallabies. Thus we have a general assembly of types which, though unrelated to them, is superficially similar to faunas in other parts of the world.

The earliest marsupial fossils in Australia come from Tasmania; small Oligocene diprodontids occur on the Derwent River at Hobart in the south, and *Wynyardia bassiana* in the Early Miocene marine limestones at Table Cape in the north. The latter is a primitive phalanger, that is, the family to which belong our possum and cuscus. From Late Miocene marine beds at Beaumaris-Melbourne bones of a kangaroo and a notothere are known, and further kangaroo remains were found in early Pliocene rocks near Hamilton in western Victoria. From the later Pliocene of the same area comes the tooth of a cuscus. From South Australia's interior, near Lake Eyre, come the remains of a bandicoot, a wallaby, and again a notothere (*Stirton, 1955*), which are of late Miocene to Pliocene age. There have been many other discoveries in recent years, but little or no systematic work has been done yet on all that material which may well solve many a puzzle concerning the history of our fauna.

So we come to the Pleistocene, which occupied the last million of the over 50 million years which this chapter described. Here there is a lot more definitive evidence and, by the abundance of the remains, the fertile plains and forests must then have teemed with marsupials as well as other forms of life.

Many of the Pleistocene marsupials were identical with or closely related to living species, but others are now extinct. Most conspicuous amongst the extinct forms was *Diprotodon* (Plate 27), of which the remains in some of the Queensland bone breccias are so abundant as to give them the name of "*Diprotodon* beds". In a muddy mass grave in Lake Callabonna in South Australia a whole herd died literally on its feet from hunger and thirst, and beautifully preserved whole skeletons have been found there. The animal was an enormous creature. As large as a rhinoceros, thick-bodied, it stood erect on four massive legs and was not unlike the South American tapir in appearance. It was a plant-feeder and probably very sluggish in movement. Somewhat smaller in size, but similar in habit, was *Nototherium*, while the curious creature known as *Euryzygoma* had extraordinary wide pouches.

Among the carnivorous animals, *Thylacoleo*, the "marsupial lion", is extinct, but the smaller forms—the Tasmanian wolf, *Thylacinus*, and the Tasmanian devil, *Sarcophilus*—though extinct upon the mainland, still survive precariously in Tasmania. Kangaroos and wallabies were abundant, some representatives of existing species, others now extinct. Another smaller type of the giant marsupial *Diprotodon*, also extinct, was *Palorchestes*. These were many wombats, some of small size, but *Phascolonus* was about twice the size of the living species. It is interesting that before the wombat was known as a living creature, it was described and named in England by the famous Professor Owen from fossil bones found in the Wellington Caves, New South Wales. The ancestral form of our native "bear", too, is known under the name *Koalemus*.

On the whole, however, it is well to keep in mind that, as Archer and Bartholomai (*1978*) have expressed it, our knowledge of the development of the Australian mammals through the Tertiary represents no more than the proverbial tip of the iceberg. While the records from the Pleistocene are obviously much better and more numerous, we are still far away from knowing whence all our mammalian families came.

BIRDS. Though some families of birds are peculiar to Australia, the general assemblage is similar to that of other regions, and is in striking contrast in that regard to the primitive mammals. In fact, the avifauna is not only thoroughly representative, but is about the most highly developed in the world, for of the four families at the head of the bird world, three, including the bower birds, the birds of paradise, and the bell magpies, are confined to Australian regions, and the fourth, that of the crows, is shared with other regions. This is not surprising, since the barriers to the migration of land animals are not so to creatures capable of flight. Fossil birds are extremely rare in beds earlier than Pleistocene in Australia although, like the marsupials, there is no reason why they should not have been plentiful. The only birds which have a well documented fossil record in Australia are the penguins, of which remains have been found in the Eocene of South Australia, in the Oligocene of the Mount Gambier area, and in the Miocene of western Victoria and Beaumaris, near Melbourne—altogether, five specimens (*Simpson, 1965*). Being essentially marine animals, it is not surprising that various marine formations have yielded their remains.

After the Oligocene the fossil record of birds increases markedly. Finds were made in inland salt lakes which include a high proportion of aquatic birds. They have naturally a better chance of preservation than those living on higher land or remote from large streams. Species identified and partly described so far are pelicans, ibises, spoonbills, ducks, pigeons, falcons, and stork, most of them Pliocene to Pleistocene. Of special interest are the more primitive and large running birds—the emu is a present-day survivor. In the Miocene to Pleistocene there were several which have since become extinct, notably the giant *Genyornis*, which was over two metres high, and the somewhat smaller genera *Dromaeus* and *Dromornis*.

REPTILES. In Australia, as elsewhere, the giant forms of the Mesozoic became extinct at the end of the Cretaceous Period, but some spectacular forms occurred in subsequent ages. Remains of crocodiles have been found in Tertiary rocks as far south as Victoria, and fossil turtles were discovered in a deep lead deposit under basalt in New South Wales which must be at least of Pliocene age.

Once again it is not until we examine the Miocene deposits that we find abundant fossil evidence. It would seem that for very long ages conditions were not so favourable for the preservation of the remains of land-living animals as they were when the large alluvial plains were being formed. On the other hand, the vast extent of these alluvials as such makes the search for the fossils a task far worse than looking for a needle in a haystack.

In the Pleistocene, monitor lizards, the goannas of the Australian forests, were abundant, one of them a gigantic extinct species which attained a length of six metres. The extinct crocodile *Palimnarchus* reached 4.5 metres. The fact that not only crocodiles, but also the Queensland lung-fish *Epiceratodus*, are found in Pleistocene deposits in the north as well as in the south of the continent shows not only that the then climate was mild and warm, but that the waterways of central eastern Australia were open to the far north.

A fine form among the reptiles was the gigantic horned turtle *Miolania* (Plate 27). This was from 1.5 to 1.8 metres in length, and the first specimens were found in old coral rock at Lord Howe Island (*Anderson, 1925*). Its discovery was followed by that of a second species on the Darling Downs in Queensland and a third under younger basalt in New South Wales (the latter older than Pleistocene). A very similar species, *Niolania*, has since been found in older Tertiary rocks in Patagonia, and another, younger one on Walpole Island near New Caledonia. These, together with additional remains from Lord Howe Island, have enabled a good reconstruction of the animal to be made. *Miolania* differed from both the existing orders of turtles in being unable to retract its head within the shell and in having a strong tail enclosed in a sheath armed with stout spines. The most striking feature was, however, the two horns which were actual parts of the skull. These were placed on either side of the head and were from seven to ten centimetres long. In the Darling Downs species the horns protruded at right angles, but in that from Lord Howe Island they were curved upwards and backwards.

MARINE LIFE. The disappearance of the ammonites and belemnites at the end of the Cretaceous Period left the general balance of marine life much as it is now, and in the sea the faunal changes were perhaps not so spectacular, though in their way equally important, as on land. Eocene fossils practically all consist of extinct species, but they can to a large extent be placed in living families or even genera, and except to the specialist they look very much alike. As mentioned before, in the Epochs following the Eocene living species began to appear, until by the Pleistocene almost all of the fauna consisted of species still inhabiting the ocean. Indeed, there is more to strike the imagination in the bizarre forms of pelagic fish and other creatures that are captured today than there is in an average assemblage of Tertiary fossils. That does not mean, of course, that Tertiary fossils are uninteresting, or that their scientific importance and usefulness is lessened by their familiarity which, in any event to the specialist, is largely superficial.

The tracing of the origin of living forms depends largely on the study of Cainozoic fossils, as does the determination of facts of distribution, migration, and oceanography generally. One or two aspects may be considered. It seems that in earlier geological periods the total number of species at any given time was much less than at the present day, even allowing for the fact that only a small proportion of the earlier forms have been preserved. There is a much finer distinction between living species, because we include the soft parts of an animal, its colouring, and even its living habits in the definition—all of which are absent from the definition of a fossil animal species. All the same, it seems the geographical range of earlier forms—partly because we cannot define them as closely—was also greater, and, especially the distribution of shallow-water species, rather more worldwide. This has been of great value in linking geological formations of the same age in regions that are far apart. At the present time, although the pelagic animals of the open ocean often have a worldwide range, those which live in shallow water or on the sea bottom close to the shore tend to be grouped within narrow geographical limits. Such a limited area, with its assembly of animals, is known as a zoogeographical province.

Such provinces are only partly the result of climatic conditions. It is probable that rapid evolution and the appearance of great numbers of new species leads to their restriction within narrow geographical limits. For instance, if the living marine faunas of South Africa, South America, New Zealand, New South Wales and eastern Tasmania, and the southern shores of Australia from Victoria to Western Australia be compared, it will be found that they have rather few species in common, although climatic and oceanic conditions in all five regions appear very similar. If there be geologists some tens of millions of years hence, they might have difficulty in correlating the rocks now being formed in these seas by their fossil content. Fortunately, simply because only the hard parts of the animal are capable of being preserved (together, very rarely, with impressions only of soft parts), the problem is not likely to be too serious. Even now, if we look only at what will be preserved of the animals of the mentioned Southern Ocean faunas, we find that they are in fact much more similar than the many names on a zoological list would suggest. Nevertheless, in principle, the problem does exist, and one must ascertain how far back in time such zoological boundaries go, and what allowance must be made for them when determining the age of geological formations in widely separated localities. In fact, modern palaeontology has shown in many cases how such problems can be solved.

It is difficulties of this kind which once made the determination of the exact age of Australian Tertiary marine rocks so uncertain and led to widely differing, often rather personal opinions among authorities dealing with the subject. But practically all these once so hotly disputed questions have now been settled.

The Tertiary rocks of southern and western regions of the continent are in many places packed with fossils of many kinds, and they have been collected from innumerable localities, many of which were mentioned in the first part of this chapter. Among the most interesting and important, though not spectacular, fossils are the foraminifera. Most of these are minute organisms but some Cretaceous and Tertiary genera were comparatively huge. Among these are the forms derived from the genus known as *Nummulites* and its relatives, which had the form of a discus, some types being about the size of a 20¢ piece, others smaller as well as larger. Nummulitids are common in the Eocene limestones in the Carnarvon Basin (the French still call the Eocene the "Nummulitic" because of the abundance of this form in that Epoch), but are absent from rocks of the same age in southern Australia probably for climatic reasons. In Egypt the Great Pyramids are built of limestone which is composed almost entirely of certain species of these organisms. Similarly large discoidal forms belonging to other genera occur also in the Miocene, both in western and southern Australia. A great many species of foraminifera have been described, and they are of great use in subdividing the formations.

Apart from the foraminifera, the total number of species found is very great, literally in their thousands. They include sponges, corals, starfish, sea-urchins, lamp-shells, molluscs, ostracods, bryozoans, and teeth of gigantic extinct sharks and toothed whales. A feature was the great abundance, especially in Miocene seas, of whales and dolphins. Their remains have been found in many places, notably in the Grange Burn area near Hamilton and at Beaumaris near Melbourne. Teeth and ear bones of extinct sperm whales are very common, and a rib found at Beaumaris and now in the National Museum in Melbourne is 1.6 metres in length.

The genera and species of shells are too numerous to mention in detail, but most belong to living families and many to living genera. The Miocene ones have a typically warm-water character. Among these is the largest cowry ever found, *Gigantocypraea*, which reached upwards of fifteen centimetres in length. In another cowry, found at Balcombe Bay, the margin of the shell was expanded so that it was not unlike the skirt of a ballet dancer. Among the bivalves was the curious

Neotrigonia, the direct descendant of the Mesozoic *Trigonia* already mentioned. Species of *Neotrigonia* are still living around the Australian coast and are especially common in southeastern waters. It is interesting to note that many other species of shells first discovered as fossils have later been found, and are still being found, alive in deeper water on the Australian continental shelf.

Sometimes the reverse is the case, and living species of shells, which are rather out of keeping with their present environment, are turning up one by one as fossils in nearby Tertiary rocks. Living in the Great Australian Bight are certain shells, of distinctly tropical character and quite anomalous in their present habitat. These include large cowries, curious small volutes, a harp shell and some others. Closely related forms then became known as fossils from later Tertiary, probably upper Miocene, marine rocks of the Nullarbor Plain just to the north. Nothing specific has been published as yet on these occurrences, but considerable collections have by now accumulated in the Adelaide museums. There is certainly a very interesting study awaiting research students, professional or amateur, of finding out the story behind such curious adaptive powers of marine animals, a story evidently linked also to the growth of reef corals in Pliocene or even later times in St. Vincent Gulf as far south as Glenelg.

MAN IN AUSTRALIA. The coming of man to Australia is one of the latest events in geological history, and this story is still far from clear. The first arrivals seem to have come during the later Pleistocene, 50–60 000 years ago, when the lowering of the sea-level either removed or greatly narrowed the sea barriers in their path.

The first indication suggesting human influences on the Australian landscape may be the sudden and considerable increase of charcoal fragments in soil horizons which are 60 000 to 100 000 years old. Had man arrived with his firesticks? In any event, the first and very crude stone tools found in Australia (e.g. on Kangaroo Island) date back some 40–50 000 years. The oldest fossil human bones, discovered in 1969 near Mildura (Lake Mungo, New South Wales), are however only 25–30 000 years old. These bones are associated with tools which are much more refined and numerous than the just mentioned older ones.

It may come as a surprise to many—but the Mungo people belong to the earliest known forms of modern *Homo sapiens* in the whole world. They were a small and gracile race of negroid character, with a round head and quite small eye-brow ridges; in other words, they resembled very much some of the more gracile types of today's humans. Being only 1.3–1.5 metres tall, they were also much smaller than the Aboriginal of our time. The negrito-like pygmies living in the area of Mount Molloy on the Atherton Tableland in the north of Queensland, as well as the extinct Tasmanian native could perhaps be suspected to be descendants of a race such as the Mungo people. Such a theory will, however, be hard to prove because, among other things, there is no reason why these Molloy and Tasmanian tribes could not have been much later immigrants.

The most confusing aspect of the problem is the fact that a much more ugly and rugged hominid with high eye-brow ridges, receding forehead, jutting face and massive jaws was also living in Australia at a much later stage than the Mungo people, namely towards the end of the last ice age, or 10–15 000 years ago. Skulls and bones of this hominid, who is related to the ancient type known as Java Man, were found long ago at Talgai on the Darling Downs in Queensland, and much more recently in the Kow Swamp in western Victoria. Could it be that the seemingly more primitive Talgai-Kow people were in fact the earliest immigrants, and that the living Aboriginal is the result of their mixing with the later arrivals, the Mungo people?

There has been much progress in such archaeological studies in Australia, but we are still very much in the dark about very important aspects of man's development here. It will be highly interesting to see what the scientists at the Australian National University under Professor John Mulvaney—the most active and determined archaeological team in Australia—are going to discover in the coming years. It is a rather difficult task because the study of man's occupation of this continent, his spreading and adaptation to the special conditions here until the arrival of the European conquerors is limited to the very simplest records of human activities, that is, the records left behind by simple nomads whose style and standard of living have hardly changed through the tens of thousands of years since they first arrived here.

When the Aboriginal Australian came in from the north he brought with him the dingo, which is closely related to the Asiatic wolf. The remains of dingoes have been found in the Wellington Caves, the Diamantina River and other localities associated with the bones of extinct marsupials such as the giant kangaroo. Whether such beasts as *Diprotodon* were hunted by man, and figured in his diet, is still an open question. Considering that modern research has pushed back the time of man's arrival in this continent by tens of thousands of years, the question may well be soon answered in the affirmative.

So we come to the present; to a world which seems physically static, but which remains in the process of inexorable change. Whether we look on the present as the span of a human life, of the life of a nation, or of the life of all humanity, it is still an infinitesimal speck in the vast panorama of all that has gone before, and of all that is yet to come.

What the future holds we know not, of that which is past no more than fragments can ever be revealed. Nevertheless, to endeavour to reconstruct even a minute portion of the past is a task which has an irresistible attraction, and though the task can never be completed, it is possible in doing it to find the secret of great contentment.

Map 25:

GEOLOGICAL MAP

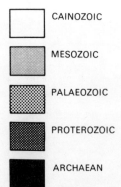
CAINOZOIC
MESOZOIC
PALAEOZOIC
PROTEROZOIC
ARCHAEAN

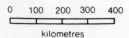
kilometres

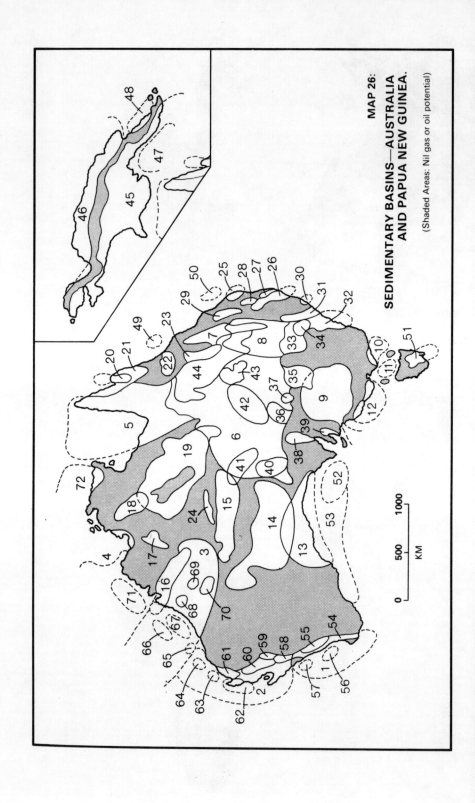

A. Major Basins known from surface mapping

- x   1—Perth Basin (WA)
- x   2—Carnarvon Basin (WA)
- x   3—Canning Basin (WA)
- ○   4—Bonaparte Basin (WA/NT)
-      5—Carpentaria Basin (QLD/NT)
- x   6—Artesian Basin (QLD/NSW/SA)
-      7—Bowen Basin (QLD)
-      8—Surat Basin (QLD)
-      9—Murray Basin (SA/VIC)
-     10—Gippsland Basin (VIC)
-     11—Bass Basin (VIC/TAS)
- x   12—Otway Basin (VIC/SA)
-     13—Eucla Basin (WA/SA)
-     14—Officer Basin (WA/SA)
- ○   15—Amadeus Basin (NT/WA)
-     17—Ord Basin (WA/NT)
-     18—Daly River Basin (NT)
-     19—Georgina Basin (NT/QLD)
-     20—Laura Basin (QLD)
-     21—Hodgkinson Basin (QLD)
-     22—Clarke River Basin (QLD)
-     23—Drummond Basin (QLD)
-     24—Ngalia Basin (NT)
-     25—Maryborough/Gympie Basin (QLD)
-     26—Clarence-Morton Basin (QLD/NSW)
-     27—Nambour Basin (QLD)
-     28—Esk Graben Trough (QLD)
-     29—Yarrol Basin (QLD)
-     30—Lorne Basin (NSW)
-     31—Tamworth Trough (NSW)
-     32—Sydney Basin (NSW)
-     33—Coonamble Basin (NSW)
-     34—Oxley Basin (NSW)
-     37—Bancannia Trough (NSW)
-     38—Pirie-Torrens Basin (SA)
-     39—St. Vincent Basin (SA)
-     45—Papuan Basin (PNG)
-     46—North New Guinea Basin (PNG)
-     47—Gulf of Papua Basin (PNG)
-     48—Cape Vogel Basin (PNG)
-     51—Tasmania Basin (TAS)

B. Concealed older Basins discovered by deep drilling

-     16—Fitzroy Trough (WA) — Palaeozoic
-     35—Darling Basin (NSW) — Ordovician to Devonian
-     36—Frome Basin (SA) — Cambrian
-     40—Arckaringa Basin — Permian
-     41—Pedirka Basin — Cambrian to Permian
- x   42—Cooper Basin — Permian and Triassic
- ○   43—Adavale Basin — Devonian and Carboniferous
-     44—Galilee Basin — Carboniferous to Triassic
-     54—Bunbury Trough (WA) — Permian to Jurassic
- x   55—Dandaragan Trough (WA) — Triassic and Jurassic
-     58—Coolcalaya Basin (WA) — Permian
-     59—Byro Basin (WA) — Permian
-     60—Merlinleigh Basin (WA) — Devonian to Permian
-     61—Ashburton Basin (WA) — Devonian to Permian
-     62—Exmouth Basin (WA) — Devonian to Jurassic
-     68—Willara Basin (WA) — Ordovician to Devonian
-     69—Joanna Springs Basin (WA) — Ordovician to Devonian
-     70—Kidson Basin (WA) — Ordovician to Devonian

C. Offshore Basins discovered by drilling and seismic surveys
(Note that a number of onshore basins extend to offshore)

-     49—Halifax Basin (QLD) — Mesozoic and Tertiary
-     50—Capricorn Basin (QLD) — Mesozoic and Tertiary
-     52—Duntroon Basin (SA) — Jurassic to Tertiary
-     53—Great Australian Bight Basin (SA/WA) — Jurassic to Tertiary
-     56—Vlaming Basin (WA) — Jurassic to Tertiary
-     57—Abrolhos Basin (WA) — Permian to Tertiary
- x   63—Barrow Basin (WA) — Permian to Jurassic/Cretaceous
- x   64—Dampier Basin (WA) — Permian to Cretaceous
-     65—Beagle Basin (WA) — Triassic and Jurassic
-     66—Rowley Basin (WA) — Triassic to Cretaceous
-     67—Bedout Basin (WA) — Triassic to Cretaceous
- x   71—Browse Basin (WA) — Permian to Cretaceous
-     72—Arafura Basin (NT) — Cambrian and Ordovician (?)

x: Hydrocarbon producing (or in the near future producing) basins
○: Basins in which significant hydrocarbon finds were made, but where production remains problematical for economic reasons

Map 27

MINERAL MAP
Major occurrences only

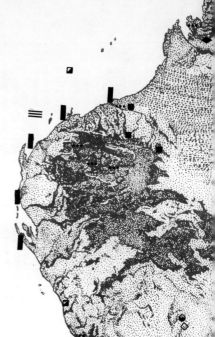

▮	Salt
...	Iron
✦	Alumina (Bauxite)
■	Manganese
⌣	Black Coal
⌢	Brown Coal
▲	Copper
○	Zinc
◆	Lead
∨	Silver
⏺	Gold
◇	Nickel
●	Heavy Mineral Sands
◆	Tungsten (Scheelite)
†	Tin
□	Uranium
﴾	Phosphate Rock
≡	Oil
▄	Natural Gas
⬦	Oil Shale
◆	Diamonds
⬦	Opal

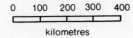

kilometres

REFERENCES

A COMPLETE bibliography of Australian geology would itself fill a large volume. It is neither possible nor necessary, in a work such as this to refer to more than a few of the sources from which information has been drawn. You will find scarcely any reference to the writings of the pioneers of Australian geology in the 19th century. Their work is now not only so incorporated in the science, as it is taught in the universities and colleges, that there is no need to specify it, but much of it has long become redundant or extensively modified in the light of more recent exploration of our continent. For those interested in the historical development of the geological science in Australia, Fairbridge (1953), below, in a brief summary in his first chapter, provides the road-signs to relevant literature. Most of the references therefore are to writings on later, and especially most recent, discoveries, some of which have drastically modified our knowledge of the structure and development of this continent. A number of references are to more generalized papers or books which have condensed and synthetized the writings of many authors; such references are emphasized in the list in special type. Even the references to more recently published works are far from complete, and apologies are extended to those living geologists whose papers have been consulted but, because of lack of space, have not been specified here.

ANDERSON, C., 1925—Notes on the extinct chelonian *Meiolania*. *Aust. Mus., J.*, 14
ANDREWS, E. C., 1938—The structural history of Australia. *Roy. Soc. N.S. Wales, J.*, 71
ARCHER, M., and BARTHOLOMAI, A., 1978—Tertiary mammals of Australia: a synoptic review. *Alcheringa*, 2, 1
ARGAND, E., 1924—La tectonique de l'Asia. *Int. Geol. Congress*, XIII, *p. 171–372*
ARKELL, W. J., and PLAYFORD, P. E., 1954—The Bajocian ammonites of Western Australia. *Roy. Soc. London, Phil. Trans. B*, 237, *No. 651*

BALLY, A. W., and SNELSON, S., 1979—Realms of Subsidence. *Canad. Soc. Petrol. Geol., Mem.*, 6

BANKS, M. R., 1957—The Middle and Upper Cambrian Series (Dundas Group and its Correlates) in Tasmania. In: The Cambrian Geology of Australia. *Bur. Min. Resour. Aust., Bull. 49*

———, 1962—Ordovician System. In: The Geology of Tasmania. *Geol. Soc. Aust., J.*, 9, 2

BARTHOLOMAI, A., and HOWIE, A., 1970—Vertebrate fauna from the lower Trias of Australia. *Nature, 225 (5273)*

BELFORD, D. J., 1960—Upper Cretaceous Foraminifera from the Toolonga Calcilutite and Gingin Chalk, Western Australia. *Bur. Min. Resour. Aust., Bull. 57*

BERRY, W. B., 1965—Monograptids from Eildon, Victoria. *Roy. Soc. Vic., Proc.*, 78

BINNS, R. A., 1964—Zones of progressive regional metamorphism in the Willyama Complex, Broken Hill District, New South Wales. *Geol. Soc. Aust., J.*, 11, 2

BLACK, L. P., MORGAN, W. R., and WHITE, M. E., 1973—Age of a Mixed *Cardiopteris-Glossopteris* Flora from Rb-Sr Measurements on the Nychum Volcanics, North Queensland. *Geol. Soc. Aust., J.*, 19, 2

BROWN, D. A., CAMPBELL, K. S. W., and CROOK, K. A. W., 1968—The Geological Evolution of Australia and New Zealand. *Pergamon, Oxford*

BROWNE, W. R., 1945—An attempted post-Tertiary chronology for Australia. *Linn. Soc. N.S. Wales, Proc.*, 70

———, 1947—A Short History of the Tasman Geosyncline of eastern Australia. *Sci. Progr.*, 35

BRUNNSCHWEILER, R. O., 1954—Mesozoic stratigraphy and history of the Canning Basin and Fitzroy Valley, Western Australia. *Geol. Soc. Aust., J.*, 1

———, 1959—New Aconeceratinae (Ammonoidea) from the Albian and Aptian of Australia. *Bur. Min. Resour. Aust., Bull.*, 54

———, 1960—Marine Fossils from the Upper Jurassic and the Lower Cretaceous of Dampier Peninsula, Western Australia. *Bur. Min. Resour. Aust., Bull.*, 59

———, 1961—Jungproterozoische (assyntische) Gebirgsbildung in Australien. *Eclog. Geol. Helvetiae*, 54, 2

———, 1966—Upper Cretaceous ammonites from Western Australia, Part I: The heteromorph Lytoceratina. *Bur. Min. Resour. Aust., Bull.*, 58

BUICK, R., 1981—Did Life exist in the Early Archaean? Evidence from shallow-water sediments, North Pole, Western Australia (Abstract). *5th Geol. Convent. Perth, Geol. Soc. Aust., Abstr.*, 3

BURNE, R. V., and CROWE, R. W. A., 1977—Aeolianites in the Jurassic Jurgurra Sandstone, Canning Basin, Western Australia. *Bur. Min. Resour. Aust., J.*, 2, 4

CAMP, C. L., and BANKS, M. R., 1978—A proterosuchian reptile from the Early Triassic of Tasmania. *Alcheringa*, 2, 1

CAMPBELL, K. S. W., and BELL, M. W., 1977—A primitive amphibian from the Late Devonian of New South Wales. *Alcheringa*, 1, 4

CAMPBELL, K. S. W., and ENGELL, B. A., 1963—The faunas of the Tournaisian Tulcumba Sandstone and its members in the Werrie and Belvue Synclines, New South Wales. *Geol. Soc. Aust., J.*, 10, 1

CAMPANA, B., 1955—Tillites and related glacial topography of South Australia. *Eclog. Geol. Helvetiae*, 48, 2

CAMPANA, B., and WILSON, R. B., 1954—The geology of the Jervis and Yankalilla Military Sheets. *Geol. Surv. S. Aust., Rep. Investig.*, 3

CAREY, S. W., 1947—Occurrence of tillite on King Island. *Australas. Ass. Adv. Sci., Rep.*, 25

———, 1977—The Expanding Earth. *Elsevier, Amsterdam*

CARTER, A., 1958—Tertiary foraminifera from the Aire District, Victoria. *Geol. Surv. Vic., Bull.*, 55

CARTER, E. K., BROOKS, J. H., and WALKER, K. R., 1961—The Precambrian mineral belt of north-western Queensland. *Bur. Min. Resour. Aust., Bull.*, 51

CAWOOD, P. A., 1976—Cambro-ordovician Strata, northern New South Wales. *Search*, 7, 7

CHAPMAN, F., 1909—On some microzoa from the Wianamatta Shales, New South Wales. *Geol. Surv. N.S. Wales, Rec.*, 8

CLARKE, E. de C., and TEICHERT, C., 1948—Cretaceous stratigraphy of Lower Murchison River area, Western Australia. *Roy. Soc. W. Aust., J.*, 32

COLEMAN, P. J., 1957—Permian productacea of Western Australia. *Bur. Min. Resour. Aust., Bull.*, 40

CONDON, M. A., 1965—The geology of the Carnarvon Basin, Western Australia, Part I: Pre-Permian stratigraphy. *Bur. Min. Resour. Aust., Bull.*, 77

CONOLLY, J. R., 1969—Models for Triassic deposition in the Sydney Basin. *Geol. Soc. Aust., Spec. Publ.*, 2

COOKSON, ISABEL C., 1935—On plant-remains from the Silurian of Victoria. *Phil. Soc. London, Trans., Ser. B*, 225

———, 1965—Cretaceous and Tertiary microplankton from south-eastern Australia. *Roy. Soc. Vic., Proc.*, 78, 1

COOKSON, ISABEL C., and MANUM, S., 1964—on *Deflandrea victoriensis* and *D. tripartita*. *Roy. Soc. Vic., Proc.*, 77, 2

COOKSON, ISABEL C., and PIKE, M., 1954—Some dicotyledonous pollen types from Cainozoic deposits in the Australian region. *Aust. J. Bot.*, 2, 2

COSGRIFF, J. W., 1974—Lower Triassic Temnospondyli of Tasmania. *Geol. Soc. Amer., Spec. Pap.*, 149

COX, L. R., 1961—The Molluscan fauna and probable Lower Cretaceous age of the Nannutara formation of Western Australia. *Bur. Min. Resour. Aust., Bull.*, 61

CRESPIN, IRENE, 1943—The occurrence of the genus *Conoclypus* in the North-west Division, Western Australia. *Roy. Soc. W. Aust., J.*, 28

———, 1957—Permian bryozoa from the Fitzroy Basin, Western Australia. *Bur. Min. Resour. Aust., Bull.*, 34

———, 1958—*Hantkenina* in Western Australia. *Micropaleontology*, 4, 3

CRICK, F., 1966—Of Molecules and Men. *Univ. Washington, Seattle*

CROCKFORD, JOAN M., 1941—Permian bryozoa of eastern Australia. *Roy. Soc. N.S. Wales, J.*, 74

CROOK, K. A. W., and COOK, P. J., 1966—Gosses Bluff—Diapir, Crypto-Volcanic Structure or Astrobleme? *Geol. Soc. Aust., J.*, 13, 2

DAILY, B., 1957—The Cambrian in South Australia. In: The Cambrian geology of Australia. *Bur. Min. Resour. Aust., Bull.*, 49

DAILY, B., and MILNES, A. R., 1972—Revision of the Stratigraphic Nomenclature of the Cambrian Kanmantoo Group, South Australia. *Geol. Soc. Aust., J.*, 19, 2

DARRAGH, T. A., 1965—Revision of the species of *Eucrassatella* and *Spissatella* in the Tertiary of Victoria and Tasmania. *Roy. Soc. Vic., Proc.*, 78, 1

DAVID, T. W. E., 1932—Explanatory notes to accompany a new geological map of the Commonwealth of Australia. *Cmwlth. Soc. Ind. Research, Sydney*

———, 1950—The Geology of the Commonwealth of Australia. *Edw. Arnold, London.*

DENMEAD, A. K., 1964—Note on marine macrofossils with Triassic affinities from the Maryborough Basin, Queensland, *Aust. J. Sci.*, 27, 4

DETTMAN, MARY, 1961—Lower Mesozoic megaspores from Tasmania and South Australia. *Micropaleontology*, 7

DICKINS, J. M., 1956—Permian pelecypods from the Carnarvon Basin, Western Australia. *Bur. Min. Resour. Aust., Bull.*, 29

———, 1957—Lower Permian pelecypods and gastropods from the Carnarvon Basin, Western Australia. *Bur. Min. Resour. Aust., Bull.*, 41

———, 1963—Permian pelecypods and gastropods from Western Australia. *Bur. Min. Resour. Aust., Bull.*, 63

———, 1978—Climate of the Permian in Australia: The Invertebrate Faunas. *Palaeogeography, 23*
DICKINS, J. M., and McTAVISH, R. A., 1963—Lower Triassic marine fossils from the Beagle Ridge (B.M.R. 10) Bore, Perth Basin, Western Australia. *Geol. Soc. Aust., J., 10, 1*
DOUGLAS, J. G., and FERGUSON, J. A., eds, 1976—Geology of Victoria. *Geol. Soc. Aust., Spec. Publ., 5*
DU TOIT, A. L., 1937—Our Wandering Continents, An Hypothesis of Continental Drifting. *Oliver Boyd, Edinburgh*
EDGELL, H. S., 1955—A Middle Devonian lichid trilobite from south-eastern Australia. *Paläont. Ztschr., 29*
———, 1964—Precambrian fossils from the Hamersley Range, Western Australia, and their use in stratigraphic correlation. *Geol. Soc. Aust., J., 11, 2*
EIGEN, M., and WINKLER, RUTHILD, 1975—DAS SPIEL. Piper, Munich
ELLIOTT, G. F., 1960—A new Mesozoic terebratellid brachiopod. *Geol. Ass., Proc., 71, 1*
ETHERIDGE, R., JUN., 1901—Additional notes on the palaeontology of Queensland, Part II. *Geol. Surv. Qld., Bull., 13*
———, 1902—The Cretaceous mollusca of South Australia. *Roy. Soc. S. Aust., Mem.*
———, 1905—Monograph of the Cretaceous invertebrate fauna of New South Wales. *Geol. Surv. N.S. Wales, Mem. (Palaeont.), 11*
ETHERIDGE, R., JUN., and JACK, R. L., 1892—The geology and palaeontology of Queensland and New Guinea. *Geol. Surv. Qld., Publ., 180*
FAIRBRIDGE, R. W., 1953—Australian Stratigraphy. *Univ. W. Aust., Textbk Bd*
FELDTMAN, F. R., 1951—Pectens of the Gingin Chalk. *Roy. Soc. W. Aust., J., 35*
FLEMING, C. A., 1959—*Buchia plicata* (Zittel) and its allies, with a description of a new species, *Buchia hochstetteri*. *N. Zeal. J. Geol. Geophys., 2, 5*
FLETCHER, H. O., 1964—New linguloid shells from Lower Ordovician and Middle Palaeozoic rocks of New South Wales. *Aust. Mus., Rec., 26, 10*
FORMAN, D. J., 1966—Regional geology of the south-west margin, Amadeus Basin, central Australia. *Bur. Min. Resour. Aust., Bull., 144*
FORMAN, D. J., and HANCOCK, 1965—Geology of southern margin of the Amadeus Basin. *Bur. Min. Resour. Aust., Rep., 87*
FORMAN, D. J., and SHAW, R. D., 1973—Deformation of the crust and mantle in central Australia. *Bur. Min. Resour. Aust., Bull., 144*

GALE, N. H., BECKINSALE, R. D., and WADGE, A. J., 1979—A Rb-Sr whole rock isochron for the Stockdale Rhyolite of the English Lake District and a revised mid-Palaeozoic time-scale. *Geol. Soc. London, J., 136, 2*

GEOL. SURV. WESTERN AUSTRALIA, 1975—The Geology of Western Australia. *Mem., 2*

GILL, E. D., 1955—Radiocarbon dates for Australian archaeological and geological samples. *Aust. J. Sci., 18, 2*

———, 1963—Evolution of the Zenatiinae. *Roy. Soc. Vic., Proc., 77, 1*

GLAESSNER, M. F., 1960—Precambrian fossils from South Australia. *21st Int. Geol. Congr. (Copenhagen), Pt. 22*

———, 1961—Precambrian animals. *Scientific American, 204*

———, 1962—Precambrian fossils. *Biolog. Review, 37*

———, 1971—Geographic distribution and time range of the Ediacara Precambrian fauna. *Geol. Soc. Amer., Bull., 82, 2*

GLAESSNER, M. F., and DAILY, B., 1959—The geology and late Precambrian fauna of the Ediacara fossil reserve. *S. Aust. Mus., Rec., 8*

GLAESSNER, M. F., and PARKIN, L. W., Ed., 1958—The geology of South Australia. *Geol. Soc. Aust., J., 5, 2*

HARRIS, W. J., and THOMAS, D. E., 1938—A revised classification and correlation of the Ordovician graptolite beds of Victoria. *Min. Geol. J. Vic., 1, 3*

HERBERT, C., 1973—A sedimentological appraisal of the Wianamatta Group. *Sympos. (8th) Adv. Study Sydney Basin, Dep. Geol. Univ. Newcastle (Abstracts)*

HILL, DOROTHY, 1934—The Lower Carboniferous corals of Australia. *Roy. Soc. Qld, Proc., 45*

———, 1954—Coral faunas from the Silurian of New South Wales and the Devonian of Western Australia. *Bur. Min. Resour. Aust., Bull., 23*

HILL, DOROTHY, and DENMEAD, A. K., Ed., 1960—The geology of Queensland. *Geol. Soc., Aust., J., 7*

HILLS, E. S., 1931—The Upper Devonian fish of Victoria. *Geol. Mag., 68*

———, 1958—A brief review of Australian fossil vertebrates. *Studies on Foss. Vertebr. (Presented to D. M. S. Watson; T. S. Westoll, ed.), Bristol*

HORWITZ, R. C., 1960—Géologie de la région de Mt Compass (Feuille Milang), Australie Méridionale. *Eclog. Geol. Helvetiae, 53, 1*

HOWCHIN, W., 1908—Glacial beds of Cambrian age in South Australia. *Geol. Soc. London, Quart. J., 54*

IRVING, E., 1964—PALAEOMAGNETISM. *Wiley and Sons, New York*

JELL, P. A., and JELL, J. S., 1976—Early Middle Cambrian corals from western New South Wales. *Alcheringa, 1, 2*
JENNINGS, J. N., 1972—The Age of the Canberra Landforms. *Geol. Soc. Aust., J., 19, 3*
JENNINGS, J. N., and BANKS, M. R., 1958—The Pleistocene glacial history of Tasmania. *J. Glaciology, 3*
JONES, B. G., 1972—Upper Devonian to Lower Carboniferous Stratigraphy of the Pertnjarra Group, Amadeus Basin, Central Australia. *Geol. Soc. Aust., J., 19, 2*
JONES, B. G., CAMPBELL, K. S. W., and ROBERTS, J., 1973—Correlation chart for the Carboniferous System in Australia. *Bur. Min. Resour. Aust., Bull., 156A*
JONES, P. T., 1962—The ostracod genus *Cryptophyllus* in the Upper Devonian and Carboniferous of Western Australia. *Bur. Min. Resour. Aust., Bull., 62, 3*
JOPLIN, GERMAINE A., 1955—A preliminary account of the petrology of the Cloncurry mineral field. *Roy. Soc. Qld, Proc., 66*
KNIGHT, C. L., Ed., 1975—Economic Geology of Australia and Papua New Guinea: 1. Metals. *Australas. Inst. Min. Met., Melbourne*
LASERON, C. F., 1918—Permo-carboniferous fenestellidae. *Roy. Soc. N.S. Wales, J., 52*
LESLIE, R. B., EVANS, H. J., and KNIGHT, C. L., Eds, 1976—Economic Geology of Australia and Papua New Guinea: 3. Petroleum. *Australas. Inst. Min. Met., Melbourne*
LEWIS, A. N., 1945—Pleistocene glaciation in Tasmania. *Roy. Soc. Tas., Pap., (for 1944)*
LONGMAN, H., 1933—A new dinosaur from the Queensland Cretaceous. *Qld. Mus., Mem., 10*
McKEOWN, K. C., 1937—New fossil insect wings. *Aust. Mus., Rec., 20, 1*
McWHAE, J. R. H., PLAYFORD, P. E., LINDNER, A. W., GLENISTER, B. F., and BALME, B. E., 1958—The stratigraphy of Western Australia. *Geol. Soc. Aust., J., 4, 2*
MARSHALL, C. G. A., MAY, I. W., and PERRET, C. J., 1964—Fossil micro-organisms: Possible presence in the Precambrian shield of Western Australia. *Science, 144*
MAWSON, D., 1949—The Elatina Glaciation. A third recurrence of glaciation evidenced in the Adelaide System. *Roy. Soc. S. Aust., Trans., 73*
MILES, R. S., 1978—Dipnoan (Lungfish) skulls and the relationships of the group: A study based on new species from the Devonian of Australia. *Linn. Soc. London (Zoo.), J., 61, 1/3*
MILLOT, J., and ANTHONY, J., 1958—Anatomie de *Latimeria*

chalumnae. Ctre Nat. Rech. Sci., Paris

MUIR, MARJORIE D., 1976—Proterozoic microfossils from the Amelia Dolomite, McArthur Basin, Northern Territory. *Alcheringa, 1*

———, 1978—Occurrence and potential uses of Archaean microfossils and organic matter. In: Glover, J. E., and Groves, D. I., Eds—Archaean cherty metasediments etc. *Geol. Dep. Univ. W. Aust., Publ., 2*

MURRAY, G. E., 1965—Cambrian and Precambrian petroleum—an appraisal. *Aust. Petrol Expl. Ass., J.*

NAYLOR, G. F. K., 1936—Note on the geology of the Goulburn district. *Roy. Soc. N.S. Wales, J., 69*

NYE, P. B., 1929—Osmiridium at Adamsfield. *Geol. Surv. Tas., Bull., 39*

OLLIER, C. D., 1978—Tectonics and geomorphology of the Eastern Highlands. In: Landform Evolution in Australasia. *Aust. Nat. Univ. Press, 5-47*

OPARIN, A. I., 1961—Life: Its Nature, Origin and Development. *Academic Press, New York*

OPIK, A. A. Ed., 1957—The Cambrian geology of Australia. *Bur. Min. Resour. Aust., Bull., 49, (Symposium)*

———, 1958a—The Cambrian trilobite *Redlichia*: Organization and generic concept. *Bur. Min. Resour. Aust., Bull., 42*

———, 1958b—The geology of the Canberra City district. *Bur. Min. Resour. Aust., Bull., 32*

———, 1961—The geology and palaeontology of the headwaters of the Burke River, Queensland. *Bur. Min. Resour. Aust., Bull., 53*

———, 1963a—Early Upper Cambrian fossils from Queensland. *Bur. Min. Resour. Aust., Bull., 64*

———, 1963b—*Nepea* and the nepeids (Trilobites, Middle Cambrian, Australia). *Geol. Soc. Aust., J., 10, 2*

———, 1967—The Mindyallan fauna of northwestern Queensland. *Bur. Min. Resour. Aust., Bull., 74*

———, 1975a—Templetonian and Ordian xystridurid trilobites of Australia. *Bur. Min. Resour. Aust., Bull., 121*

———, 1975b—Cymbric Vale fauna of New South Wales and early Cambrian biostratigraphy. *Bur. Min. Resour. Aust., Bull., 159*

OSBORNE, G. D., 1948—A review of some aspects of the stratigraphy, structure and physiography of the Sydney Basin. *Linn. Soc. N.S. Wales, Proc., 73*

PACKHAM, G. H., Ed., 1969—The geology of New South Wales. *Geol. Soc. Aust., J., 16*

PARKIN, L. W., Ed., 1969—Handbook of South Australian Geology. *Govt. Printer, Adelaide*

PERCIVAL, I. G., 1979—Ordovician plectambonitacean brachiopods from New South Wales. *Alcheringa, 3, 2*
PHILIP, G. M., 1965a—Lower Devonian conodonts from the Tyers area, Gippsland, Victoria. *Roy. Soc. Vic., Proc., 79*
———, 1965b—The Tertiary echinoids of South-Eastern Australia—III: Stirodonta, Aulodonta, and Camarodonta (1). *Roy. Soc. Vic., Proc., 78, 2*
———, 1966—The occurrence and palaeogeographic significance of Ordovician strata in northern New South Wales. *Aust. J. Sci., 29, 4*
PLAYFORD, G., 1965—Plant microfossils from Triassic sediments near Poatina, Tasmania. *Geol. Soc. Aust., J., 12, 2*
———, 1979—Floral evolution in the late Palaeozoic and early Mesozoic of Australia. *Aust. Geologist, Newsl., 26*
PLAYFORD, G., JONES, B. G., and KEMP, E. M., 1976—Palynological evidence for the age of the synorogenic Brewer Conglomerate, Amadeus Basin, central Australia. *Alcheringa, 1, 2*
PLAYFORD, P. E., 1980—Devonian "Great Barrier Reef" of Canning Basin, Western Australia. *Amer. Ass. Petrol. Geol., Bull., 64, 6*
PLUMB, K. A., 1979—The tectonic evolution of Australia. *Earth Sci. Reviews, 14, Elsevier, Amsterdam*
PONNAMPERUMA, C., 1972—The Origins of Life. *Dutton, London*
PRICHARD, C. E., and QUINLAN, T., 1962—The geology of the southern part of the Hermannsburg 1:250,000 Sheet. *Bur. Min. Resour. Aust., Rep., 61*
PRIDER, R. T., 1952—South-west Yilgarnia. In: Sir Douglas Mawson Anniv. Vol., *Univ. Adelaide Publ.*
RANFORD, L. C., COOK, P. J., and WELLS, A. T., 1965—The geology of the central part of the Amadeus Basin, Northern Territory. *Bur. Min. Resour. Aust., Rep., 86*
RANFORD, L. C., FORMAN, D. J., and WELLS, A. T., 1965—The geology of the northwestern part of the Amadeus Basin, Northern Territory. *Bur. Min. Resour. Aust., Rep., 85*
RETALLACK, G., 1975—The life and times of a Triassic lycopod. *Alcheringa, 1, 1*
REYMENT, R. A., 1963—Albian ammonites from Fossil Creek near Oodnadatta. *Roy. Soc. S. Aust., Trans., 88*
RIEK, E. F., 1970—Lower Cretaceous fleas. *Nature, 227, 5259*
———, 1976—Neosecoptera, a new insect suborder based on specimens discovered in the Late Carboniferous of Tasmania. *Alcheringa, 1, 2*
RITCHIE, A., and GILBERT-TOMLINSON, J. 1977—First Ordovician vertebrates from the Southern Hemisphere. *Alcheringa, 1, 4*

ROBERTS, J., 1964—Lower Carboniferous brachiopods from Greenhills, New South Wales. *Geol. Soc. Aust., J., 11, 2*

ROSS, JUNE P., 1961—Ordovician, Silurian, and Devonian Bryozoa of Australia. *Bur. Min. Resour. Aust., Bull., 50*

RUNNEGAR, B., and JELL, P. A., 1976—Australian Middle Cambrian molluscs and their bearing on early molluscan evolution. *Alcheringa, 1, 2*

SIMPSON, G. G., 1965—New record of a fossil penguin in Australia. *Roy. Soc. Vic., Proc., 79, 1*

SKWARKO, S. K., 1963a—New Mesozoic fossil occurrences in New Guinea and their stratigraphical significance. *Aust. J. Sci., 26, 1*

———, 1963b—Australian Mesozoic trigoniids. *Bur. Min. Resour. Aust., Bull., 67*

SMITH, K. G., 1972—Stratigraphy of the Georgina Basin. *Bur. Min. Resour. Aust., Bull., 111*

SMITH-WOODWARD, A., 1890—Triassic fishes, Railway Quarry, Gosford. *Geol. Surv. N.S. Wales, Mem. (Palaeont.), 2*

———, 1908—The fossil fishes of the Hawkesbury series at St. Peters. *Geol. Surv. N.S. Wales, Mem. (Palaeont.), 10*

SPATH, L. F., 1926—Note on two ammonites from the Gin Gin Chalk. *Roy. Soc. W. Aust., J., 12*

———, 1940—On Upper Cretaceous (Maestrichtian) ammonoidea from Western Australia. *Roy. Soc. W. Aust., J., 26*

SPRIGG, R. C., 1947—Early Cambrian (?) jellyfishes from the Flinders Ranges. *Roy. Soc. S. Aust., Trans., 71, 2*

———, 1949—Early Cambrian "jellyfishes" of Ediacara, South Australia, and Mount John, Kimberley district, Western Australia. *Roy. Soc. S. Aust., Trans., 73*

SPRY, A., and BANKS, M. R., Ed., 1962—The geology of Tasmania. *Geol. Soc. Aust., J., 9, 2*

STAUB, R., 1927—Der Bewegungsmechanismus der Erde. *Gebr. Borntraeger, Berlin*

SUESSMILCH, C. A., 1935—The Carboniferous period in eastern Australia. *Aust. N. Zeal. Ass. Adv. Sci., 22* (Melbourne Congr.)

TALENT, J. A., 1965—The stratigraphic and diastrophic evolution of Central and Eastern Victoria in Middle Palaeozoic times. *Roy. Soc. Vic., Proc., 79, 1*

TALENT, J. A., and PHILIP, G. M., 1956—Siluro-Devonian mollusca from Marble Creek, Thomson River, Victoria. *Roy. Soc. Vic., Proc., 68*

TAYLOR, T. G., 1910—Archaeocyathinae from the Cambrian of South Australia. *Roy. Soc. Aust., Mem., 2, 2*

TEICHERT, C., 1940a—*Helicoprion* in the Permian of Western Australia. *J. Palaeont., 14*

———, 1940b—Fossil nautiloid faunas from Australia. *Roy. Soc. W. Aust., J.*, 26

———, 1941—Upper Palaeozoic of Western Australia, correlation and palaeogeography. *Amer. Ass. Petrol. Geol., Bull.*, 25

———, 1944a—Permian trilobites from Western Australia. *J. Palaeont.*, 19

———, 1944b—Two new ammonoids from the Permian of Western Australia. *J. Palaeont.*, 18

———, 1953—A new ammonoid from the eastern Australian Permian province. *Roy. Soc. N. S. Wales, J.*, 87

TEICHERT, C., and GLENISTER, B. F., 1952—Fossil nautiloid faunas from Australia. *J. Palaeont.*, 26

———, 1954—Early Ordovician cephalopod fauna from north-western Australia. *Bull. Amer. Palaeont.*, 35

THOMAS, D. E., and KEBLE, R. A., 1933—The Ordovician and Silurian rocks of the Bulla-Sunbury area, and a discussion of the sequence of the Melbourne area. *Roy. Soc. Vic., Proc., NS*, 45

THOMAS, G. A., 1957—Lower Carboniferous Deposits in the Fitzroy Basin, Western Australia. *Aust. J. Sci.*, 19, 4

———, 1958—The Permian Orthotetacea of Western Australia. *Bur. Min. Resour. Aust., Bull.*, 39

———, 1965—An Echinoid from the Lower Carboniferous of the Fitzroy Basin, Western Australia. *Roy. Soc. Vic., Proc.*, 79, 1

THOMSON, B. P., 1969—Precambrian crystalline basement; Precambrian basement cover (The Adelaide System); the Kanmantoo Group and early Palaeozoic tectonics. In: Handbook of South Australian Geology, Parkin, L. W., Ed. *Geol. Surv. S. Aust. Publ.*

TILLYARD, R. J., 1923—Mesozoic insects of Queensland. *Geol. Surv. Qld, Publ.*, 273

———, 1925—Upper Permian insects of New South Wales. *Linn. Soc. N.S. Wales, Proc.*, 51, 60

TRAVES, D. M., and KING, D., Ed., 1975—Economic Geology of Australia and Papua New Guinea. 2. Coal. *Australas. Inst. Min. Met., Melbourne*

UREY, H. C., 1952—The Planets, Their Origin and Development. *Yale University Press*

VALENTINE, J. W., 1965—Quaternary mollusca from Port Fairy, Victoria, Australia, and their palaeoecologic implications. *Roy. Soc. Vic., Proc.*, 78

VEEVERS, J. J., 1959—Devonian and Carboniferous brachiopods from north-western Australia. *Bur. Min. Resour. Aust., Bull.*, 45 and 51

VEEVERS, J. J., and WELLS, A. T., 1961—The geology of the

Canning Basin, Western Australia. *Bur. Min. Resour. Aust., Bull.*, 60

VOISEY, A. H., 1959—Australian Geosynclines. *Presid. Address Aust. N. Zeal. Ass. Adv. Sci. Aust. J. Sci.*, 22, 5

WADE, R. T., 1935—The Triassic fishes of Brookvale. *Brit. Mus. Nat. Hist., Rec.*

WALKOM, A. B., 1921—On *Nummulospermum*, the probable megasporangium of *Glossopteris*. *Geol. Soc. London, Quart. J.*, 77, 4

———, 1944—The succession of Carboniferous and Permian floras in Australia. *Roy. Soc. N. S. Wales, Proc.*, 78

WALTER, M. R., 1972—Stromatolites and the biostratigraphy of the Australian Precambrian and Cambrian. *Palaeont. Ass. London, Spec. Pap.*, 11

WARREN, J. W., and WAKEFIELD, N. A., 1972—Trackways of tetrapod vertebrates from the Upper Devonian of Victoria, Australia. *Nature*, 238, 469–470

WEBB, A. W., and THOMSON, B. P., 1977—Archaean Basement rocks in the Gawler Craton, South Australia. *Search*, 8, 1/2

WEBBY, D. A., 1972—Devonian geological history of the Lachlan Geosyncline. *Geol. Soc. Aust., J.*, 19, 1

———, 1978—History of the Ordovician Continental Platform Shelf Margin of Australia. *Geol. Soc. Aust., J.*, 25, 1

WEGENER, A., 1924—The Origin of Continents and Oceans. (*Engl. Transl.*), *W. A. Skerl, London*

WELLMAN, P., 1973—Early Miocene Potassium-argon Age for the Fitzroy Lamproites of Western Australia. *Geol. Soc. Aust., J.*, 19, 4

———, 1979—On the Cainozoic uplift of the southeastern Australian highland. *Geol. Soc. Aust., J.*, 26, 1

WELLS, A. T., FORMAN, D. J., and RANFORD, L. C., 1965—Geological reconnaissance of the Rawlinson-Macdonald area, Western Australia. *Bur. Min. Resour. Aust., Rep.*, 65

WELLS, A. T., FORMAN, D. J., RANFORD, L. C., and COOK, P. J., 1970—Geology of the Amadeus Basin, Central Australia. *Bur. Min. Resour. Aust., Bull.*, 100

WILLIAMS, G. E., 1964—Geology of Kinglake District. *Roy. Soc. Vic., Proc.*, 77

WHITE, D. A., 1959—New stratigraphic units in north Queensland geology. *Qld. Gvt. J.*, 60

———, 1965—The geology of the Georgetown/Clarke River area, Queensland. *Bur. Min. Resour. Aust., Bull.*, 71

WHITEHOUSE, F. W., 1926/28—The Cretaceous ammonoidea of eastern Australia. *Qld. Mus., Mem.*, 8 and 9, (Parts 1, 2 3)

———, 1936/39—The Cambrian faunas of North-eastern Australia. *Qld. Mus., Mem.*, 11, (Parts 1, 2, 3)

———, 1941—Early Cambrian echinoderms. *Qld Mus., Mem., 12*, (*Parts 4, 5*)

———, 1941/45—The Cambrian faunas of northeastern Australia. *Qld. Mus., Mem., 12A, (Parts 4, 5)*

———, 1953—The Mesozoic environments of Queensland. *Aust. N. Zeal. Ass. Adv. Sci., Rep., 29*

WILSON, A. F., COMPSTON, W., JEFFERY, P. M., and RILEY, G. H., 1960—Radioactive ages from the Precambrian rocks in Australia. *Geol. Soc. Aust., J., 6, 2*

WOODS, J. T., 1962—A new species of *Hatchericeras* (ammonoidea) from North Queensland. *Geol. Soc. Aust., J., 8, 2*

WOODWARD, A. S., (see also Smith-Woodward), 1906—On a tooth of *Ceratodus* and dinosaurian claw from the Lower Jurassic of Victoria. *Ann. Mag. Nat. Hist. (Ser. 7), 18*

WOPFNER, H., 1972—Depositional history and tectonics of South Australian sedimentary basins. *Min. Resour. Rev. S. Aust., 133*

WOPFNER, H., CALLEN, R. A., and HARRIS, W. K., 1974—The lower Tertiary Eyre Formation of the southwestern Great Artesian Basin. *Geol. Soc. Aust., J., 21, 1*

WRIGHT, C. W., 1963—Cretaceous ammonites from Bathurst Island, North Australia. *Palaeontology, 6, 4*

WYATT, D. H., and WHITE, D. A., 1960—Carboniferous of the upper Burdekin River Valley. *Geol. Soc. Aust., J., 7*

INDEX

The following abbreviations may be used:
B	for Basin		**Gp**	for Group
Cgl	for Conglomerate		**Lst**	for Limestone
Dol	for Dolomite		**Sh**	for Shale
F	for Formation		**Ss**	for Sandstone

Names of States and Territories are abbreviated as follows:
W.A.	Western Australia		N.S.W.	New South Wales
S.A.	South Australia		A.C.T.	Australian Capital Territory
N.T.	Northern Territory		VIC.	Victoria
QLD.	Queensland		TAS.	Tasmania

Geographical coordinates show first latitude, then longitude, e.g.: Yarrol Qld. 24°58′–151°21′. Localities for which no geographical coordinates are given in the Index can be found on the Geographical Map.

There is no special glossary for the simple reason that the meaning of geological terms is sure to be explained somewhere in the text; such parts of the text are marked in the Index by an asterisk (*).

Aboriginals 286–7
absence of polar ice-caps 218
acacias 244, 279
Acado-Baltic fauna (trilobites) 85
Acanthoceras 247
Aconeceras 247
Actinoceras 122
Actinocystis 138
ADAMSFIELD TAS. 102, 42°43′– 146°20′
adaptation 77, 280
ADELAIDE S.A. 87, 165, 167, 269, 286
Adelaide Geosyncline 58, 61, 62, 63, 84, 87, 108
ADELAIDE HILLS S.A. 213
Adelaidean Era (Erathem) 57, 58, 60

aeolian deposits 5
Aetheolepis 222
AFRICA 38, 276
Agathiceras 189
age and metamorphism of rocks 39
age of continents 38
age of opal 241
age of rocks (relative) 21, 22, 23, 226
age of rocks (years, or absolute) 23, 24
agglomerate 18
Agicondian System 48
agnostids 86, 87, 97
AIRLY COAL MINE, LITHGOW N.S.W. 182, 183, 210
AJAX MINE, COPLEY S.A. 94

ALASKA 62, 93, 158, 174, 226
ALBANY W.A. 37, 50, 56
Albany-Fraser sequence (zone) 50, 56, 58
Albian Epoch 233–48
ALCURAH QLD. 217, 26°06'–148°22'
ALDINGA S.A. 262, 35°16'–138°29'
ALEXANDRA VIC. 131, 136, 37°12'–145°43'
algae 26, 69, 70, 79, 135, 150, 214
ALICE SPRINGS N.T. 28, 106, 107, 197, 239
Alice Springs Orogeny 106, 126
alkaline extrusives 216, 265
Allagecrinidae 189
ALLANDALE N.S.W. 29, 171, 186, 32°44'–151°25'
alluvials 6, 103, 258, 270*, 283
Alpine Orogeny 235, 257*
alpinotype geosynclines 52
ALPS 10, 14, 179, 193, 257, 271
Amadeus **B** 106, 108, 109, 112, 126, 217
amber 26
ammonites 193, 211, 212*, 214, 226, 227, 236, 242, 246, 247, 248, 278, 283
amphibians 73, 138, 142, 156, 183, 184, 185, 201, 209, 210, 223, 225
Amphioxus 141
Anagaudryceras 247
Anaspides 207
ancient river systems 258, 259, 266, 267, 269, 270, 271, 275, 283
ANDAMOOKA S.A. 241, 30°27'–137°10'
ANDES 193, 257
andesite 16, 150
ANGARA 181
angiosperms 70, 71*, 244
anhydrite 145
animal kingdom 71
ANTARCTICA 38, 61, 66, 93, 158, 166, 167, 180, 187, 200, 247, 271–4
anticline and syncline 6
ant-lions 207
apatite 243

Aphanaia 187
Aphnelepis 222
Aphrophyllum 155
appalachiotype geosyncline 50
Aptian Epoch 231–48
Arab scientists 20
ARAFURA SEA 48
aragonite 111
ARAL SEA 202
Arandaspis 112, 141
Araucaria 71, 206, 221
Araucarites 244
Archaean bastions 38
Archaeocyathus 88, 92, 93
Archaeomaene 222
Archaeopteris 137
Archaeopterix 225
Archaeozoic (Archaean) Eon (Eonothem) 23, 31–45
Archaeozoic (Archaean) orogenies 44
Archelone 246
Archeozoon 79
Arckaringa **B** 167
ARCKARINGA CREEK 194, 28°02'–134°55'
ARDROSSAN S.A. 94, 24°26'–137°55'
Areyonga **F** 63
ARGENTINA 220
Argyllian trough 48
arid topography 4, 64
aridity 64, 218, 275, 276, 277
ARIZONA 206
Arkell, W. J., Dr. 226
ark-shell 111
ARNHEM LAND N.T. 48, 52, 66, 83
artesian basins 160, 204, 217, 236–40*, 261, 262
artesian bores 160, 217, 237–9, 261
arthropods 73, 94, 96, 122
ARTHUR RANGE TAS. 273, 43°09'–146°15'
ARTHUR RIVER W.A. 161, 33°10'–117°15'
articulated vertebrae (fish) 156
Arumbera **Ss** 66, 80
Arunta System 39
asaphids 111
Asaphus 111
ascidians 73

INDEX 313

Ashburton **F** 58
ASHBURTON RIVER W.A. 37, 47, 58
ASHFORD N.S.W. 175, 29°19′–151°06′
ash showers 130, 174, 205, 219, 268
ASIA 15, 42, 94, 158, 167, 181, 210, 213, 233, 244, 271, 280
asiatic wolf 287
Astartila 187
asymmetry of folded mountain ranges 9
Atavus Time 86, 87
ATHERTON TABLELAND QLD. 268, 287
Athyris 117
atmosphere 76, 78
Atrypa 122
ATTUNGA N.S.W. 132, 30°56′–150°50′
augite 17
AUSTRALASIA 257
Australasian Antarctic Expedition (1913) 166
AUSTRALIAN ALPS 84, 265, 266, 267
AUSTRALIAN CAPITAL TERRITORY (A.C.T.) 104, 129
AUSTRALIAN MUSEUM, SYDNEY N.S.W. 27, 183, 189, 211
Australiarcula 248
Australiceras 247
Austropelor 225
Austrosaurus 245
avalanches 4
Avicenna 20
Aviculopecten 187
avifauna 282
AVON RIVER VIC. 137, 37°48′–146°53′

BACCHUS MARSH VIC. 197, 258, 37°41′–144°26′
back-reef deposits 125
bacteria 26, 69, 79
Baculites 247
Badgeradda **Gp** 53, 56
Badgeradda-Moore-Yandanooka series 56
Bajocian Epoch 226
Baker **F** 183
Bakevellia 212
Balanoconcha 148

Balboa 28
BALCOMBE BAY VIC. 261, 285, 38°15′–145°01′
BALD HILL VIC. 197, 37°39′–144°25′
BALD HILLS N.S.W. 258, 33°29′–149°32′
BALGO MISSION W.A. 125, 20°09′–127°58′
BALGOWAN COLLIERY QLD. 223, 27°26′–151°43′
BALLARAT VIC. 103
BALLINA N.S.W. 241, 28°52′–153°34′
BALMAIN COLLIERY, SYDNEY N.S.W. 170
bandicoot 281
Bangemall **B** 64
Bangemall **Gp** 58
Banksia 279
Baragwanathia 135
BARKLY TABLELAND N.T. 52, 97
BARN BLUFF TAS. 220, 41°44′–145°55′
barnacles 73
Barossa Metamorphics 53
BARRABA N.S.W. 132, 134, 135, 152
Barraba **Gp** 134
Barraba Mudstone 134
BARRABOOL HILLS VIC. 220, 38°10′–144°
BARRIER RANGE N.S.W. 50
BARROW ISLAND W.A. 162, 214, 242
basalt 16, 57, 83, 84, 92, 216, 220, 241, 258, 259, 264–9
base level of erosion 6
Bass **B** 233, 261, 265
BASS STRAIT 129, 168, 257, 259, 269–74
Bassian Isthmus (Pleistocene) 286
Batavia Deep Lead Qld. 219
BATEMAN'S BAY N.S.W. 173
BATHURST N.S.W. 119, 128, 134, 175, 258
BATHURST ISLAND N.T. 243, 247
bauxite 277
BAVARIA 255, 227
Baywulla **F** 149
BEACON HILL BRICK PIT, SYDNEY N.S.W. 201, 206
bear 278

BEAUDESERT QLD. 218, 27°59′–153°
BEAUMARIS, MELBOURNE VIC. 281, 282, 285
beaver 280
bêche-de-mer 72, 94
BEETLE CREEK QLD. 97, 13°39′–143°11′
beetles 183, 206
belemnites 227, 228, 248, 278, 283
Belemnopsis 228
Bell magpies 282
Bell Point **Lst** 133
BELLENDEN KER RANGE QLD. 266, 17°12′–145°51′
Bellerophon 122, 140
Bellerophontidae 188
BELMONT N.S.W. 182, 183, 33°02′–151°40′
BEN LOMOND TAS. 220, 41°33′–147°40′
BENAMBRA HIGHLANDS VIC. 104, 129, 36°57′–147°42′
Benambran Orogeny 104, 106, 113, 125, 129
BENDIGO VIC. 103
Bendigonian Stage (Age) 103
bennettitales 70, 71
Beringer, John Bartholomew, Prof. 20
BEROWRA FLAGSTONE QUARRY, SYDNEY N.S.W. 211
Beudanticeras 247
biblical flood (Diluvium) 20
BIGGENDEN QLD. 269, 25°31′–152°03′
billabong 200
BILOELA GOVERNMENT DOCKS, SYDNEY N.S.W. 209
BINDI VIC. 133, 37°07′–147°49′
BINGERA N.S.W. 132, 135, 146, 155, 24°54′–152°12′
biochemical action in rocks 5
biosphere 5
BIRD ROCK BLUFF VIC. 261
birds 73, 210, 225, 246, 277, 282
birds of paradise 282
bird's-nest ferns 205
birth of Australia 33–42
Bitter Springs **Dol** 56, 58
bivalves 72, 96, 111, 122, 131, 139
black soil country Qld. 270

BLAIR ATHOL QLD. 178, 22°42′–147°33′
BLANCHE POINT, ALDINGA S.A. 262
blastoids 72, 121
Blina **Sh** 196, 208, 209
Blinasaurus 209
blind trilobites 96, 97
block-faulting 266, 267
blue asbestos 47, 48
BLUE HILLS VIC. 262, 35°13′–142°34′
blue metal 174
BLUE MOUNTAINS N.S.W. 7, 70, 137, 174, 175, 198, 199, 201, 258, 267, 268
blue-bottle 108
Blythesdale **Gp** 238
BOGANTUNGAN QLD. 156, 23°39′–147°18′
Bolindian Stage (Age) 103
Boliteceras 247
BOMBO N.S.W. 174, 34°40′–150°51′
BONANG HIGHWAY VIC. 116, 37°11′–148°44′
BONAPARTE GULF (JOSEPH BONAPARTE GULF) 83, 105, 124, 144, 145, 164, 196
Bonaparte Gulf **B** 160, 164, 196
bone breccia (bone-bed) 270, 281
bony fish 156, 220, 221, 222, 245
bony pike 143
BOONAH QLD. 218, 28°–152°41′
BORDER RIVERS DISTRICT QLD. 175
BORENORE N.S.W. 30, 120, 122, 33°15′–148°59′
boring mussels 236, 244
BORNEO 48
BOTANY BAY, SYDNEY N.S.W. 274
Bothriceps 210
Bothriolepis 142
bottlebrush 279
boulder beds 102, 161, 165, 168, 173, 242, 262
boulder clays 60–3
BOULIA QLD. 237
BOWEN QLD. 176, 180
Bowen **B** 167, 176–9, 204, 237, 239
BOWEN RIVER QLD. 176
Bowen "series" 176
bower birds 282

INDEX

BOWNING N.S.W. 122, 34°46′–148°49′
Bowning Orogeny 129
brachiae of brachiopods 95
brachiopod limestone 139
brachiopods 72, 95, 109, 111, 117, 125, 129, 139, 144, 148, 149, 155, 173, 186, 187, 248
Brachythyris 148
bracken 205
brackish waters 133, 134, 145, 196, 198, 201, 208, 209, 218, 264
Braeside Tillite 164
Brahmaites 247
BRAMBLE CAY QLD. 269, 9°07′–143°52′
BRANXTON N.S.W. 175, 186, 32°40′–151°21′
BRAZIL 38
break (interruption) in the deposition of sediments 99, 237, 254
break-up of continental crust 40
breathing of earth 179
Bresnahan **Gp** 58
Brewer **Cgl** 126
brick shales 117, 208
BRIDGE CREEK, MANSFIELD VIC. 146
BRINGO W.A. 225, 28°45°–114°50′
BRISBANE QLD. 103, 116, 153, 177, 202, 203, 206, 217, 231, 239, 255, 256, 258
Brisbane Schists 66, 84, 104, 203
BRITISH COLUMBIA 24, 94, 226
BRITISH ISLES 271
BRITISH MUSEUM 209
brittle stars 227, 248
Brockman Iron **F** 47
BROKEN HILL N.S.W. 38, 52
Bronteus 140
Brontosaurus 223
BROOKVALE, SYDNEY N.S.W. 27, 201, 203, 206, 207, 209
BROOME W.A. 105, 125, 231
brown coal (lignite) 257, 258, 259
BRUNSWICK, MELBOURNE VIC. 121
BRUNY ISLAND TAS. 169, 221, 43°15′–147°20′
bryophytes 70
bryozoa 72, 109, 122, 125, 186, 257, 285

BUCHAN VIC. 30, 130–2, 37°30′–148°11′
Buchan **Gp** 133
Buchan **Lst** 130, 132
BUCHAN RIVER VIC. 116
Buchanosteus 142
Buchia 228
bugs 206
Bulgadoo **Sh** 183
Bumastus 112
BUNBURY W.A. 216
BUNDABERG QLD. 176, 219
Bundamba **Gp** 203, 219
BUNDANOON N.S.W. 174, 34°39′–150°18′
Bunga beds 141
bunya-bunya 221
BURDEKIN RIVER QLD. 116, 130, 137, 138
BUREAU OF MINERAL RESOURCES, GEOLOGY AND GEOPHYSICS, CANBERRA (BMR) 102, 107
Burindi "series" 148
Burra **Gp** 60
burrawang 71, 179, 221
BURRINJUCK DAM N.S.W. 141, 34°58′–148°42′
Burrum Coal Measures Qld. 235
BURT RANGE W.A. 125, 15°48′–28°57′
butterflies 206
buttes 163

caddis-flies 183
Cainozoic Era 23, 251–88
CAIRNS QLD. 116, 135, 178
Calamites 154, 181
Calceola 138
Calceolispongia 189
Callanna beds 55, 57
Callide Coal Measures 203
Callistemon 279
Callythara **F** 186
Calpionella 228
Calymene 117, 122
calyx of crinoids 72, 121*
Camarotoechia pleurodon 139
Cambria 113
Cambrian Period 23, 64, 66, 83–98, 104, 105, 217
camouflage 79

CANADA 15, 35, 38, 62, 94, 141, 271
CANBERRA A.C.T. 84, 103, 119, 129, 139, 266
CANIA QLD. 146, 149, 24°38′–150°58′
CANNINDAH QLD. 149, 24°54′–151°13′
Canning **B** 132, 160, 163, 164, 196, 214–6, 226–8, 239, 242
CANNING (OR GREAT WESTERN) DESERT W.A. 47, 105, 113, 125, 227, 276
CANNING STOCK ROUTE W.A. 163
CANTERBURY, NEW ZEALAND 207
CAPE JERVIS S.A. 94, 35°36′–138°08′
CAPE PATERSON VIC. 224, 38°40′–145°37′
CAPE RANGE W.A. 214, 22°07′–113°57′
CAPE RAOUL TAS. 220, 43°15′–147°47′
CAPE YORK QLD. 236, 10°41′–142°32′
CAPE YORK PENINSULA QLD. 38, 50, 104, 105, 135, 146, 150, 154, 204, 219, 236
CAPERTEE N.S.W. 129, 33°09′–149°59′
"Capertee Islands" (Ordovician) 129
CAPITOL HILL, CANBERRA A.C.T. 104
Capitosaurus 209
captorhinomorphs 210
carapace of arthropods 122*, 207
CARAPOOK VIC. 222, 36°40′–143°21′
carbon 76
carbonate of lime (calcium carbonate) 6
carbonatite 216
Carboniferous Period 23, 144–56
Carborough **Ss** 204
CARDABIA RANGE W.A. 242, 24°35′–113°43′
Cardiopteris 154
Carey, W. S., Prof. 10, 11
CARNARVON W.A. 160, 162, 256
Carnarvon **B** 126, 139, 145, 154, 160–2, 187, 196, 214, 226, 227, 242, 243, 247, 248, 249, 256, 261, 263, 276, 285
Carpentaria **B** 218, 231, 237
Carpentarian Era (Erathem) 23, 50
CARPENTARIAN REGION 39, 44, 57, 58, 63, 66, 84, 217, 228, 233
cartilaginous fish 156, 222
Cascades **Gp** 169
CASPIAN SEA 202
Castlemainian Stage (Age) 103
casts and moulds of fossils 19, 111
casuarinas 244
catalyst enzymes 77
CATHEDRAL RANGE VIC. 131, 37°24′–145°45′
Catillocrinidae 189
cats 278
Cave Hill beds 131
caves 26, 263, 270, 277, 281
CAWARRAL QLD. 132, 23°15′–150°41′
cedar 206
Cedroxylon 206
CELEBES 235
celery-top pine 71, 244
Cenomanian Epoch 243
CENTRAL AUSTRALIA 39, 81, 107, 108, 124, 217, 270
CENTRAL NEW SOUTH WALES 103, 130, 133, 217
CENTRAL QUEENSLAND 113, 133, 153, 155, 244, 270
CENTRAL VICTORIA 103, 130, 131
CENTRALIAN RAILWAY 276
cephalopods 73, 96, 109–10*, 117, 125, 140, 155, 188, 211, 212, 225, 228, 246–8
Ceratites 212
Ceratodus 222
chain coral 120
chalcedony 180, 203
chalk 213, 242*, 249
chambers of cephalopod shell 109, 110, 140, 227
CHARLEVILLE QLD. 218
charophytes 69
CHARTERS TOWERS QLD. 217, 269
chemical erosion 5
chemical solution and reconcentration 48
chert 92*, 116, 148
Chewtonian Stage (Age) 103

INDEX

CHILE 247
CHILLAGOE QLD. 116, 17°09′–144°31′
Chillagoe **F** 116
CHINA 62, 94
CHINCHILLA QLD. 218, 26°44°–150°38′
chitinous shells 208
chitons 72, 122*
Choanichthyes 142, 143
chocolate shales 63
Chondrichthyes 143
Chonetidae 155
Chonetes 122, 148, 155
cicadas 206
Cinnamomum 279
circumpacific volcanism 14
Cladochonus 186
Cladophlebis 181, 205, 221, 243
cladoxyles 70
Claraia 212
CLARE S.A. 57
Clarence River **B** 201, 202
CLARENCETOWN N.S.W. 154, 32°35′–151°47′
CLARKE RIVER QLD. 116, 19°31′–144°57′
classification of fossil plants 278, 279
Clatrotitan 207
Cleithrolepis 208
Clematis **Ss** 204
Cleobis 187
CLERMONT QLD. 135
CLIFTON QLD. 218, 270, 27°56′–151°54′
Climacograptus 109
climate 62, 64, 88, 93, 118, 158, 162, 176, 179, 180, 182, 200, 205, 207, 218, 225, 239, 263, 264, 270, 271, 275, 283–6
climatic belts 88, 118, 119
CLONCURRY QLD. 217, 237
club-mosses 70, 135, 137, 153, 154
CLUTHA STATION QLD. 245, 20°22′–142°31′
Clyde Coal Measures 170
CLYDE RIVER N.S.W. 134, 172, 35°36′–150°10′
clymenioids 125, 155
coal measures (deposits) 144, 153, 157, 162, 165, 166, 169–78*, 196, 197, 202, 217, 221, 223, 235, 236, 255, 257, 259
coal swamps 153, 157, 171, 172, 202, 203, 218, 235, 257
coal-charcoal (fossil charcoal) 286
COALCLIFF, ULLADULLA N.S.W. 170
COALSTOUN LAKES QLD. 269, 25°37′–151°53′
Coastal **Lst** 276
coastal plains 275
coat-of-mail shell 72, 122*
COBAR N.S.W. 104, 130, 146
COBARGO N.S.W. 27, 36°23′–149°54′
Coccolepis 222
Coccosteus 142
COCKBUNDOON RIVER N.S.W. 134
cockle 72
cockroach 206
coelacanths 142, 222, 227
coelenterates 71, 138
COFFS HARBOUR N.S.W. 202
cold-water marine faunas 185, 186, 187, 239, 248
collecting fossils 27–30
Collenia 79
Collie Coal **B** 162
Collinsville Coal Measures Qld. 176
collision of continents 10, 53
comparative anatomy 74
compass directions 42
competition between races 77, 280, 286, 287
compound animals 108
compound eyes 111, 122, 140
conglomerate 6*, 56, 57, 63, 102, 106, 133, 134, 149, 161, 173, 176, 198, 202, 241
conifers (pines) 70, 71, 153, 179, 206, 244
conodonts 106, 109, 112*, 126, 144
Conophyton 79
continent 56
continental crust 41, 53, 144
continental drift 10, 11, 180
continental outline 56, 153, 167, 168
continental shelf 56, 164
contradictory evidence 200, 218, 256

Conularia 117, 188
convection currents 9
COOBER PEDY S.A. 241
COOKTOWN QLD. 146, 178, 219, 236
COOMA N.S.W. 27
COONABARABRAN N.S.W. 175, 198, 31°16′–149°17′
Cooper (Coopers Creek) **B** 107, 115, 118, 158, 167, 178, 237
COOPERS CREEK QLD./S.A. 264
COORDEWANDY HOMESTEAD W.A. 162, 25°36′–115°58′
COPLEY S.A. 94, 30°33′–138°25′
copper 47, 50, 198
CORAL SEA 195
corallian faunas 102, 118–21, 131, 132
corallian limestone 116, 118, 119, 120, 121, 134, 138, 149, 263
corals 27, 71, 109, 119–21, 125, 126, 129, 132, 144, 148, 149, 155, 186, 239, 248, 256, 263, 285
Corbicula 236
cordaitales 70, 153, 179
CORNWALL QLD. 217, 26°15′– 143°33′
CORNWALL COLLIERY TAS. 197
Corunna **Cgl** 56
cosmic radiation 78
COSTERFIELD VIC. 117, 36°52′– 144°48′
Cothonion 93
Cotylosaurus 210
cowry shell 285
crabs 73, 94
CRADLE MOUNTAIN TAS. 220, 41°41′–145°57′
Crania 95
Cratochelone 246
crescent-shaped dunes (barchans) 276
Cretaceous Period 23, 229–49, 280
Cretaceous System 229–49
crinoidal limestones 121, 139
crinoids 72, 95, 121, 129, 139, 155, 189, 248
Crioceras 247
crocodiles 277, 282, 283
cross-bedding 169, 199*
Crossopterygii 142, 156, 184

crows 282
crustal imbalance 7–14, 84, 124, 157, 193, 219, 239, 264, 265
Cryptozoon 79
crystallization (igneous) 16, 17, 18
CUDAL N.S.W. 141, 33°17′–148°44′
cuestas 163
CUMBERLAND PLAINS N.S.W. 199, 33°45′–150°45′
cunjevoi 73
CUNNAMULLA QLD. 241, 28°04′– 145°41′
CURDAMUDA WELL W.A. 214, 23°45′–115°15′
CURRABUBULA N.S.W. 152, 31°16′– 150°44′
current-bedding 131, 199
cuscus 281
cuttlefish 73, 109, 227
Cyathophyllum 120
cycads 70, 71, 179, 180, 205, 213, 243
cyclical mountain-building 10
cycloid fish scales 222
Cyclotosaurus 209
Cygnet Coal Measures 169
Cymbionites 95
Cyrtograptus 117
cystoids 72, 121
Cyzicus 198, 201, 208, 209

Dadoxylon 179
DALBY QLD. 218, 27°11′–151°15′
Dalmanites 117, 122, 140
Dalwood **Gp** 170
DALY RIVER N.T. 97
DAMPIER PENINSULA W.A. 231, 17°–123°
DANDARAGAN W.A. 163, 30°40′– 115°42′
Dandaragan Gasfield 163
Dargile beds 122
DARLING DOWNS QLD. 217, 223, 270, 283, 30°–147°
DARLING RIVER QLD. 167, 217, 233, 262, 264
DARLING RANGE W.A. 271, 32°30′– 116°
DARNLEY ISLAND QLD. 269, 9°35′– 143°46′
DARRA QLD. 255, 27°34′–152°57′

DARRINAL VIC. 165, 36°55′–144°42′
Darriwillian Stage (Age) 103
DARWIN N.T. 63, 164, 233, 243, 247
Darwin, Charles 74
dasyures 280
DAVENPORT RANGE N.T. 48–52
David, T. W. E., Prof. 269
DAWSON RIVER QLD. 258
DAYLESFORD VIC. 103, 37°21′–144°09′
DE GREY RIVER W.A. 163
DEAD SEA 7
deep leads 26, 209, 258*, 282
deep-sea deposits 8, 52, 84, 85, 102, 129, 135, 148
deep-sea sounding 269
deep-sea trenches 8, 40
DEEPWATER N.S.W. 258, 29°26′–151°51′
deer 278
deformation of rocks 14
Delamerian Orogeny 88
DELEGATE N.S.W. 116, 37°03′–158°57′
delta sedimentation 6, 161, 199, 209, 264
Deltasaurus 209
Deltopecten 187
DENMARK HILL, IPSWICH QLD. 203, 206
DEPOT CREEK S.A. 57, 32°14′–137°55′
DERBY W.A. 228, 231
Derbyia 187
DERRENGULLEN CREEK N.S.W. 119, 34°51′–148°53′
DERWENT RIVER, HOBART TAS. 169, 274, 280
Desert (Canning) **B** 163
desert conditions 88, 229, 275, 276
detritus 6, 102–5, 152
DEVIL'S COACHHOUSE TAS. 168
devil's fingers (belemnites) 227
Devonian Period and System 23, 124–43, 173
DEVONSHIRE 113
DIAMANTINA RIVER QLD. 264, 287
diamonds 216, 258
diatomaceous earth 70, 265

diatoms 69, 70, 265
Dicellograptus 109
dicotyledons 71
Dicroidium 193, 205
Dictyoclostus 148
Dielasma 187
DIGLUM QLD. 146, 149, 24°14′–151°07′
diluvian theory 20
Diluvium 267
Dimegelasma 148
Dimitopyge 189
Dinantian Epoch (series) 144
dingo 287
dinosaurs 222–3, 245, 277, 278
diorite 17, 219
diphycercal fishtail 142
Diplograptus 109
Dipnoi 142, 143, 156, 184, 185, 208, 245
Dipnorhynchus 143
Diprotodon 280, 281, 287
Diptera 183
disconformity 58
Disphyllum 132, 138
dissection of tablelands 267
diversion of drainage 258
Dogger Epoch (series) 214
dogs 278, 280
dolerite 220, 273
DOLODROOK RIVER VIC. 92, 37°34′–146°43′
dolomite 17, 56, 57, 88*
dolphins 285
domal structures (opposite: basinal) 162
DONNYBROOK W.A. 223, 35°35′–115°49′
double helix 76
dragonflies 206
DRAYTON QLD. 270, 27°36′–151°55′
drift coal 180
drift of continents 10, 11, 166
Dromaeus 282
Dromornis 282
drowned coastal features 274, 275
DRUMMOND RANGES QLD. 135, 150, 155, 156, 177, 24°43′–146°41′
Du Toit, A. L., Prof. 10, 11
DUARINGA QLD. 258, 23°43′–149°40′

DUBBO N.S.W. 104, 175, 217
Duchess Land (Cambrian) 85, 92
ducks 282
Dundas **Gp** 92
Dundas Peninsula (Miocene) 262
dune alignments 276
dunes 5, 163, 198, 201, 275, 276
Dunkleosteus 142
DUNOLLY VIC. 103, 36°52′–143°44′
Dunstan, **B** 203
DURHAM DOWNS QLD. 223, 26°10′–149°02′
duricrust (silcrete) 277
dwarfed fauna 182
dyke rocks 16

EAGLEHAWK NECK TAS. 168, 43°01′–147°55′
ear bones of whales 285
earthquake 14, 152, 179
earthworm 72
EAST KIMBERLEY W.A. 42
EAST LONDON, SOUTH AFRICA 142
EASTERN NEW SOUTH WALES 103, 118, 177, 239, 258
EASTERN QUEENSLAND 104, 115, 148, 153, 176, 235, 236, 244, 255
EASTERN TASMANIA 168, 169
EASTERN VICTORIA 103, 115, 116, 117, 130, 131
Eastmanosteus 142
Eastonian Stage (Age) 103
echidna 279
echinoderms 72, 80, 95, 109, 121, 139, 146, 155, 248, 263
ecology of marine environment 247, 248
economic significance of geological systems 157
EDEN N.S.W. 198
Ediacara fauna 60, 80
egg-stone 148
EGYPT 285
Eifelian Stage (Age) 124
EILDON DAM VIC. 130, 131, 37°07′–145°58′
Eildonian Stage (Age) 116
EINASLEIGH QLD. 50, 52, 104
Eldon **Gp** Tas. 117, 122
elephant 278

elm 71
Emsian Stage (Age) 124
emu 282
Encrinurus 117
enigma of evolution of monotremes 279
Eocene Epoch (series) 23, 253–7, 278, 279, 282, 283–5
eocrinoidea 95
Eon (Eonothem) 34
Eoses 206
epeirogenetic movements of the crust 144, 179, 258, 264–7
Epiceratodus 245, 283
epicontinental seas 8, 125
EQUATOR, position in geological past 88, 158
equisetales 70
Era (Erathem) 36
EROMANGA QLD. 241
Eromanga **B** 237
erosion 3–6, 44, 169, 173, 177, 213, 229, 236, 255, 263, 266, 267, 268
erratics 4*, 161–75, 177, 185, 187, 239, 271
Erskine **Ss** 196
Erythrobatrachus 209
ESK RIVER QLD. 202, 27°14′–152°28′
Esk Trough 202
ESPERANCE W.A. 37
Estheria (now *Cyzicus*) 198
ETON VALE QLD. 270, 26°07′–146°54′
Eubaculites 247
eucalyptol 279
Eucalyptus 279
Euomphalidae 188
Euomphaloceras 247
EURASIA 167
Euroka Ridge 237
EUROPE 15, 93, 94, 123, 135, 143, 166, 167, 181, 182, 184, 207, 211, 226, 229, 271, 273, 280
Eurydesma 187
Euryphyllum 186
eurypterids 123
Euryzygoma 281
Evans, J. W., Dr. 183
evaporites 145
Evergreen **Sh** 221

INDEX 321

evolution of Australian mammals 281
evolution (growth) of continents 40
evolution of species 21, 22, 23, 73, 74, 75, 110, 136, 153, 184, 193, 205, 226, 247, 248, 277, 280, 283, 284
evolutionary peaks 111
excess of lime (calcium) 78
excursions 28
expansion of earth 8
external shell of cephalopods 110
extinct volcanoes 216, 258, 265, 268, 269
extinction of species 21, 22, 23, 154, 189, 193, 247, 275, 277, 278, 280, 283, 287
extraction of fossils 27, 30
extrusive rocks 16, 18, 129, 178, 216, 219, 220, 258, 259
Eyre **F** 255
EYRE PENINSULA S.A. 38, 47, 50, 52

faceted eyes 111, 122, 140
Falciferella 247
falcon 282
Famenian Stage (Age) 124, 144
fanglomerate 102, 126
FAR EAST 71, 206
fault corridor (graben) 266, 267
faulting 178*, 266, 267
faunal differences between continents 22
Favosites 120, 138
feathers of birds 210, 225
feldspar 17
Fenestella 186
Fenestellidae 186
FENNELL BAY N.S.W. 180, 33°05′–151°33′
FERNBROOK N.S.W. 30, 119, 30°23′–152°35′
ferns 70, 137, 153, 154, 181, 182, 205
Ferntree **Gp** 169
FERNVALE QLD. 116, 27°27′– 152°39′
Fickling beds 56
FIELD, BRITISH COLUMBIA 94
FIJI ISLANDS 195
filicales 70

Fingal Coal Measures Tas. 197
Finke **Gp** 126
FINKE RIVER 166, 167
fiords 273, 274
first forests 153, 179–82
first higher flowering plants 244
first land animals 136, 137, 138
first plants on land 132, 136, 137
first tetrapod vertebrates 137, 138, 183
first vertebrates 112, 141
fish 73, 134, 141–3, 155, 156, 183, 201, 208, 209, 219, 222, 245, 255, 258, 285
fissile (fissility) 18
Fitzroy **B** (or Trough) W.A. 125, 140, 144, 160, 163, 164, 196, 208, 209
FITZROY RIVER QLD. 105
FITZROY RIVER W.A. 105, 163, 216, 228
flagellates 69
FLINDERS ISLAND TAS. 129, 262
FLINDERS RANGES S.A. 38, 50, 53, 60, 87, 94, 196, 217, 218
flood deposits 152, 161, 200, 256
flow texture of volcanic rocks 16
flying reptiles 224, 277
folded mountain ranges 7–14
foliated rock 18
footprints and other animal tracks (Palichnology) 198, 211, 223, 224
foraminifera 71, 201, 209, 225, 242, 249, 256, 263, 285
FORBES N.S.W. 138
forces from the earth's interior 11
fore-reef deposits 125
fore-set bedding 199
forms of life 68
FORREST W.A. 263
FORSTER N.S.W. 122, 32°11′– 152°21′
Fortesque **Gp** 47
FORTESQUE RIVER W.A. 47
Fossil Cliff **F** 162
fossil wood 174, 179, 180, 203, 205, 206, 239, 244, 259, 278
fossilized suncracks 133, 198, 200
fossils 19
FOWLER'S BAY S.A. 262
Franklin **F** 57

322   INDEX

Fraser Range Metamorphics 53
FRASER RIVER W.A. 145, 17°19′–123°07′
Frasnian Stage (Age) 124
Frenchman Orogeny 57
freshwater bivalves 209, 220, 259
friction at underside of earth's crust 9, 10
fringing reefs 125
frogs 73, 74, 184
fungi 69, 137

gabbro 17
Galilee **B** 178
Gangamopteris 180
ganoids 143, 156*, 183, 208, 222
garden slater 207
garnet 17, 258
gas 107, 157, 162, 168
GASCOYNE PROVINCE W.A. 53, 56, 58
GASCOYNE RIVER W.A. 160
gastropods 72, 96, 111, 125, 140, 188, 242, 248, 255, 259, 285
Gawler Range Volcanics 57
Gedinnian Stage (Age) 124
GEELONG VIC. 220
genetic code and alphabet 76
GENOA RIVER VIC. 138, 37°27′–149°33′
Genyornis 282
geochronology 23, 24
geographic coordinates 42
geological correlation 21, 22, 24, 34, 35, 131*, 158, 213
geological record 19–24, 253, 254
Geological Society of Australia 126, 149
geology, a rational natural science 21, 22
geophysical methods 42
GEORGETOWN QLD. 50, 52, 237
Georgina **B** 107, 217
GEORGINA RIVER 97
geo-suction 97
geosyncline (versus geanticline) 45, 48, 50*, 51, 115, 116
geotectonic depression 50
geo-tumor 9
GERALDTON W.A. 160, 162, 214, 222, 225, 226
GERMANY 271, 279

GERRINGONG N.S.W. 29, 174, 34°45′–150°50′
giant kangaroo 270, 287
giant period of Australian continent 153*, 157, 163, 195
giant species 110, 207, 270, 283, 287
gibber plains 277
Gibbus Time 86, 88
GIDGEALPA GASFIELD S.A. 167, 27°50′–140°09′
Gigantocypraea 285
GIGOOMGAN QLD. 177, 25°44′–152°15′
GILBERTON QLD. 135, 137
Giles Complex 47
Gillatia 117
gills 184
GINGIN W.A. 214, 242, 245, 247
gingkos 70, 206, 244
GIPPSLAND VIC. 92, 129, 133, 197, 220, 233, 257, 259, 261
Gippsland **B** 233, 259, 265
GIRALIA RANGE W.A. 242, 22°44′–114°20′
Gisbornian Stage (Age) 103
Givetian Stage (Age) 124
glacial erosion 4, 15, 158, 161, 162, 164–7, 175, 271–4
glacial lakes (tarns) 273
glacial topography 4, 15, 165, 271–4
glaciation and glaciers 4, 15, 60, 61, 62, 118, 150–2, 158–66, 171, 174, 177, 271–4
GLADSTONE QLD. 116, 132, 255, 258
GLASSHOUSE MOUNTAINS QLD. 265, 26°54′–152°57′
GLENELG S.A. 286, 34°58′–138°32′
GLENELG RIVER VIC. 92, 220, 38°03′–141°11′
Globigerina 249
Globotruncana 249
Glossopteris 167, 180, 181
Glossopteris flora 166, 167, 169, 180, 181, 193, 205, 210
GLOUCESTER N.S.W. 146, 152, 32°01′–152°
Glyptophiceras 212
Glyptoxoceras 247
gneiss 17, 33, 177

Gneudna **F** 126
goanna 283
GOBI DESERT 229, 278, 280
Gogo **F** 141
gold-bearing quartz veins and pockets 103
goldfield districts and mines 37; 103, 179, 219, 258
Gondwanaland (Southland) 61-7, 166, 167, 168, 180, 210, 233
goniatites 125, 140, 155, 188, 193
goose barnacles 96
Gordon **Lst** 102, 117
GORE QLD. 270, 28°18′–151°29′
gorges 268
GOSFORD QUARRIES N.S.W. 208, 210
Gosfordia 208
GOSSES BLUFF N.T. 107, 23°48–132°19′
GOULBURN N.S.W. 104
graben tectonics 102, 116, 163, 266, 267
graded bedding 131
GRAFTON N.S.W. 202, 219, 269
Graham Creek **F** 235
GRAMPIAN MOUNTAINS VIC. 145, 268
Grampian series 146
GRANGE BURN VIC. 285, 37°44′–141°57′
granite decomposition 5
granite rock 17, 48, 92, 115, 129, 130, 152, 154, 177, 178, 219, 239
Grant **F** 164
GRANT RANGE W.A. 145, 18°03′–124°03′
graphite 219
graptolites 27, 71, 106, 108*, 109, 112, 117, 119, 131, 138
graptolitic faunas 108, 109, 138
grasses 277
gravitational movements in crust 9
grazing animals 280
Great Artesian Basin of Eastern Australia 88, 107, 115, 132, 167, 202, 204, 217, 221, 233, 240, 247, 256
GREAT AUSTRALIAN BIGHT 63, 153, 241, 262, 286
GREAT BARRIER REEF QLD. 261, 269, 277

GREAT BRITAIN 94, 116, 226, 242, 279
great Cretaceous submergence 229, 233, 236, 237
GREAT DIVIDING RANGE 130, 146, 198, 202, 217, 218, 240
Great Ice Age 15, 16, 271–4
Great Southland 61, 62
GREAT WESTERN (CANNING) DESERT 42, 276
Greek (ancient) scientists 19, 20
GREENLAND 62, 138, 158, 184, 271, 274
greensand (glauconitic sandstone) 105, 237, 249
GREAT, NEWCASTLE N.S.W. 171
Greta Coal Measures 170, 171, 173, 175
greywacke (or graywacke) 149
Groenlandaspis 142
guide fossils 85, 139
GULF OF CARPENTARIA (GULF COUNTRY) 50, 213, 216, 236, 247, 269
Gulfstream 185
GULGONG N.S.W. 28, 219, 258, 32°22′–149°32′
gums 279
Gunnarites 247
GUNNEDAH N.S.W. 146, 198
GUYRA N.S.W. 258, 30°13′–151°40′
Gymnosolen 79
gymnosperms 70
GYMPIE QLD. 177, 219, 265
gypsum 56, 275

HALL'S CREEK N.S.W. 155
Halls Creek beds W.A. 48
Halysites 120, 138
HAMERSLEY W.A. 39
Hamersley **B** 58
Hamersley **Gp** 47, 80
HAMERSLEY RANGE W.A. 39
HAMILTON VIC. 262, 281, 285
Hamites 247
HAMPTON RANGE W.A. 263, 33°–126°40′
HANNAM'S GAP QLD. 156, 23°39′–147°12′
Hardmanoceras lobatum 110
hardness of water 5
harp shell 286

HARPER'S HILL N.S.W. 171, 32°44'–151°25'
Harpes 140
HARRISVILLE QLD. 218, 27°49'–152°40'
HARTS RANGE N.T. 38, 23°10'–135°09'
Hatches Creek **Gp** 50
HATTON'S CORNER N.S.W. 121, 34°51'–148°55'
Hauericeras 247
Hawkesbury **B** 199
HAWKESBURY RIVER N.S.W. 199, 201
Hawkesbury **Ss** 199, 201, 206–9, 267
headwater stealing 267
HEALESVILLE VIC. 133, 37°39'–145°32'
heat convection in earth's mantle 9
HEATHCOTE VIC. 92, 116, 117, 121, 122, 165, 36°55'–144°42'
heather 71
Helicoprion 183
helicopter geologists 163, 276
Heliocrinus 121
Heliolites 120, 138
Hemiptera 183
HERCULANEUM 18
Herodotus 20
heterocercal fishtail 142, 156, 222
Heysen, Hans 107
higher flowering plants 243, 244, 278
higher vertebrates 156
HILL RIVER W.A. 214, 30°23'–115°31'
HIMALAYA 10, 14, 179, 193, 211, 227, 257
HINCHINBROOK PASSAGE QLD. 266
hinge-line of brachiopod shell 111
HOBART TAS. 169, 197, 210
Hodgkinson beds 135
Holostei 222
holothurians 94
Homo sapiens 286
homocercal fishtail 156, 222
hoop pine 244
horizontal crustal movements 9–14
hornblende 17

horned turtles 283
horny shells 95, 111
horse 75, 278
horse-tails 70, 153, 154, 181*, 205
horst-and-graben topography 266
Hostimella 135
hot spring deposits 203, 259, 265
house-flies 183
HOWARD QLD. 235, 25°19'–152°34'
HUCKITTA-MARQUA REGION N.T. 97, 100, 22°35'–135°35'
human immigration 286
HUNGARY 240
HUNGERFORD QLD. 241, 29°–144°25'
Hunter-Bowen Orogeny 178, 179, 193, 198, 202
HUNTER RIVER N.S.W. 152, 154, 170, 174, 178, 274, 31°56'–151°10'
Hutton **Ss** 217
hydrology 240
hydrozoa 92, 138
Hyolithes 96, 188
hypercycle 77
hypercyclic macromolecules 78
Hystrichosphaeridae 233

ibis 282
ice ages 60–2, 158, 161, 162, 164–7, 173–7, 271–4
iceberg erratics 4, 61, 62, 175, 177
icebergs 4, 61, 62, 161
ICELAND 151
ice-sheets and caps 158, 165, 271
Ichthyosauridae 245
Ichthyostega 138
igneous rocks (see also volcanics) 16, 17, 220
Iguanodon 223
illaenids 112
"*Illaenus* band" 117
ILLAWARRA DISTRICT N.S.W. 170, 174, 175, 34°30'–150°45'
Illawarra Coal Measures 170
imagination and intuition in geology 101
impact craters 107
inclination of earth's rotational axis 119
independence of evolution from local geological events 99, 205

INDEX

INDI RIVER VIC. 116, 36°30′–148°10′
INDIA 62, 153, 158, 166, 180, 247
INDIAN OCEAN 42, 123, 164
INDIANA 155
INDONESIA 44, 123, 211, 227, 274
influence of sedimentary environment on life 116, 237
Infusoria 228
inland seas and lakes 125, 133, 134, 145, 150, 156, 164, 171, 175, 178, 195, 196–204, 210, 213, 216–20, 229–33, 235, 239, 240, 258–61, 265, 269, 270
INMAN VALLEY S.A. 165, 35°30′–138°28′
Inoceramus 227, 243, 248
Inoruni **Ss** 217
insectivores 280
insects 73, 94, 138, 182, 183, 201, 203, 206, 207, 255, 258
insect's metamorphosis 183
intake beds of Great Artesian Basin 240
interfingering of formations 151, 169
interglacial periods 61, 158, 271–4
internal shell of molluscs 110
interpretation of geological observations 101
intracontinental basin 85
intrusive rocks 17, 173, 179, 218, 239
IOWA 155
IPSWICH QLD. 202, 203, 216–8, 258
Ipswich beds ("series") 206, 255
iridium 102
iron carbonates 235
iron ore 47, 48
ironbark 279
ironstone 201, 209
IRWIN RIVER W.A. 162, 28°58′–115°29′
island arcs 52, 115
island groups 42, 50, 118, 177, 231, 235, 239
Isoarca 111
isolation of faunas 156, 183, 244, 246, 280
isopods 207

JAMIESON VIC. 131, 37°18′–146°08′
JAPAN 206
Jarlemai Siltstone 226
Java man 287
jellyfish 60, 71, 80, 94, 108, 119
JEMALONG RANGE N.S.W. 138, 33°28′–147°45′
JENOLAN CAVES N.S.W. 119, 35°49′–150°02′
JERVIS BAY N.S.W. 274
joining of islands to coast 275
JOSEPH BONAPARTE GULF (**B**) 160, 164, 196
Jukesian Orogeny 88
JUNDAH QLD. 241, 17°33′–145°18′
Junee **Gp** 102
JURA MOUNTAINS 213
Jurassic fish beds 28, 219
Jurassic Period (System) 23, 213–28
Jurusania 79

KALBAR QLD. 218, 24°56′–152°14′
KALGOORLIE W.A. 47
KANGAROO ISLAND S.A. 87, 165, 287
KANGAROO RIVER N.S.W. 173, 34°45′–150°21′
kangaroos 281
Kanimblan Orogeny 150
Kanmantoo **Gp** 87
KATHERINE-DARWIN REGION N.T. 38, 44, 47, 83
Katherine River **Gp** 50
KATOOMBA N.S.W. 70, 137, 175, 199
Keilorian Stage (Age) 116
KEMPSEY N.S.W. 152, 153
KENNEDY RANGE W.A. 256, 23°50′–116°58′
kerosene shales 168, 169, 183
KIAMA N.S.W. 174, 34°40′–150°51′
KIANDRA N.S.W. 258
Kiandra beds 104
Kiandra deep leads 258
KIDSTON QLD. 137, 18°53′–144°10′
Kimban Orogeny 48
Kimberley **B** 53
Kimberley Block 105
Kimberley **Gp** 53

INDEX

KIMBERLEY (EAST AND WEST) REGION W.A. 38, 47, 48, 50, 53, 57, 66, 83, 90, 105, 106, 109, 125, 139, 154, 158, 164, 196, 205, 231, 232
king crabs 123
KING ISLAND TAS. 63, 262
KING LEOPOLD RANGE W.A. 125
KINGSCOTE S.A. 165
koala 281
Koala Creek series 131
Koalemus 281
Kockatea **Sh** 196, 209
KOONWARRA VIC. 210, 38°33′–145°57′
Kosciusko "Uplift" (or upwarp) 265, 266
Kossmatia 226, 228
KOW SWAMP VIC. 287, 35°57′–144°17′
Kronosaurus 245
KROOMBIT QLD. 132, 24°27′–150°48′
Kuttung **Gp** ("series") 152–4
Kuttung Time 167
KYNUNA QLD. 241, 21°35′–141°55′

LA TROBE TAS. 169, 41°14′–146°24′
La Trobe **Gp** 259
LA TROBE RIVER VIC. 259
La Trobe Valley Brown Coal Measures 259, 37°58′–146°03′
labelling of fossils 30
labyrinthodonts 184*, 185, 209, 225
lace-corals 72, 122, 186
lace-wings 183, 206
LACHLAN RIVER N.S.W. 262
Laevigata Time 86, 87
lagoonal environments 169
LAKE BARRINE QLD. 289, 17°15′–145°38′
LAKE BATHURST N.S.W. 27, 140, 35°–149°39′
LAKE BLANCHE S.A. 264, 269
LAKE CALLABONNA S.A. 264, 269, 270, 281
Lake Callabonna *Diprotodon* beds 281
LAKE DIERI 269, 270, 275

LAKE DRUMMOND 150
LAKE EACHAM QLD. 269, 17°18′–145°37′
LAKE EYRE S.A. 56, 58, 63, 107, 133, 167, 264, 269, 270, 275, 279, 281
LAKE FROME S.A. 87, 264, 269, 279
LAKE GEORGE N.S.W. 7, 35°05′–149°25′
LAKE GREGORY S.A. 264, 269, 29°–139°
LAKE MACQUARIE N.S.W. 180, 182, 35°05′–151°33′
LAKE MUNGO (MILDURA AREA) N.S.W. 286
LAKE NARRABEEN N.S.W. 198
LAKE ST. CLAIR TAS. 169, 273, 42°05′–146°10′
LAKE TALBRAGAR N.S.W. 219
LAKE TIBERIAS TAS. 197, 42°25′–147°20′
LAKE TORRENS S.A. 63, 241, 264, 269
LAKE WALLOON 195, 196, 204, 216–9, 229
LAKE WINTON 233, 254, 255
LAKES ENTRANCE VIC. 259, 37°53′–148°
Lamboo Complex 47, 48
laminae in shales 17
lamp-shells 72, 80, 95, 119, 121, 138, 139, 145, 148, 149, 155, 161, 186, 237, 256, 285
LANCEFIELD VIC. 92, 37°17′–144°44′
Lancefieldian Stage (Age) 103
lancelet 141
land bridges 166, 195, 233, 244, 274, 286
land surfaces (fossil) 85, 92, 129, 266
Landsborough **Ss** 203
LANEFIELD EXTENDED COLLIERY QLD. 223, 27°39′–152°26′
LANGEY CROSSING W.A. 228, 17°40′–123°34′
large extinct animals 30
larval stages and states 74, 95, 210
laterite 277
Latimeria 142, 227
LAUNCESTON TAS. 129, 169

INDEX 327

LAURA QLD. 219, 15°34′–144°27′
Laura **B** 219, 235
laurel 71
Laurel beds 145
lava 5, 15, 83, 117, 133, 152, 171, 220, 257, 258, 265, 268, 269
LAVERTON W.A. 242
lead 50, 52, 97
leaf scars of lycopods 137
Leigh Creek Coal Measures S.A. 196
Leiopyge laevigata 86, 87
Leonardo da Vinci 20
LEONORA W.A. 47, 242
Lepidodendron 137, 154
Leptolepis 222
Leptophloeum 137
Levipustula 149, 155
Liassic Epoch (series) 214
LIBYAN DESERT 276
lichens 69
life in test tubes 77
LIGHTNING RIDGE N.S.W. 240, 241, 29°26′–147°59′
lignites (brown coals) 236, 257, 259
LILYDALE VIC. 117, 131, 37°46′–145°21′
Lilydale **Lst** 131
limestone 17
limestone decomposition 5
Limmen Geosyncline 52
Lingula 80, 95, 145, 212
LION CREEK QLD. 149, 23°23′–150°22′
LISMORE N.S.W. 218
LITHGOW N.S.W. 134, 170, 175
Lithgow Coal Measures 170, 175, 183, 210
lithographic limestone 225, 227
Lithostrotion 155
Liveringa **F** 164
living fossils 71, 142, 279, 280, 286
living tissues 68
lizards 73, 277
Llandoverian Stage (Age) 116
load on earth's crust 9
lobe-fins 142, 156, 184, 222
lobsters 73
LOCHINVAR (HUNTER VALLEY) N.S.W. 171
locust 206

LONG REEF N.S.W. 199, 23°45′–151°19′
LONGREACH QLD. 237
LORD HOWE ISLAND 283
lowering of highland surfaces 266
LOWMEAD QLD. 255, 24°32′–151°45′
LOWOOD QLD. 225, 27°28′–152°35′
Loxonema 140
Ludlovian Stage (Age) 116
lung-fish 143, 208, 209, 222, 245, 283
lycopods 70, 135, 137, 179
LYNDON RIVER W.A. 160, 23°18′–114°55′
Lyons **Gp** 161, 162, 187

MACDONNELL RANGES N.T. 97, 106, 107, 111, 264, 276
MACKAY QLD. 219
Macrocephalites 226
macrofossils 249, 256
Macrotaeniopteris 205
Macrozamia 71, 221
MADAGASCAR 142, 153, 247
Madariscus 22
Madiganella magna 110
Madiganian Orogeny 58, 63
MADURA W.A. 241, 31°54′–126°
magnesia 88
magnesian limestone (dolomite) 259
magnesite 60
magnetic poles 88, 120
maidenhair plant 182
maidenhair tree (gingko) 70, 206
MAITLAND N.S.W. 152, 171, 175, 190
MAITLAND DISTRICT N.S.W. 28, 189
Maitland **Gp** 170
MALAYA 207
Malbina **Ss** 169
MALDON VIC. 103, 37°–144°04′
Mallacoota-Wagonga phyllites 104
MALLEE COUNTRY VIC. 262, 264
mammals 73, 74, 210, 225, 245, 246, 279–82
MANILLA N.S.W. 132, 135, 30°45′–150°43′
MANN RANGE S.A./N.T. 38, 58, 105

INDEX

MANNING RIVER N.S.W. 170, 201, 31°49′–151°38′
MANN-MUSGRAVE RANGE S.A./N.T. 58
Mann-Musgrave System 47, 53, 58
MANSFIELD VIC. 145, 146, 155, 156
MANTUAN DOWNS QLD. 177, 24°25′–147°15′
MANY PEAKS QLD. 149, 24°32′–151°23′
Maorites 247
map-making 99, 101
marble 17*, 52, 132, 139
Marburg **Ss** 217, 218, 225
Marginirugus 155
MARIA ISLAND TAS. 168, 42°40′–148°05′
marine currents 185
marine erosion 5, 6, 173
marine erosion platforms 173, 263, 275
marine straits 50, 56, 64, 83, 87, 113, 167, 236
Marinoan series 62, 64
marker floras 182
marker horizons 149
MARMOR QLD. 270, 23°41′–150°42′
marshes 26
marsupials 246, 270, 280, 281
Martiniopsis 187
MARULAN N.S.W. 174 34°43′–150°
MARYBOROUGH QLD. 195, 219, 235, 239
Maryborough **B** 196, 202, 203, 235, 236
Maryborough **F** 235
Maryburian Orogeny 239, 265
Mastodonsaurus 209
Mathinna beds 117, 135
matrix of rocks 17, 29
Mawson, Sir Douglas 57
McArthur **Gp** 50, 56, 57
MCARTHUR RIVER N.T. 52, 15°35′–136°35′
McIvor beds 117
Mecoptera 183
MEDITERRANEAN 94
Medusae 71, 80, 119
Megadesmus 187
Melaleuca 279

MELBOURNE VIC. 92, 103, 116, 117, 119, 121, 122, 135, 141, 145, 160, 168, 259, 262, 268, 281, 282, 285
Melbournian Stage (Age) 116, 131
MEREWEATHER N.S.W. 182, 32°57′–151°45′
Meridional Divide in Middle Cambrian times 84, 85, 88
Merimbula **F** 134
Mersey Coal Measures Tas. 169
Mersey **Gp** 169
mesa topography 255
Mesotitan 207
Mesozoic Era (Erathem) 23, 191–249
Metalegoceras jacksoni 189
metalliferous sediments 50
metamorphic grade 52, 53
metamorphic rocks 17*, 18, 52, 53, 84, 87, 104, 105, 116, 154
metaquartzite (metamorphic quartzite) 17
Metaxygnathus 138
mica 17
micaschist 52
MICHELAGO N.S.W. 139, 35°43′–149°10′
microfloras 221
microfossils 112, 249, 256
Microsaurus 210
Middle Cambrian 84–8
Middleback **Gp** 48
mid-Eocene transgression 256, 259
migration barriers 84, 185, 210, 280, 282
migration of floras 181, 210
Miles, Campbell 97
MILTON N.S.W. 172, 35°19′–150°26′
Milton Quartzporphyry 172
Minchin Siltstone 196
mineral deposits 47, 50, 52, 116, 170, 179
mineral solutions 53
MINILYA RIVER W.A. 160, 161, 162, 214
Miocene Epoch (series) 23, 253, 254, 259–64, 280, 281, 285, 286
Miolania 282

Miria Marl 242
missing links in evolution 78, 93, 225, 279
Mitchell, J. 187
MITCHELL RIVER VIC. 133, 37°38′–147°22′
molecular ring systems 77
molecular structure 76, 77
molluscs 72, 73, 122, 139, 155, 161, 248, 283–6
MOLUCCAS 153, 226, 231, 235
MONARO (TABLELAND) N.S.W. 118, 129, 258
monkey puzzle tree 221
monocotyledons 71
Monograptus 109, 117, 119
monolith 265
MONTO QLD. 148, 149, 24°52′–151°07′
MOOGOORIE HOMESTEAD W.A. 161, 24°04′–115°12′
Moogoorie **Lst** 145
Moolayember **Sh** 204
MOOMBA GASFIELD S.A. 167, 28°07′–140°11′
MOONBI RANGE N.S.W. 134, 30°50′–151°05′
Moonlight Valley Tillites 63
Moora **Gp** 53, 56
MOORE CREEK N.S.W. 132, 30°58′–150°59′
Moore Creek **F** 124
Moorlands Brown Coal Measures S.A. 257, 35°16′–139°52′
Mootwingee beds 146
moraines 4, 60, 61, 62, 161, 164–7
MORINISH QLD. 132, 23°15′–150°12′
MORNINGTON VIC. 261, 38°14′–145°02′
MORUYA N.S.W. 104
MOSS VALE N.S.W. 201, 34°33′–150°23′
mosses 70
moulds and casts of fossils 111
mountain moss 70, 137
mountain-building (orogenesis) 7–14, 101, 104, 106, 113, 129, 133, 134, 144, 150, 178, 179, 233, 235, 239, 257, 269
Mt. Barren beds 56

MT. BOPPLE QLD. 218, 25°50′–152°35′
MT. BRITTON QLD. 176, 21°24′–148°32′
MT. CANOBOLAS N.S.W. 265, 33°21′–148°59′
MT. CROSBY QLD 203, 206, 27°32′–152°48′
MT. DEVLIN QLD. 176, 20°20′–147°42′
Mt. Devlin Coal Measures 176
MT. ELEPHANT VIC. 268, 37°54′–143°12′
MT. EMU QLD. 269, 20°07′–144°32′
MT. ETNA QLD. 132, 23°07′–150°59′
MT. GAMBIER S.A. 268, 282, 37°50′–140°47′
MT. GRENFELL STATION N.S.W. 141, 31°20′–145°19′
MT. GRIM QLD. 149, 23°41′–150°20′
MT. HOWITT VIC. 141
MT. ISA QLD. 38, 50, 52, 53, 63, 84, 85, 90, 97, 217
MT. KOSCIUSKO N.S.W. 265, 266, 271
MT. LAMBIE N.S.W. 134, 175, 33°27′–149°59′
MT. LARCOM QLD. 132, 23°49′–150°59′
MT. LAURA VIC. 268, 38°15′–143°15′
MT. LE BRUN QLD. 269, 25°37′–151°53′
MT. LOFTY RANGE, ADELAIDE S.A. 38, 57, 87, 256, 262, 34°58′–138°40′
MT. LYELL TAS. 102, 41°40′–145°15′
MT. MATLOCK VIC. 131, 37°35′–146°06′
MT. MORGAN QLD. 146, 148, 149, 23°39′–150°23′
MT. MULLIGAN QLD. 176, 16°52′–144°57′
MT. MURCHISON TAS. 102, 41°50′–145°36′
MT. NOORAT VIC. 268, 38°11′–142°54′
MT. PAINTER S.A. 57, 30°15′–139°21′

Mt. Painter Complex 53
MT. PIERRE W.A. 139, 18°22′–125°59′
MT. PLEASANT VIC. 135, 37°35′–143°51′
Mt. Rigg **Gp** 53, 57
MT. ROLAND TAS. 102, 41°58′–145°34′
MT. SCHANK S.A. 268, 37°56′–140°44′
MT. TOUSSAINT QLD. 176, 20°30′–147°49′
MT. VINCENT N.S.W. 189, 32°55′–151°29′
MT. WAPTA, CANADA 94
MT. WELLINGTON, HOBART TAS. 220
MT. WELLINGTON VIC. 92, 37°31′–146°50′
Mucophyllum 121
mud volcano 107
mud-cake polygons 200
MUDDY CREEK VIC. 261, 37°46′–142°02′
MUDGEE N.S.W. 130, 135, 174
Mulga Downs **F** 141, 146
multiplication of cells 74, 76, 77
Mulvaney, J., Prof. 287
MUNDUBBERA QLD. 146, 149, 25°35′–151°18′
Mungo people 286
MURCHISON RIVER W.A. 36, 126, 160, 243
Murchisonia 122
Murray **B** 167, 233, 255, 257, 261
Murray Gulf 233, 261, 264
MURRAY ISLAND QLD. 269, 15°06′–145°16′
MURRAY RIVER 116, 133, 165, 262, 264
MURRUMBIDGEE RIVER N.S.W. 130, 139, 141, 143, 262, 264
MURRURUNDI N.S.W. 146, 152, 171, 198, 31°46′–150°49′
MUSGRAVE RANGE S.A. 47, 53, 58, 64
mussels 187, 212
MUSWELLBROOK N.S.W. 171, 174, 198, 32°16′–150°54′
MYALL RIVER N.S.W. 172, 32°18′–152°11′
Myonia 187

Myophoria 227
Myriolepis 208

Nabberu **B** 58
Namambu Complex 38, 44
Namatjira, Albert 107
Namurian Stage (Age) 149
NANDEWAR HILLS N.S.W. 265, 30°18′–150°32′
Nangetty **F** 161, 162
nannofossils 70
NARRABEEN, SYDNEY N.S.W. 275
Narrabeen **Gp** 198, 205, 208, 219
NARRABRI N.S.W. 174, 198
Nathorsti Time 86, 87
NATIONAL MUSEUM, MELBOURNE VIC. 28, 119, 121, 285
native bear (koala) 281
native cat 281
natural selection 74
natural self-organization 77
NAUGUDA VIC. 137
Nautilus 73, 106, 110, 122, 125, 131, 140, 155, 188, 278
Nebine Ridge 237
Neerkol **F** 149
negritos 286, 287
Neil's Creek beds 149
NEMINGHA N.S.W. 139, 31°07′–150°59′
Neocomian Epoch (series) 235, 236
Neocretaceous 23
Neogene 23
Neohamites 247
Neotrigonia 227, 286
NERANLEIGH QLD. 116, 27°27′–152°40′
Neritacea 188
NEVADA 93
NEW CALEDONIA 195, 283
NEW ENGLAND N.S.W. 90, 130, 132, 146
New England Geosyncline 148
NEW GUINEA (PAPUA-NIUGINI) 66, 153, 195, 239, 271, 274
New South Wales Permian sequence 169–76, 178, 179
NEW ZEALAND 5, 101, 129, 153, 195, 207, 227, 239, 240, 247, 273, 274, 284
NEWCASTLE N.S.W. 146, 150, 152,

153, 175, 177, 178, 198
Newcastle Coal Measures 170, 175, 182, 210
NEWCASTLE HARBOUR N.S.W. 274
Newlandia 79
Newmarracarra **Lst** 214, 222, 226
newt 73, 184
Ngalia **B** 107, 197
NHILL VIC. 165
nickel 47
NILE RIVER, AFRICA 20, 143
NINETY MILE BEACH VIC. 220, 38°27′–147°
Niolania 283
nitrogen 68, 76
Noeggerathiopsis 179
non-marine sediments 115
non-permanence of continents and oceans 19
Noonkanbah **F** 164
NORMANVILLE S.A. 94, 35°27′–138°19′
NORSEMAN W.A. 256
NORTH AMERICA 50, 110, 123, 143, 167, 179, 182, 226, 271, 273, 280
North Arm Volcanics 202
NORTH VIETNAM 180
NORTHEASTERN TASMANIA 117, 135
NORTHERN HEMISPHERE 42, 144, 154, 166, 167, 181, 213, 246
NORTHERN QUEENSLAND 104, 137, 178, 203, 210, 219, 229, 239
Northwest (Carnarvon) **B** 160
NORTHWEST CAPE W.A. 105, 124, 126, 145, 214, 231, 264
NORTHWEST SHELF 264
Norton Gully **Ss** 131
NORWAY 62, 273
Nostoceras 247
Notochelone 246
notochord 73, 141
Notothemium 281
NOWRA N.S.W. 173, 186
nuclei of continents 42
nucleotic acids 76
Nuculana 187
Nuculopsis 212
Nullaginian Era (Erathem) 23, 45, 48
Nullarbor Embayment 56

Nullarbor Gulf 264
Nullarbor **Lst** 263
NULLARBOR PLAIN 42, 47, 56, 163, 166, 167, 216, 262, 263
Nummulites 285
Nummulitic (time term) 285
Nummulospermum 180
Nura Nura Member (of Poole **Ss**) 164

OAKEY QLD. 223, 27°26′–151°43′
OAKOVER RIVER W.A. 163, 164
Obdurodon 279
obsidian 16
oceanic crust 40, 41, 44
octopus 73, 96, 109, 212
Officer **B** 105, 108
Ogygites 111
oil accumulation 165, 196, 204, 242, 259
oil and gas resources 157, 163, 164, 168, 169, 183, 196, 204, 214, 242, 259
oil reservoir rocks 162, 163, 204, 242, 259
oil source formations 162, 204, 242, 259
Olarian Orogeny 56
OLARY S.A. 56
Old Red **Ss** (England) 138
oldest Australian rocks 37, 79
oldest fossils 78, 79, 80
oldest preserved land surfaces 38, 85
Oligocene Epoch (and series) 16, 253, 254, 257, 258, 259, 282
olivine 17
Omphalotrochus 122
onchophoroids 94
ONSLOW W.A. 214
ontogeny 74
OODNADATTA S.A. 245, 247
oolitic rocks 148–50
oozes 249
opal 203, 240*, 241, 245
OPALTON QLD. 241, 23°15′–142°46′
open-cut (open-cast) mining 178*, 259
operculum 96
Ophiceras 212
Ophiura 227

Opik, A. A., Dr. 84, 85, 97
opossum (American) 280, 281
ORANGE N.S.W. 104, 265
ORD-VICTORIA RIVER REGION
 W.A. 63, 66, 123
Ordovices 113
Ordovician Period (System) 23,
 27, 99–112, 172, 278
ore deposits (chemical-
 sedimentary) 5
organic limestone 25, 88, 259,
 261, 262
organ-pipe coral 120
origin of coal 171–2*, 181, 257,
 259
origin of life 68, 69, 76, 77, 78
origins of geologic nomenclature
 113, 144, 213, 229, 253, 254
Ormoceras 122
orogenesis 7–14, 44, 52, 101, 113,
 178, 179, 257
orogenetic framework 52
orogenetic metamorphism 39, 52
Orthis 111, 117
Orthoceras 140, 188
Osagia 79
osmiridium 102
osmium 102
Osteolepis 142
ostracods 112, 117, 125, 154, 222,
 225, 255, 285
Ostracodermata 141
Ostrea 227
Otoceratidae 212
Otozamites 221, 244
Ottokaria 181
OTWAY RANGES VIC. 220, 262,
 38°27°–154°
OUSE RIVER TAS. 169, 42°03′–
 146°42′
outwash (glacial) gravels 161, 171
Owen, R., Prof. 281
Owen **Cgl** 102
oxen 278
oxidizing bottom environments
 169
OXLEY QLD. 255, 27°33′–152°59′
oxygen 68, 76, 77, 78
oxygen-carbon dioxide balance
 78
oysters 72, 227, 248, 256
ozone layer 78

Pachydiscus 247
Pachymyonia 187
PACIFIC OCEAN 13, 14, 44, 115,
 146, 226, 235, 269
Pacoota **Ss** 111
Palaeacis 149
Palaeaster 189
Palaeocene Epoch (series) 23,
 254–7
Palaeocretaceous Subperiod 23
Palaeogene Subperiod 23
palaeogeographic maps 40
palaeomagnetism 11, 39, 88, 118,
 132, 134, 200, 207, 218
Palaeoniscidae 156, 222
Palaeozoic Era (Erathem) 23,
 81–189
Palimnarchus 283
PALMER RIVER QLD. 116
palms 71
Palorchestes 281
Palynology 182, 195, 221, 222
Pander Greensand 105
Panenka 131
PAPUA 44
Paramecoptera 183
parasitic worms 72
Parvifrons Time 87
Pascoe River beds 146
PATAGONIA 283
PATERSON N.S.W. 152, 154,
 32°36′–151°37′
PATERSON RANGE W.A. 58,
 21°50′–122°
Paterson Range conglomerates
 164
PATERSON RIVER N.S.W. 152,
 30°29′–151°32′
PEAK DOWNS QLD. 270, 22°56′–
 148°05′
Peake Metamorphics 53
PEAKE-DENISON RANGE S.A. 57, 165,
 28°30′–136°
pearl oyster 139
peat 169, 181
Pecten 227
Pedirka **B** 107, 133
pelagic organisms 96, 102, 108,
 249
pelecypods 72, 96, 111, 117, 129,
 139
pelican 282

peneplain 6, 255
penguin 281
Pentamerus 121
peppermint oil 279
Peridionites 95
Peripatus 94
Perisphinctes 226
permanence of continent positions 8
Permian Period (System) 23, 157-89
Permian sequence of New South Wales 170-6, 179, 180
Pertaknurra **Gp** 56, 58
Pertatataka **GP** 58, 63
PERTH W.A. 105, 160, 214, 242, 271
Perth **B** 157, 160, 161, 162, 196, 209, 214, 226, 231, 256, 261
Pertnjarra **Gp** 126
PETRIE QLD. 258, 27°16′-152°59′
Phacops 122, 140
phalangers 281
Phanerozoic Eon (Eonothem) 60
Phascolonus 281
Phialocrinus 189
phosphatic rocks 85, 208, 243
Phreatoicus 207
phyla 69-73, 79
phyllite 52
Phylloceras 247
Phyllograptus 109
Phyllolepis 142
Phyllopachyceras 247
Phyllotheca 181, 205
pigeon 282
Pikaia 141
Pilbaran Era (Erathem) 24, 37, 42
PILBARA REGION W.A. 48
Pilbara Shield 37, 38, 40, 44, 47, 53, 56, 58, 79, 231
Pilbara-West Kimberley Block 53
Pilpah **Ss** 63
PINE CREEK N.T. 47
Pine Creek Geosyncline 47, 48
pines 70, 71, 153, 179, 206, 244
Pittman **F** 104
PITTWATER, SYDNEY N.S.W. 275
placoderms 141
plankton 248
Planorbis 259

plant kingdom 68-71
plant-bug 183
plantless landscape 136
plasticity of hard rocks 11
plate tectonics 11
plateau 7, 266-8
plateau basalts 83
Platyceps 209, 210
platyceratids 188
platypus 279
Pleistocene Epoch (series) 15, 16, 23, 253, 254, 267-74
Pleospongia 93
Pleuracanthus 209
Pleuromeia 205
Pleurotomariidae 188
PLEVNA QLD. 258, 23°30′-149°30′
Pliny (Plinius) 188
Pliocene Epoch (series) 23, 253, 254, 264-7
pliosaurs 224, 245
plutonic rocks (see also volcanics) 17
Podozamites 221, 244
POINT CULVER W.A. 262, 32°54′-124°45′
polar ice-caps 15, 16, 218*
polar regions 4, 15, 200
pollen 112, 182, 244, 254, 278
polymetamorphism 53
polyp 120
Polypora 186
Polypterus 143
POMPEII 18
Pond Argillite (Mudstone) 149
Pool **Ss** 164
porcellanite 243
Porophoraspis 112
porphyry 17, 130, 154, 173, 174, 177, 178
PORT AUGUSTA S.A. 57
PORT DAVEY TAS. 273, 43°17′-145°56′
PORT HEDLAND W.A. 47, 163, 231
PORT JACKSON, SYDNEY N.S.W. 15, 274
PORT KEATS N.T. 164, 14°14′-129°31′
PORT MACQUARIE N.S.W. 18, 201
PORT PHILLIP BAY, MELBOURNE VIC. 168, 261
PORT PIRIE S.A. 57

PORT STEPHENS N.S.W. 17, 172, 174
PORTLAND VIC. 261
Portuguese man-of-war (jellyfish) 108
POSEIDON VIC. 103
possum (Australian) 280
potassic extrusives 216, 265
Poteriocrinitidae 189
Pound **Ss** (or Quartzite) 60, 66, 80
praying mantis 206
Precambrian eons 33–67
Precambrian fossils 60, 68, 78, 79, 80
Precambrian shields of Australia 83
Precipice **Ss** 217, 221
preservation of fossils 28–30, 85, 208, 282
PRICES CREEK W.A. 110, 18°41′–125°53′
primitive mammals 224, 225, 279, 280, 281
probability laws 76
Productidae 155
Productus 149, 155, 187
Proetus 117
progress in geology 45
Propinacoceras 189
Proterozoic Eon (Eonothem) 23, 45–67, 79, 80, 217
Proterozoic-Phanerozoic boundary 60
Protophiceras 212
Protophyllocladus 244
Protoretopora 149, 186
protozoa 71
provincial faunas and floras 153, 185, 239, 284
Pseudogastrioceras 189
Pseudophyllites 247
Pseudorhacopteris 154
Pseudotoites 226
psilophytes 70, 135, 136, 137
Pteria 139
pteridophytes 70
pteridosperms 180
pterodactyls 224, 277
pteropods 72, 96, 131, 188
pterosaurs 224
Pterygotus 123
Ptychagnostus atavus 86, 87

PURGA QLD. 218, 27°43′–152°44′
pyroxenite 17
pygmies 286

QATTARA DEPRESSION, AFRICA 7
Quambi **Gp** 168
quartz 17
Quaternary Period (System) 23, 253, 254, 267–77
Queensland lung-fish 143, 184, 209, 283
QUEENSLAND MUSEUM, BRISBANE 156
Queensland-South Australia seaway 165, 176, 178
QUEENSTOWN TAS. 117, 273
Quidongan Orogeny 129
QUITA CREEK QLD. 86, 87, 22°03′–138°55′
QUORN S.A. 57

radioactive minerals 23, 24, 36
radioactive properties 24, 36
radiocarbon 268
radiolaria 71, 92, 139, 148, 243, 249
radiolarian chert 134, 139, 249
radiometry (absolute age determination) 23, 24, 34, 35, 45, 48, 268
radium 23
RAGLAN QLD. 132, 23°43′–150°49′
rain forest 279
RAINBOW HILL N.S.W. 122, 35°05′–149°16′
raindrop impressions 133, 198
raised beaches 273, 274, 275
raised fault blocks (horsts) 266
Raphistoma 111
rate of sedimentation 115, 116
rate of subsidence of depositional area 116
RAVENSFIELD QUARRY N.S.W. 171, 189, 32°44′–151°34′
RAVENSWOOD-ANAKIE AREA QLD. 88
Recent Epoch (series) 23, 267, 275–7
Receptaculites 139
REDBANK PLAINS QLD. 255, 27°39′–152°51′
Redbank Plains **F** 259
red-bed formations 64, 88, 131,

134, 135, 218
Redlichia 84, 86, 87
reducing bottom environments 168, 169
reef-building corals 118–21, 263, 277, 286
regional metamorphism 39, 40, 52, 53
rejuvenation of topography 6
Remigolepis 142
removal of eroded material from Australian Alps 266
Renaissance scientists 20
reptiles 73, 156, 185, 210, 211, 222, 223, 224, 245, 246, 277, 283
reptilian footprints (trackways) 137, 138, 211, 223, 224
REPULSE BAY QLD. 146
resin of pines 26
Rhacopteris 154
Rhacopteris flora 154, 181
Rhaetosaurus 223, 224
rhinoceroses 278
rhizome 180
Rhizophyllum 121
Rhynchonella 139
Rhynie Chert (Scotland) 138
rhyolite 16
Riek, E. F., Dr. 183
Rocellaria 236
ROCKHAMPTON QLD. 116, 130, 132, 146, 149, 151, 153, 155, 176–8, 203, 231, 235, 236, 258
Rockhampton **Gp** ("series") 148, 149, 150
ROCKLEY N.S.W. 30, 121, 33°42′–149°34′
ROCKY MOUNTAINS U.S.A. 257
Rodingan Deformation (Orogeny) 106
ROLLING DOWNS AREA QLD. 245, 27°–149°
Rolling Downs **Gp** 237
ROMA QLD. 223
Roma **F** 237, 244, 245
Roopena Lavas 57
roots of plants 180
Roper **Gp** 57
ROSEWOOD QLD. 223, 27°39′–152°36′
ROSS SEA, ANTARCTICA 200

rostroconchs 96
rotation of earth 9, 10
rotational pole of earth 88
ROUGH RANGE W.A. 161, 242, 22°25′–114°05′
rudimentary organs 74
RUM JUNGLE N.T. 38, 44, 13°02′–131°
Rum Jungle Archipelago 42
Rum Jungle Orogeny 44
running birds 282
RUSSIA 94, 181, 202
RYDAL N.S.W. 134, 33°29′–150°02′
RYLSTONE N.S.W. 189, 32°48′–149°58′

saddle reefs of Bendigo Goldfields 103
Saghalinites 247
SALE VIC. 259
salinity 237
salt (and rocksalt) 56, 90, 107, 113, 132, 275
salt-flats and salt-pans 163, 275
sand-ridge desert 275, 276
Sanidophyllum 138
Sanmartinoceras 247
sapphire 258
Sarcophilus 281
sardine 222
SARDINIA 93
Saurichthys 208
sauropods 224
scale-trees 137, 179
scallops 187, 212, 227
SCANDINAVIA 38, 85
Scaphites 247
scaphopods 73
scarcity of fossils 25, 26, 134
schists 17, 52, 177
Schizodus 227
Schrödinger, Erwin, Prof. 76
science and arts 101
scoriae 152, 202
scorpion-flies 183, 206
scorpions 73, 136, 138
SCOTLAND 138
scree 102
sea-anemones 71, 120
sea-cucumbers 72, 94
SEAHAM N.S.W. 151, 32°40′–151°44′

sea-level changes (general) 14, 271, 274, 275, 286
sea-lice 73
sea-lilies 72, 95, 121
seals 278
sea-pens 60, 80
sea-spider 146
Seaspray **Gp** 261
sea-squirt 73
sea-urchins 72, 95, 121, 248, 256, 285
seaways (straits) 50, 52, 58, 60, 157, 165
seaweeds 70, 120, 135, 177
secondary nerve centres 223
sedimentary rocks 16
sedimentation 6
seed-ferns 70, 153, 180, 205
seeds 180
seismograph 14
self-catalytic activity 77
SELLICKS BEACH S.A. 262, 35°20'–138°27'
SELLICKS HILL S.A. 94, 35°20'–138°29'
SELWYN RANGE N.T. 86, 87, 21°28'–140°21'
Seneca 20
SENEGAL, AFRICA 143
Serpula 212
serpulids 72
shallow-water environments 198, 199, 239, 256
sharks 143, 145, 156, 208, 245, 257, 285
shark's teeth 145, 157
shelly animals 25
shelly marine faunas 102, 106, 109, 131, 161, 236, 261
shields (Precambrian) 37, 38, 83
ship-worms 244
Shoalhaven **Gp** 170
SHOALHAVEN RIVER N.S.W. 104, 134, 173, 35°02'–150°02'
shrimps 73, 94, 207
shrinking of earth 8
SIBERIA 26, 38, 93, 181
Siegenian Stage (Age) 124, 131
Silkstone **F** 259
Silures 117
Silurian Period (System) 23, 113–23, 172

silver 50, 52, 179
Simmons, T. H. 203
SIMPSON DESERT 105, 108, 276
simulated erratics 239
single corals 186, 248, 256
SINGLETON N.S.W. 176
siphuncle of cephalopods 110, 227
size of continent 56, 153
slate 17, 52
slope creep (submarine) 103
sloths 278
slugs 72
slumping (submarine) 103, 239
Smith, William 21, 227
snails 72, 122, 136, 140, 188
snakes 73
snow-line 4, 150
SNOWY MOUNTAINS N.S.W. 66, 104, 129, 36°40'–148°17'
SNOWY MOUNTAINS SCHEME (AUTHORITY) 129, 266
soft-bodied animals 24, 25, 94
solar radiation 78
SOMERSBY, GOSFORD N.S.W. 208
SOMERTON N.S.W. 154, 30°56'–150°39'
SORRENTO VIC. 261, 38°20'–144°45'
SOUTH AFRICA 62, 142, 158, 166, 180, 207, 210, 279, 284
SOUTH AMERICA 94, 166, 207, 279, 284
SOUTH COAST N.S.W. 104, 118, 134, 141, 170, 172
SOUTH EASK RIVER QLD. 197
SOUTH GIPPSLAND HILLS VIC. 220, 38°25'–146°40'
South Nicholson **Gp** 63
SOUTH POLE 166, 167
SOUTHEAST ASIA 66, 181, 211
SOUTHEASTERN QUEENSLAND 202, 265, 269
SOUTHERN HEMISPHERE 62, 153, 158, 166, 181
SOUTHERN KIMBERLEYS W.A. 141
SOUTHERN NEW SOUTH WALES 115, 133, 140, 145, 265
SOUTHERN OCEAN 42, 84, 118, 165, 168
SOUTHERN TABLELANDS N.S.W. 129
SOUTHERN VICTORIA 239, 261

Southland 61, 62, 64, 67, 118, 166
SOUTHPORT QLD. 269
Spanodonta 111
specialization of organisms 110, 201, 237, 239, 244, 279, 280
species (living and fossil) 284, 285
species definition 284
Speewak **Gp** 53
SPENCER GULF S.A. 56, 165, 218, 261, 262, 264, 270
spermatophytes 70, 71
sperm whale 245, 285
Sphenopteris 182, 243
spicules of sponges 71
spiders 73, 94, 136, 138
spines of brachiopods 173
spiral processes of brachiopods 95, 111, 186, 187
Spirifer 122, 139, 155, 186, 187
Spirifer disjunctus 139
Spirit Hill **Lst** 144–5
sponges 71, 285
Spongophyllum 120
spoonbills 282
spores 112, 169, 182, 195, 221, 222, 244, 254, 278
spreading centres between crustal plates 11
SPRING HILL N.S.W. 30, 122, 33°24′–149°09′
SPRINGS CREEK VIC. 261, 38°20′–144°23′
SPRINGSURE QLD. 167
squids 73, 109, 227
ST. PETERS, SYDNEY N.S.W. 201, 206, 208, 209
ST. VALENTINE'S PEAK TAS. 102, 41°21′–145°45′
ST. VINCENT GULF S.A. 256, 261, 262, 264, 286
Stairway **Ss** 112
stalagmite (and stalactite) 26
STANWELL QLD. 146, 148, 229, 235, 236, 23°29′–150°19′
starfish 72, 95, 121, 171, 189, 285
STATION CREEK QLD. 149, 24°27′–150°10′
Staub, Rudolf, Prof. 10, 11
Stegocephalia 184
STEIGLITZ VIC. 103, 37°53′–144°11′
Stenopora 186

STIRLING RANGE W.A. 37, 34°23′–117°35′
Stirling Range beds 56
Stirling-Mt. Barren beds 56
stone implements and tools 286
stork 282
Strabo 20
stratosphere 5
Streptorhynchus 187
striated rock surfaces 162, 165, 168, 273
stringybark 279
Stromatopora 139
Strophalosia 173, 187
structural nuclei 37
STUART RANGE S.A. 239, 28°47′–134°32′
sturgeon 143
Sturtian series 60, 61, 62
Stutchbury, Samuel 227
Styliolina 131
Styx River Coal Measures Qld. 236, 244
sub-basins 237
subcrustal rocks 9, 257, 258
subduction 11
Subinyoites 212
submarine lavas 84, 148, 174
submarine mountains, ridges, spurs 166, 237, 269
submarine mud flows 239
subpolar regions 4
subsidence, intermittent 8, 106, 179, 196, 233, 237, 262, 270–5
subsidence, strong and persistent 146, 164, 172, 179, 214, 219, 235
subsurface geology 167, 170, 204
subulitids 188
sunken continents or parts thereof 167
sunklands 262
Surat **B** 237
survival of the fittest 73, 74, 75
survival of the mediocre 75
suture line of cephalopods 140, 212
swamp environments 136, 163, 169, 171, 172, 175, 178, 182, 184, 195, 197, 198, 205, 213, 216, 233, 255, 259, 270
SWAZILAND, AFRICA 37

Swiss geology schools 10
SWITZERLAND 213, 279
SYDNEY N.S.W. 170, 174, 175, 197–201, 205, 210, 219, 267, 269, 275
Sydney **B** 165, 195, 195–201, 203
SYDNEY HARBOUR N.S.W. 227, 274
syenite 17
symbiosis 36, 69
symbiotic biochemical action 5
symmetrical orogenic tectonics 9
Syncaridae 207
syncline 8
Syringopora 120, 138, 155

TABBERABBERA VIC. 133, 37°33′–147°21′
Tabberabberan Orogeny 131, 133, 134
TABLE CAPE TAS. 262, 281, 40°57′–145°44′
TABLE MOUNTAIN, SOUTH AFRICA 207
table of Eras, Periods and Epochs 23
tablelands 7, 258, 266, 267, 268, 273
Tachylasma 186
tachylite 16
tadpole 74, 184, 209, 210
Taeniopteris 221, 244
Taeniopteris spatulata 221
TAGGERTY VIC. 141, 143, 37°19′–145°43′
TALBRAGAR N.S.W. 219, 220, 221, 222, 31°53′–149°51′
Talgai tribes 287
Talgai-Kow people 287
TALLONG N.S.W. 104, 34°43′–150°05′
TAMAR RIVER TAS. 169, 41°20′–147°02′
TAMBO QLD. 217
Tambo **F** 237, 238, 245, 248
TAMWORTH N.S.W. 132, 134, 139
TANAMI N.T. 38
TANIMBAR ISLANDS 164
Tanjil **Gp** 131
Tanjilian Stage (Age) 131
tapir 281
TARAGO N.S.W. 27, 35°04′–149°39′

TARCOOLA-EYRE PENINSULA REGION S.A. 44
TAREE N.S.W. 155
Tasman Fold Belt 258
Tasman Geosyncline 115
TASMAN SEA 5, 269
Tasmanian devil 281
Tasmanian negritos 287
Tasmanian wolf 280, 281
tasmanite (oil shale) 169
Tasmanoceras zeehanense 110
TASMAN'S ARCH TAS. 168, 43°03′–147°56′
Tawallah **Gp** 50
taxodont bivalves (zip-type teeth) 212
tea-tree 244, 279
tea-tree swamp 172
TECHNOLOGICAL MUSEUM, SYDNEY N.S.W. 27
teeth of worms and molluscs 112
Teichert, Curt, Prof. 126
Teleostei 222
Tellebang **F** 148
TEMPLE BAY QLD. 105, 12°16′–143°09′
TEMPLETON RIVER N.T. 97, 21°10′–138°17′
TENNANT CREEK N.T. 97
Tentaculites 96
Tenticospirifer 148
Terebratellidae 248
teredo (ship-worm) 244
TERRIGAL N.S.W. 198, 33°27′–151°27′
Tertiary Period (System) 23, 253–67, 277–86
TESSELLATED PAVEMENT TAS. 168, 43°–147°57′
Tetragraptus 109
texture of rocks 16, 17
thallophytes 69
Thamnopora 186
THE NARROWS QLD. 258, 23°40′–151°05′
THE WALL QLD. 177, 18°04′–144°05′
Thecodontosaurus 210, 211
thelodonts 141
theropods 224
thickness of continental crust 40
Thinnfeldia 205, 221

INDEX 339

Thinnfeldia flora 193, 205
third eye 184
Thomas, David E., Dr. 108, 109
THORNTON VIC. 135, 37°15'–145°48'
THORNTON RIVER QLD. 95, 19°25'–138°59'
THORNTONIA QLD. 87, 95, 19°30'–138°56'
THREEMILE CREEK QLD. 180, 20°–148°
thresholds in sedimentary basins 103
thunder lizard 223
Thylacinus 281
Thylacoceras kimberleyense 110
Thylacoleo 281
Tiaro Coal Measures Qld. 218
tidal mud flats 198
tillite (including marine pseudotillite) 60*, 63, 151, 152, 161, 164, 165, 166, 167
Tillyard, R. J., Dr. 183
tilted tablelands 266
time lags in faunal migration 183, 184, 185, 210
time-scale in geology 23, 85, 86, 87
TIMOR ISLAND 153, 231, 235, 274
TIMOR TROUGH 42
tin mining 179, 258
TINTENBAR N.S.W. 241, 28°48'–153°31'
toads 184
TOKO RANGE N.T. 110, 22°52'–138°07'
Tolmer **Gp** 63
Tomago Coal Measures N.S.W. 170, 175
Toolonga Chalk 242
TOONGABBIE VIC. 220, 38°04'–146°38'
TOOWOOMBA QLD. 132, 218
topaz 258
TOP-END N.T. 42
TORBANLEA QLD. 235, 25°21'–152°36'
TORQUAI VIC. 261, 38°20'–144°19'
Torquai **Gp** 261
TORRENS RIVER S.A. 213, 43°53'–138°40'

Torrensian Period (System) 58
Torrensian "series" 58, 60, 63
TORRES STRAIT QLD. 15, 274
Torrowangee beds 63, 66
Torrowangee embayment 66
TOWER HILL VIC. 268, 38°20'–142°21'
TOWNSVILLE QLD. 105, 116, 130, 146, 178
trachyte 16
TRANSCONTINENTAL RAILWAY 241, 263
transgressive seas 8, 236
transitional series 100, 102, 135*, 197, 237
tree stumps in coal beds 180, 203
tree-ferns 181, 205
Triassic Period (System) 23, 193–212
Triassoblatta 206
Trigonia 227, 248, 286
Trigonucula 212
trilobites 22, 96, 97, 107, 109, 111, 116, 117, 122, 125, 129, 138, 140, 154, 189
Tropaeum imperator 247
TROPIC OF CAPRICORN 239
tropical latitudes in the past 88, 93
Tryplasma 120
tuff 18, 130, 134, 170, 203, 220, 235, 268, 269
Tumblagooda **Ss** 105
TUMUT N.S.W. 130, 35°18'–148°13'
tundra 26
Tungussia 79
tunnelling 130
Turonian Epoch (series) 233
Turrilites 247
turtles 73, 246, 277, 283
tusk shells 73
TYENNAN RANGE TAS. 102, 42°42'–146°35'
Tyrannosaurus 245

Uabryichthys 222
Uintatherium 278
ULA (ULAM) QLD. 132, 23°53'–150°37'
ULLADULLA N.S.W. 174, 190, 35°21'–150°29'
Umberatana **Gp** 60

unconformity 45*, 101, 102, 104, 131, 133, 134, 150, 193
underground drainage 263
underground water reserves 204, 240
Undilla **B** 86, 87
unicellular plants 69
Unio 209
UNITED STATES OF AMERICA 15, 62, 85, 94, 155, 158, 271
Upper Permian marine fauna 164
upwarping 48, 265, 266
Uraloceras irwinense 189
uranium 24, 38
U-valleys 4, 165, 273

valley glaciers 271, 272, 273
value of microfossils 112
varve 161
varved shales and clays 161, 167, 273
vascular plants 70, 71, 123, 135
Verbeekiella 186
Vertebraria 180
vertebrates 73, 109, 141–3, 183–5
vertical movements of earth's crust 7, 8, 257, 258, 262, 264, 265, 266, 274
VICTORIA DESERT (GREAT VICTORIA DESERT) 263, 276, 28°–128°
VICTORIA RIVER N.T. 83, 97, 16°–131°
Victoria River **Gp** 53, 57
Viséan Epoch (series) 150
Voisey, A., Prof. 148
volcanic brecciation 18, 151
volcanic cones, plugs and necks 216, 265, 268, 269
volcanics 16, 18, 26, 47, 48, 53, 57, 58, 63, 83, 92, 104, 116, 119, 120, 123, 133, 134, 148, 150, 151, 163, 170, 171, 174, 178, 198, 202, 216, 218–20, 235, 257, 258, 264, 265, 268, 269
volutes 286

Walcott, Charles Doolittle, Prof. 94

WALES 113
WALHALLA VIC. 116, 131, 37°57′–146°27′
wallabies 280, 281
Walloon Coal Measures Qld. 218, 221, 223
WALPOLE ISLAND, NEW CALEDONIA 283
Walsh Tillite 63
WANDAGEE HILL W.A. 162, 23°50′–114°27′
wandering dunes 275
Wapentake beds 117
waratah 279
WARATAH BAY VIC. 132, 133, 38°51′–146°
WARBURTON VIC. 133, 37°45′–145°42′
Warburton **B** 108
WARBURTON RANGE W.A. 57, 166, 26°06′–126°40′
WARDEN HEAD N.S.W. 173, 35°22′–150°29′
WARIALDA N.S.W. 132, 29°32′–150°35′
warm water faunas 88, 120, 263, 285, 286
WARNER'S BAY N.S.W. 182, 32°58′–151°39′
warping of crust 48, 144, 255, 264–7, 269
Warramunga Geosyncline 48
Warramunga **Gp** 48
Warrego River Qld. 218
WARRNAMBOOL VIC. 262, 268
WARWICK QLD. 146, 177, 270
water beetles 183, 203
WATERHOUSE RANGE N.T. 106, 107, 24°02′–133°22′
WATERPARK QLD. 258, 23°30′–149°30′
Watson, James, Dr. 76
weathering depth 266
Wegener, Alfred, Prof. 10, 11
welding of continental blocks 44, 48
WELLINGTON N.S.W. 119, 122, 130, 270
WELLINGTON CAVES 270, 281, 287
welt 44
Wenlockian Epoch (series) 116
Wentworth **Gp** 133

INDEX

WERRIBEE VIC. 160, 37°05′–144°40′
WERRIBEE GORGE VIC. 165, 37°51′–144°37′
WERRIS CREEK N.S.W. 152, 31°21′–150°39′
West Australian Shield 37, 38
WEST IRIAN 44, 48
WESTBROOK QLD. 270
WESTERN NEW SOUTH WALES 109, 113, 115, 118, 130–4, 141, 143, 146, 240, 244
WESTERN PORT BAY VIC. 220, 258
WESTERN QUEENSLAND 107, 109, 146, 150, 176
WESTERN TASMANIA 102, 117, 129, 168, 169
WESTERN TIERS TAS. 266, 41°45′–148°50′
WESTERN VICTORIA 101, 103, 133, 146, 165, 233, 256, 261, 262, 268, 276, 281, 282
whale ribs 285
whales 254, 278, 285
wheat 71
WHITE CLIFFS N.S.W. 240, 241, 245
WHITE CLIFFS OF DOVER 229
white man's arrival 287
WHITSUNDAY ISLAND QLD. 146
WHITSUNDAY PASSAGE QLD. 266
Wianamatta **Gp** 201, 208, 209
Wilkinson, C. S. 170
WILLIAMBURY W.A. 161, 23°52′–115°09′
WILLOUGHBY, SYDNEY N.S.W. 208
Willouran Period (System) 57, 58
WILLOWTREE N.S.W. 171, 31°39′–150°44′
Willyama 57
Willyama Block 84, 85, 90
Willyama Complex 53
Wilpena **Gp** 62, 63
WILPENA POUND S.A. 60, 31°35′–138°32′
WILSON, PORT AUGUSTA AREA S.A. 94
Wilson Bluff **Lst** 256
WILUNA W.A. 47
wind erosion 4, 5, 256, 275, 276
winged snails 72, 96, 188
WINTON QLD. 244
Winton **F** 255

WIRREALPA S.A. 94, 31°08′–138°58′
WITTAGOONA STATION N.S.W. 141, 31°09′–145°23′
Wittagoonaspis 142
wombat 280, 281
WONTHAGGI VIC. 220, 38°37′–145°36′
Wooramel **Gp** (Sandstone) 160
worms 60, 72, 80, 94, 120, 136, 198
WRECK ISLAND QLD. 261, 23°20′–151°57′
Würm Stage of Pleistocene glaciation 271, 275
Wyloo **Gp** 50, 58
WYNDHAM W.A. 125, 15°28′–128°06′
WYNDHAM RIVER W.A. 161, 25°03′–115°30′
Wynyard Tillites 168
Wynyardia bassiana 280

Xanthus 19
Xenacanthus 209
Xenophanes 19
Xenostegium 112
Xystridura 97
Xystridura Time 86, 87

YALLOURN VIC. 220, 38°12′–146°21′
YALWAL CREEK N.S.W. 134, 173, 34°51′–159°22′
Yalwal **Gp** 134
Yapeenian Stage (Age) 103
YARRA RIVER VIC. 117, 133
Yarravia 135
YARROL QLD. 132, 148, 149, 24°58′–151°21′
Yarrol **B** 132, 146, 148, 150, 152, 157, 177, 203
YASS N.S.W. 119–22, 137, 140, 143
Yering **Gp** 131
Yeringian Stage (Age) 131
Yilgarn Shield 37, 38, 42, 44, 47, 50, 53, 56, 58, 160, 231
Yilgarnia 56
Yilgarnian Era (Erathem) 23, 37, 42, 44, 50
YORKE PENINSULA S.A. 38, 47, 94, 165, 167
YOUNG N.S.W. 130

Zaphrentis 138, 155
Zemistephanus 226
Zenophila 120
Zeuchthiscus 208
zinc 50, 52
zip-type teeth of pelecypods 111
zircon 258

zoning of rock sequences by fossils (biostratigraphy) 86, 109, 140, 155, 188, 226, 246, 249, 253, 254, 284, 285
zoogeographical provinces 284, 285